W9-CBY-583

Asterisk™: The Future of Telephony

Other resources from O'Reilly

Related titles

Ethernet: The Definitive
 Guide
Switching to VoIP
T1: A Survival Guide

TCP/IP Network
 Administration
VoIP Hacks™

oreilly.com

oreilly.com is more than a complete catalog of O'Reilly books. You'll also find links to news, events, articles, weblogs, sample chapters, and code examples.

oreillynet.com is the essential portal for developers interested in open and emerging technologies, including new platforms, programming languages, and operating systems.

Conferences

O'Reilly brings diverse innovators together to nurture the ideas that spark revolutionary industries. We specialize in documenting the latest tools and systems, translating the innovator's knowledge into useful skills for those in the trenches. Please visit *conferences.oreilly.com* for our upcoming events.

Safari Bookshelf (*safari.oreilly.com*) is the premier online reference library for programmers and IT professionals. Conduct searches across more than 1,000 books. Subscribers can zero in on answers to time-critical questions in a matter of seconds. Read the books on your Bookshelf from cover to cover or simply flip to the page you need. Try it today for free.

SECOND EDITION

Asterisk™: The Future of Telephony

Jim Van Meggelen, Leif Madsen, and Jared Smith

O'REILLY®

Beijing · Cambridge · Farnham · Köln · Paris · Sebastopol · Taipei · Tokyo

Asterisk™: The Future of Telephony, Second Edition
by Jim Van Meggelen, Leif Madsen, and Jared Smith

Copyright © 2007, 2005 O'Reilly Media, Inc. All rights reserved.
Printed in the United States of America.

Published by O'Reilly Media, Inc., 1005 Gravenstein Highway North, Sebastopol, CA 95472

O'Reilly books may be purchased for educational, business, or sales promotional use. Online editions are also available for most titles (*http://safari.oreilly.com*). For more information, contact our corporate/institutional sales department: (800) 998-9938 or *corporate@oreilly.com*.

Editor: Mike Loukides
Copy Editor: Sanders Kleinfeld
Production Editor: Laurel R.T. Ruma
Proofreader: Tolman Creek Design

Indexer: Joe Wizda
Cover Designer: Karen Montgomery
Interior Designer: David Futato
Illustrators: Robert Romano and Jessamyn Read

Printing History:

June 2005:	First Edition.
August 2007:	Second Edition.

Nutshell Handbook, the Nutshell Handbook logo, and the O'Reilly logo are registered trademarks of O'Reilly Media, Inc. *Asterisk™: The Future of Telephony*, the image of starfish, and related trade dress are trademarks of O'Reilly Media, Inc. Asterisk™ is a trademark of Digium, Inc. *Asterisk: The Future of Telephony* is published under the Creative Commons "Commons Deed" license (*http://creativecommons.org/licenses/by-nc-nd/2.5/ca/*).

While every precaution has been taken in the preparation of this book, the publisher and authors assume no responsibility for errors or omissions, or for damages resulting from the use of the information contained herein.

This book uses RepKover™, a durable and flexible lay-flat binding.

ISBN-10: 0-596-51048-9
ISBN-13: 978-0-596-51048-0

[M]

This book is dedicated to Rich Adamson (1947–2006).

Thanks for showing us the meaning of community.

Table of Contents

Foreword

Once upon a time, there was a boy

...with a computer

...and a phone.

This simple beginning begat much trouble!

It wasn't that long ago that telecommunications, both voice and data, as well as software, were all proprietary products and services, controlled by one select club of companies that created the technologies, and another select club of companies who used the products to provide services. By the late 1990s, data telecommunications had been opened by the expansion of the Internet. Prices plummeted. New and innovative technologies, services, and companies emerged. Meanwhile, the work of free software pioneers like Richard Stallman, Linus Torvalds, and countless others was culminating in the creation of a truly open software platform called Linux (or GNU/Linux). However, voice communications, ubiquitous as they were, remained proprietary. Why? Perhaps it was because voice on the old public telephone network lacked the glamor and promise of the shiny new World Wide Web. Or, perhaps it was because a telephone just wasn't as effective at supplying adult entertainment. Whatever the reason, one thing was clear. Open source voice communications was about as widespread as open source copy protection software.

Necessity (and in some cases simply being cheap) is truly the mother of invention. In 1999, having started Linux Support Services to offer free and commercial technical support for Linux, I found myself in need (or at least in perceived need) of a phone system to assist me in providing 24-hour technical support. The idea was that people would be able to call in, enter their customer identity, and leave a message. The system would in turn page a technician to respond to the customer's request in short order. Since I had started the company with about $4,000 of capital, I was in no position to be able to afford a phone system of the sort that I needed to implement this scenario. Having already been a Linux user since 1994, and having already gotten my feet wet in open source software development by starting l2tpd, Gaim, and cheops, and in the complete absence of anyone having explained the complexity of such a task, I decided that I would simply make my own phone system using hardware borrowed from

Adtran, where I had worked as a co-op student. Once I got a call into a PC, I fantasized, I could do *anything* with it. In fact, it is from this conjecture that the official Asterisk motto (which any sizable, effective project must have) is derived:

It's only software!

For better or worse, I rarely think small. Right from the start, it was my intent that Asterisk would do *everything* related to telephony. The name "Asterisk" was chosen because it was both a key on a standard telephone and also the wildcard symbol in Linux (e.g., `rm -rf *`).

So, in 1999, I had a free telephony platform I'd put out on the Web and I went about my business trying to eke out a living at providing Linux technical support. However, by 2001, as the economy was tanking, it became apparent that Linux Support Services might do better by pursuing Asterisk than general-purpose Linux technical support. That year, we would make contact with Jim "Dude" Dixon of the Zapata Telephony project. Dude's exciting work was a fantastic companion to Asterisk and provided a business model for us to start pursuing Asterisk with more focus. After creating our first PCI telephony interface card in conjunction with Dude, it became clear that "Linux Support Services" was not the best name for a telephony company, and so we changed the name to "Digium," which is a whole other story that cannot be effectively conveyed in writing. Enter the expansion of Voice over IP (VoIP) with its disruptive transition of voice from the old, circuit-switched networks to new IP-based networks, and things really started to take hold.

Now, as we've already covered, clearly most people don't get very excited about telephones. Certainly, few people could share my excitement the moment I heard a dial tone coming from a phone connected to my PC. However, those who *do* get excited about telephones get *really* excited about telephones. And facilitated by the Internet, this small group of people were now able to unite and apply our bizarre passions to a common, practical project for the betterment of many.

To say that telecom was ripe for an open source solution would be an immeasurable understatement. Telecom is an enormous market due to the ubiquity of telephones in work and personal life. The direct market for telecom products has a highly technical audience that is willing and able to contribute. People demand their telecom solutions be infinitely customizable. Proprietary telecom is very expensive. Creating Asterisk was simply the spark in this fuel-rich backdrop.

Asterisk sits at the apex of a variety of transitions (proprietary → open source; circuit switched → VoIP; voice only → voice, video, and data; digital signal processing → host media processing; centralized directory → peer to peer) while easing those transitions by providing bridges back to the older ways of doing things. Asterisk can talk to anything from a 1960s-era pulse-dial phone to the latest wireless VoIP devices, and provide features from simple tandem switching all the way to Bluetooth presence and DUNDi.

Most important of all, though, Asterisk demonstrates how a community of motivated people and companies can work together to create a project with a scope so significant that no one person or company could have possibly created it on its own. In making Asterisk possible, I particularly would like to thank Linus Torvalds, Richard Stallman, the entire Asterisk community, and whoever invented Red Bull.

So where is Asterisk going from here? Think about the history of the PC. When it was first introduced in 1980, it had fairly limited capabilities. Maybe you could do a spreadsheet, maybe do some word processing, but in the end, not much. Over time, however, its open architecture led to price reductions and new products allowing it to slowly expand its applications, eventually displacing the mini computer, then the mainframe. Now, even Cray supercomputers are built using Linux-based x86 architectures. I anticipate that Asterisk's future will look very similar. Today, there is a large subset of telephony that is served by Asterisk. Tomorrow, who knows what the limit might be?

So, what are you waiting for? Read, learn, and participate in the future of open telecommunications by joining the Asterisk revolution!

—*Mark Spencer*

Preface

This is a book for anyone who is new to Asterisk™.

Asterisk is an open source, converged telephony platform, which is designed primarily to run on Linux. Asterisk combines more than 100 years of telephony knowledge into a robust suite of tightly integrated telecommunications applications. The power of Asterisk lies in its customizable nature, complemented by unmatched standards compliance. No other PBX can be deployed in so many creative ways.

Applications such as voicemail, hosted conferencing, call queuing and agents, music on hold, and call parking are all standard features built right into the software. Moreover, Asterisk can integrate with other business technologies in ways that closed, proprietary PBXes can scarcely dream of.

Asterisk can appear quite daunting and complex to a new user, which is why documentation is so important to its growth. Documentation lowers the barrier to entry and helps people contemplate the possibilities.

Produced with the generous support of O'Reilly Media, *Asterisk: The Future of Telephony* was inspired by the work started by the Asterisk Documentation Project. We have come a long way, and this book is the realization of a desire to deliver documentation that introduces the most fundamental elements of Asterisk—the things someone new to Asterisk needs to know. It is the first volume in what we are certain will become a huge library of knowledge relating to Asterisk.

This book was written for, and by, the Asterisk community.

Audience

This book is for those new to Asterisk, but we assume that you're familiar with basic Linux administration, networking, and other IT disciplines. If not, we encourage you to explore the vast and wonderful library of books that O'Reilly publishes on these subjects. We also assume you're fairly new to telecommunications, both traditional switched telephony and the new world of Voice over IP.

Organization

The book is organized into these chapters:

Chapter 1, A Telephony Revolution
This is where we chop up the kindling and light the fire. Asterisk is going to change the world of telecom, and this is where we discuss our reasons for that belief.

Chapter 2, Preparing a System for Asterisk
Covers some of the engineering considerations you should have in mind when designing a telecommunications system. Much of this material can be skipped if you want to get right to installing, but these are important concepts to understand, should you ever plan on putting an Asterisk system into production.

Chapter 3, Installing Asterisk
Covers the obtaining, compiling, and installation of Asterisk.

Chapter 4, Initial Configuration of Asterisk
Describes the initial configuration of Asterisk. Here we will cover the important configuration files that must exist to define the channels and features available to your system.

Chapter 5, Dialplan Basics
Introduces the heart of Asterisk, the dialplan.

Chapter 6, More Dialplan Concepts
Goes over some more advanced dialplan concepts.

Chapter 7, Understanding Telephony
Taking a break from Asterisk, this chapter discusses some of the more important technologies in use in the Public Telephone Network.

Chapter 8, Protocols for VoIP
Following the discussion of legacy telephony, this chapter discusses Voice over Internet Protocol.

Chapter 9, The Asterisk Gateway Interface (AGI)
Introduces one of the more amazing components, the Asterisk Gateway Interface. Using Perl, PHP, and Python, we demonstrate how external programs can be used to add nearly limitless functionality to your PBX.

Chapter 10, Asterisk Manager Interface (AMI) and Adhearsion
Describes how external applications can connect to Asterisk to manipulate or monitor various aspects of the system. Also included in this chapter is a gentle introduction to the Adhearsion framework.

Chapter 11, The Asterisk GUI Framework
The Asterisk GUI Framework, new in Asterisk 1.4, is a framework system that allows web developers to create graphical interfaces with minimal interference to the standard configuration files.

Chapter 12, Relational Database Integration
> Walks you through setting up Asterisk to work with ODBC databases.

Chapter 13, Managing Your Asterisk System
> Discusses issues regarding how to best manage your Asterisk phone system, including CDR, logs, and prompts.

Chapter 14, Potpourri
> Briefly covers what is, in fact, a rich and varied cornucopia of incredible features and functions—all part of the Asterisk phenomenon.

Chapter 15, Asterisk: The Future of Telephony
> Predicts a future where open source telephony completely transforms an industry desperately in need of a revolution.

Appendix A, VoIP Channels

Appendix B, Application Reference

Appendix C, AGI Reference

Appendix D, Configuration Files

Appendix E, Asterisk Dialplan Functions

Appendix F, Asterisk Manager Interface Actions

Appendix G, An Example of func_odbc

Software

This book is focused on documenting Asterisk Version 1.4; however, many of the conventions and information in this book are version-agnostic. Linux is the operating system we have run and tested Asterisk on, with a leaning toward Red Hat syntax. We decided that while Red Hat–based distributions may not be the preferred choice of everyone, their layout and utilities are nevertheless familiar to many experienced Linux administrators.

Conventions Used in This Book

The following typographical conventions are used in this book:

Italic
> Indicates new terms, URLs, email addresses, filenames, file extensions, pathnames, directories, and Unix utilities.

`Constant width`
> Indicates commands, options, parameters, and arguments that must be substituted into commands.

Constant width bold

> Shows commands or other text that should be typed literally by the user. Also used for emphasis in code.

Constant width italic

> Shows text that should be replaced with user-supplied values.

[Keywords and other stuff]

> Indicates optional keywords and arguments.

{ choice-1 | choice-2 }

> Signifies either *choice-1* or *choice-2*.

 This icon signifies a tip, suggestion, or general note.

 This icon indicates a warning or caution.

Using Code Examples

This book is here to help you get your job done. In general, you may use the code in this book in your programs and documentation. You do not need to contact us for permission unless you're reproducing a significant portion of the code. For example, writing a program that uses several chunks of code from this book does not require permission. Selling or distributing a CD-ROM of examples from O'Reilly books does require permission. Answering a question by citing this book and quoting example code does not require permission. Incorporating a significant amount of example code from this book into your product's documentation does require permission.

We appreciate, but do not require, attribution. An attribution usually includes the title, author, publisher, and ISBN. For example: "*Asterisk: The Future of Telephony*, Second Edition, by Jim Van Meggelen, Leif Madsen, and Jared Smith. Copyright 2007 O'Reilly Media, Inc., 978-0-596-51048-0."

If you feel your use of code examples falls outside fair use or the permission given above, feel free to contact us at *permissions@oreilly.com*.

Safari® Books Online

 When you see a Safari® Books Online icon on the cover of your favorite technology book, that means the book is available online through the O'Reilly Network Safari Bookshelf.

Safari offers a solution that's better than e-books. It's a virtual library that lets you easily search thousands of top tech books, cut and paste code samples, download chapters, and find quick answers when you need the most accurate, current information. Try it for free at *http://safari.oreilly.com*.

How to Contact Us

Please address comments and questions concerning this book to the publisher:

O'Reilly Media, Inc.
1005 Gravenstein Highway North
Sebastopol, CA 95472
(800) 998-9938 (in the United States or Canada)
(707) 829-0515 (international or local)
(707) 829-0104 (fax)

We have a web page for this book, where we list errata, examples, and any additional information. You can access this page at:

http://www.oreilly.com/catalog/9780596510480

To comment or ask technical questions about this book, send email to:

bookquestions@oreilly.com

For more information about our books, conferences, Resource Centers, and the O'Reilly Network, see our web site at:

http://www.oreilly.com

Acknowledgments

Firstly, we have to thank our fantastic editor Michael Loukides, who offered invaluable feedback and found incredibly tactful ways to tell us to rewrite a section (or chapter) when it was needed, and have us think it was our idea. Mike built us up when we were down, and brought us back to earth when we got uppity. You are a master, Mike, and seeing how many books have received your editorial oversight contributes to an understanding of why O'Reilly Media is the success that it is.

Thanks also to Sanders Kleinfeld, our copy editor, Laurel Ruma, our production editor, and the rest of the unsung heroes in O'Reilly's production department. These are the folks that take our book and make it an *O'Reilly book*.

Everyone in the Asterisk community needs to thank Jim Dixon for creating the first open source telephony hardware interfaces, starting the revolution, and giving his creations to the community at large.

Thanks to Tim O'Reilly, for giving us a chance to write this book.

To our most generous and merciless review team:

- Rich Adamson, President of Network Partners Inc., for your encyclopedic knowledge of the PSTN, and your tireless willingness to share your experience. Your generosity, even in the face of daunting challenge, is inspiring to us all.*

- Tilghman Lesher, for an incredibly thorough review of our book, contributing some much needed time toward Appendixes B and F, in addition to some amazing new Asterisk applications and functions.

- Andrew Kohlsmith, for helping to write the IMAP voicemail storage section in Chapter 14.

- David Troy, for providing a technical review, for AstManProxy, and for porting Asterisk to the Roomba (first PBX to run on a vacuum cleaner!).

- Matthew Gast, fellow O'Reilly author, for reading our book from cover to cover, and then giving us a comprehensive review, and also for *T1, The Definitive Guide*.

- Dr. Edward Guy III, for your comprehensive and razor-sharp evaluation of each and every chapter of the first edition, and for your championing of Asterisk.

- Kristian Kielhofner, President, KrisCompanies, and creator of AstLinux, for the most excellent AstLinux distribution.

- Russell Bryant, for your rapid and helpful responses to our questions.

- Joshua Colp, for helping us with performance tweaking, and still more questions.

- Kevin Fleming, for raising the bar, and for being a class act, respected (dare we say loved) by all.

- Brian Capouch, for talking about what is possible, and then going out there and doing it.

- Stephen Uhler, for championing the port of Zaptel to Solaris, and for giving us some golden examples.

- Jason Parker, for not being a newb.

- Ekke Loo, for beating up the database chapter.

- Ian Darwin, for tweaking some of the verbiage for us, and for the cherry-red rotary dial phone (that works with Asterisk!).

- Joel Sisko, CEO, iConverged, for your comprehensive telecom and wiring knowledge.

Finally, and most importantly, thanks go to Mark Spencer for Gaim (recently renamed Pidgin, *www.pidgin.im*), Asterisk, and DUNDi, and for contributing his creations to the open source community.

---

* In December of 2006, Rich passed away, as his two-year battle with cancer came to an unfortunate end. Rich was posting on the Asterisk Users mailing list as late as November of that year. He was giving to the community right up until the end, which is why we dedicated this book to him.

Jim Van Meggelen

For me, it all started in the spring of 2004, sitting at my desk in the technical support department of the telecom company I'd worked at for nearly 15 years. With no challenges to properly exercise the skills I had developed, I spent my time trying to figure out what the rest of my career was going to look like. The telecommunications industry had fallen from the pedestal of being a darling of investors to being a joke known to even the most uninformed. I was supposed to feel fortunate to be one of the few who still had work, but what thankless, purposeless work it was. We knew why our industry had collapsed: the products we sold could not hope to deliver the solutions our customers required—even though the industry promised that they could. They lacked flexibility, and were priced totally out of step with the functionality they were delivering (or, more to the point, were failing to deliver). Nowhere in the industry were there any signs this was going to change any time soon.

I had been dreaming of an open source PBX for many long years, but I really didn't know how such a thing could ever come to be—I'd given up on the idea several years before. I knew that to be successful, an open source PBX would need to effectively bridge the worlds of legacy and network-based telecom. I always failed to find anything that seemed ready.

Then, one fine day in spring, I half-heartedly seeded a Google search with the phrase "open source telephony," and discovered a bright new future for telecom: Asterisk, the open source Linux PBX.[†]

There it was: the very thing I'd been dreaming of for so many years. I had no idea how I was going to contribute, but I knew this: open source telephony was going to cause a necessary and beneficial revolution in the telecom industry, and one way or another, I was going to be a part of it.

For me, more of a systems integrator than developer, I needed a way to contribute to the community. There didn't seem to be a shortage of developers, but there sure was a shortage of documentation. This sounded like something I could do. I knew how to write, I knew PBXes, and I desperately needed to talk about this phenomenon that suddenly made telecom fun again.

If I contribute only one thing to this book, I hope you will catch some of my enthusiasm for the subject of open source telephony. This is an incredible gift we have been given, but also an incredible responsibility. What a wonderful challenge. What a cosmic opportunity. What delicious fun!

[†] To get a sense of how big the Asterisk phenomenon is, type "PBX" into Google. As you look at the results, bear in mind that the traditional PBX industry represents billions of dollars. The big players are companies such as Avaya, Nortel, Siemens, Mitel, Cisco, NEC, and many, many more. It is somewhat telling that they don't seem to be concerned about how they rank in a Google search. As a cultural barometer, we're pretty sure this matters.

First of all, I need to thank Leif and Jared for inviting me to join the Asterisk Documentation Project. I have immensely enjoyed working with both of you, and I am constantly amazed at how well our personalities and skills complement each other. A truly balanced team, are we. Also, thanks goes to Figment for all the typing.

To my wife Killi, and my children Kaara, Joonas, and Joosep (who always remember to visit me when I disappear into my underground lair for too long): you are a source of inspiration to me. Your love is the fuel that feeds my fire, and I thank you.

Obviously, I need to thank my parents, Jack and Martiny, for always believing in me, no matter how many rules I broke. In a few years, I'll have my own teenagers, and it'll be your turn to laugh!

To Mark Spencer: thanks for all of the things that everybody else thanks you for, but also, personally, thanks for giving generously of your time to the Asterisk community. The Toronto Asterisk Users' Group (*http://www.taug.ca*) made a quantum leap forward as a result of your taking the time to speak to us, and that event will forever form a part of our history. Oh yeah, and thanks for the beers, too. :-)

Finally, thanks to the Asterisk Community. This book is our gift to you. We hope you enjoy reading it as much as we've enjoyed writing it.

Leif Madsen

The road to this book is a long one—nearly three years in the making. Back when I started using Asterisk, possibly much like you, I didn't know anything about Asterisk, very little about traditional telephony, and even less about Voice over IP. I delved right into this new and very exciting world and took in all I could. For two months during a co-op term, for which I couldn't immediately find work, I absorbed as much as I could, asking questions, trying things and seeing what the system could do. Unfortunately very little to no documentation existed for Asterisk, aside from some dialplan examples I was able to find by John Todd, and having questions answered by Brian K. West on IRC. Of course, this method wasn't going to scale.

Not being much of a coder, I wanted to contribute something back to the community, and what do coders hate doing more than anything? Documentation! So I started The Asterisk Documentation Assignment (TADA), a basic outline with some information for the beginnings of a book.

Shortly after releasing it on my web site, an intelligent fellow by the name of Jared Smith introduced himself. He had similar aspirations for creating a "dead-tree" format book for the community, and we humbly started the Asterisk Documentation Project. Jared set up a simple web site at *http://www.asteriskdocs.org*, a CVS server, and the very first DocBook-formatted version of a book for Asterisk. From there we started filling in information, and soon had information submitted by a number of members of the community.

In June of 2004, an animated chap by the name of Jim Van Meggelen started showing up on the mailing lists, and contributing lots of information and documentation—this was definitely a guy we wanted on our team! Jim had the vision and the drive to really get Jared's and my butts in gear and to work on something grander. Jim brought us years of experience and a writing flair that we could have hardly imagined.

With the core documentation team established, we embarked on a plan for the creation of volumes of Asterisk knowledge, eventually to lead to a complete library and a wealth of information. This book is essentially the beginning of that dream.

Firstly and mostly, I have to thank my parents, Rick and Carol, for always supporting my efforts, allowing me to realize my dreams, and always putting my needs ahead of theirs. Without their vision, understanding, and insight into the future, it would have been impossible to have accomplished what I have. I love you both very much!

I'd like to thank Felix Carapaica and Bill Farkas of the Sheridan Institute of Technology for their dedication to the advancement of knowledge. Their teaching has complemented my prior learning, and has allowed me to expand my understanding of routing and telecommunications exponentially.

There are far too many people to thank individually, but of particular importance, the following people were, and are, the most influential to my understanding of Asterisk: Joshua Colp, Tilghman Lesher, Russell Bryant, Steve Murphy, Olle Johansson, Steven Sokol, Brian K. West, John Todd, and William Suffill, for my very first VoIP phone (which I use to this day!). And for those who I said I'd mention in the book...thanks!

And of course, I must thank Jared Smith and Jim Van Meggelen for having the vision and understanding of how important documentation really is—all of this would have been impossible without you.

Jared Smith

I first started working with Asterisk in the spring of 2002. I had recently started a new job with a market research company, and ended up taking a long road trip to a remote call center with the CIO. On the long drive home we talked about innovation in telephony, and he mentioned a little open source telephony project he had heard of called Asterisk. Over the next few months, I was able to talk the company into buying a developer's kit from Digium and started playing with Asterisk on company time.

During the next few months, I became more and more involved with the Asterisk community. I read the mailing lists. I scoured the archives. I hung out in the IRC channel, just hoping to find nuggets of Asterisk knowledge. As time went on, I was finally able to figure out enough to get Asterisk up and running.

That's when the real fun began.

With the help of the CIO and the approval of the CEO, we moved forward with plans to move our entire telecom infrastructure to Asterisk, including our corporate office and all of our remote call centers. Along the way, we ran into a lot of uncharted territory, and I began thinking about creating a good repository of Asterisk knowledge. Over the course of the project, we were able to do some really innovative things, such as invent IAX trunking!

When all was said and done, we ended up with around forty Asterisk servers spread across many different geographical locations, all communicating with each other to provide a cohesive enterprise-class VoIP phone system. This system currently handles approximately 1 million minutes of calls per month, serves several hundred employees, connects to 27 voice T1s, and saves the company around $20,000 (USD) per month on their telecom costs. In short, our Asterisk project was a resounding success!

While in the middle of implementing this project, I met Leif in one of the Asterisk IRC channels. We talked about ways we could help out new Asterisk users and lower the barrier to entry, and we decided to push ahead with plans to more fully document Asterisk. I really wanted some good documentation in "dead-tree" format —basically a book that a new user could pick up and learn the basics of Asterisk. About that same time, the number of new users on the Asterisk mailing lists and in the IRC channels grew tremendously, and we felt that writing an Asterisk book would greatly improve the signal-to-noise ratio. The Asterisk Documentation Project was born! The rest, they say, is history.

Since then, we've been writing Asterisk documentation. I never thought it would be this arduous, yet rewarding. (I joked with Leif and Jim that it might be easier and less controversial to write an in-depth tome called *Religion, Gun Control, and Sushi* than cover everything that Asterisk has to offer in sufficient detail!) What you see here is a direct result of a lot of late nights and long weekends spent helping the Asterisk community—after all, it's the least we could do, considering what Asterisk has given to us. We hope it will inspire other members of the Asterisk community to help document changes and new features for the benefit of all involved.

Now to thank some people:

First of all, I'd like to thank my beautiful wife. She's put up with a lot of lonely nights while I've been slaving away at the keyboard, and I'd like her to know how much I appreciate her and her endless support. I'd also like to thank my kids for doing their best to remind me of the important things in life. I love you!

To my parents: thanks for everything you've done to help me stretch and grow and learn over the years. You're the best parents a person could ask for.

To Dave Carr and Michael Lundberg: thanks for letting me learn Asterisk on company time. Working with both of you was truly a pleasure. May God smile upon you and grant you success and joy in all you do.

To Leif and Jim: thanks for putting up with my stupid jokes, my insistence that we do things "the right way," and my crazy schedule. Thanks for pushing me along, and making me a better writer. I've really enjoyed working with you two, and hope to collaborate with you on future projects!

To Mark Spencer: thank you for your continued support and dedication and friendship. You've been an invaluable resource to our effort, and I truly believe that you've started a revolution in the world of telephony. You're always welcome in my home and at my dinner table!

To the other great people at Digium: thank you for your help and support. We're especially thankful for your willingness to give us more insight into the Asterisk code, and for donating hardware so that we can better document the Asterisk Developer's Kit.

To Steven Sokol, Steven Critchfield, Olle E. Johansson, and all the others who have contributed to the Asterisk Documentation Project and to this book: thank you! We couldn't have done it without your help and suggestions.

## CHAPTER 1

A Telephony Revolution

*It does not require a majority to prevail, but rather an
irate, tireless minority keen to set brush fires
in people's minds.*

—Samuel Adams

An incredible revolution is under way. It has been a long time in coming, but now that
it has started, there will be no stopping it. It is taking place in an area of technology
that has lapsed embarrassingly far behind every other industry that calls itself high-
tech. The industry is telecommunications, and the revolution is being fueled by an open
source Private Branch eXchange (PBX) called *Asterisk™*.

Telecommunications is arguably the last major electronics industry that has remained
untouched by the open source revolution.[*] Major telecommunications manufacturers
still build ridiculously expensive, incompatible systems, running complicated, ancient
code on impressively engineered yet obsolete hardware.

As an example, Nortel's Business Communications Manager kludges together a 15
year-old Key Telephone Switch and a 1.2 GHz Celeron PC.[†] All this can be yours for
between $5,000 and $15,000, not including telephones. If you want it to actually do
anything interesting, you'll have to pay extra licensing fees for closed, limited-
functionality, shrink-wrapped applications. Customization? Forget it—it's not in the
plan. Future technology and standards compliance? Give them a year or two—they're
working on it.

All of the major telecommunications manufacturers offer similar-minded products.
They don't want you to have flexibility or choice; they want you to be locked in to their
product cycles.

---

[*] Until now.

[†] To its credit, Nortel finally got rid of Windows NT 4.0 and installed Linux. Technically a good idea, but
rather odd, given that Nortel and Microsoft recently announced a partnership to develop enterprise telecom
applications together.

Asterisk changes all of that. With Asterisk, no one is telling you how your phone system should work, or what technology you are limited to. If you want it, you can have it. Asterisk lovingly embraces the concept of standards compliance, while also enjoying the freedom to develop its own innovations. What you choose to implement is up to you—Asterisk imposes no limits.

Naturally, this incredible flexibility comes with a price: Asterisk is not a simple system to configure. This is not because it's illogical, confusing, or cryptic; to the contrary, it is very sensible and practical. People's eyes light up when they first see an Asterisk dialplan and begin to contemplate the possibilities. But when there are literally thousands of ways to achieve a result, the process naturally requires extra effort. Perhaps it can be compared to building a house: the components are relatively easy to understand, but a person contemplating such a task must either a) enlist competent help or b) develop the required skills through instruction, practice, and a good book on the subject.

VoIP: Bridging the Gap Between Traditional and Network Telephony

While Voice over IP (VoIP) is often thought of as little more than a method of obtaining free long-distance calling, the real value (and—let's be honest—challenge as well) of VoIP is that it allows voice to become nothing more than another application in the data network.

It sometimes seems that we've forgotten that the purpose of the telephone is to allow people to communicate. It is a simple goal, really, and it should be possible for us to make it happen in far more flexible and creative ways than are currently available to us. Since the industry has demonstrated an unwillingness to pursue this goal, a large community of passionate people have taken on the task.

The challenge comes from the fact that an industry that has changed very little in the last century shows little interest in starting now.

The Zapata Telephony Project

The Zapata Telephony Project was conceived of by Jim Dixon, a telecommunications consulting engineer who was inspired by the incredible advances in CPU speeds that the computer industry has now come to take for granted. Dixon's belief was that far more economical telephony systems could be created if a card existed that had nothing more on it than the basic electronic components required to interface with a telephone circuit. Rather than having expensive components on the card, Digital Signal Processing (DSP)[‡] would be handled in the CPU by software. While this would impose a tremendous load on the CPU, Dixon was certain that the low cost of CPUs relative to their performance made them far more attractive than expensive DSPs, and, more importantly, that this price/performance ratio would continue to improve as CPUs continued to increase in power.

Like so many visionaries, Dixon believed that many others would see this opportunity, and that he merely had to wait for someone else to create what to him was an obvious improvement. After a few years, he noticed that not only had no one created these cards, but it seemed unlikely that anyone was ever going to. At that point it was clear that if he wanted a revolution, he was going to have to start it himself. And so the Zapata Telephony Project was born:

> Since this concept was so revolutionary, and was certain to make a lot of waves in the industry, I decided on the Mexican revolutionary motif, and named the technology and organization after the famous Mexican revolutionary Emiliano Zapata. I decided to call the card the "tormenta" which, in Spanish, means "storm," but contextually is usually used to imply a big storm, like a hurricane or such.[§]

Perhaps we should be calling ourselves Asteristas. Regardless, we owe Jim Dixon a debt of thanks, partly for thinking this up and partly for seeing it through, but mostly for giving the results of his efforts to the open source community. As a result of Jim's contribution, Asterisk's Public Switched Telephone Network (PSTN) engine came to be.

Massive Change Requires Flexible Technology

The most successful key telephone system in the world has a design limitation that has survived 15 years of users begging for what appears to be a simple change: when you determine the number of times your phone will ring before it forwards to voicemail, you can choose from 2, 3, 4, 6, or 10 ring cycles. Have you any idea how many times people ask for five rings? Plead as the customers might, the manufacturers of this system cannot get their head around the idea that this is a problem. That's the way it works, they say, and users need to get over it.

Another example from the same system is that the name you program on your set can only be seven characters in length.[||] Back in the late 1980s, when this particular system was designed, RAM was very expensive, and storing those seven characters for dozens of sets represented a huge hardware expense. So what's the excuse today? None. Are there any plans to change it? Hardly—the issue is not even officially acknowledged as a problem.

[‡] The term DSP also means Digital Signal Processor, which is a device (usually a chip) that is capable of interpreting and modifying signals of various sorts. In a voice network, DSPs are primarily responsible for encoding, decoding, and transcoding audio information. This can require a lot of computational effort.

[§] Jim Dixon, "The History of Zapata Telephony and How It Relates to the Asterisk PBX" (*http://www.asteriskdocs.org/ modules/tinycontent/index.php?id=10*).

[||] If your name is Elizabeth, for example, you will have to figure something else out like elizbth, or elizabe, or perhaps lizabth. OK, so liz might serve as well, but you get the point.

Those are just two examples; the industry is rife with them.

Now, it's all very well and good to pick on one system, but the reality is that every PBX in existence suffers shortcomings. No matter how fully featured it is, something will always be left out, because even the most feature-rich PBX will always fail to anticipate the creativity of the customer. A small group of users will desire an odd little feature that the design team either did not think of or could not justify the cost of building, and, since the system is closed, the users will not be able to build it themselves.

If the Internet had been thusly hampered by regulation and commercial interests, it is doubtful that it would have developed the wide acceptance it currently enjoys. The openness of the Internet meant that anyone could afford to get involved. So, everyone did. The tens of thousands of minds that collaborated on the creation of the Internet delivered something that no corporation ever could have.

As with many other open source projects, such as Linux and the Internet, the development of Asterisk was fueled by the dreams of folks who knew that there had to be something more than what the industry was producing. The strength of the community is that it is composed not of employees assigned to specific tasks, but rather of folks from all sorts of industries, with all sorts of experiences, and all sorts of ideas about what flexibility means, and what openness means. These people knew that if one could take the best parts of various PBXes and separate them into interconnecting components—akin to a boxful of LEGO bricks—one could begin to conceive of things that would not survive a traditional corporate risk-analysis process. While no one can seriously claim to have a complete picture of what this thing should look like, there is no shortage of opinions and ideas.[#]

Many people new to Asterisk see it as unfinished. Perhaps these people can be likened to visitors to an art studio, looking to obtain a signed, numbered print. They often leave disappointed, because they discover that Asterisk is the blank canvas, the tubes of paint, the unused brushes waiting.[*]

Even at this early stage in its success, Asterisk is nurtured by a greater number of artists than any other PBX. Most manufacturers dedicate no more than a few developers to any one product; Asterisk has scores. Most proprietary PBXes have a worldwide support team comprised of a few dozen real experts; Asterisk has hundreds.

The depth and breadth of the expertise that surrounds this product is unmatched in the telecom industry. Asterisk enjoys the loving attention of old Telco guys who

---

[#] From the release of Asterisk 1.2 to Asterisk 1.4, there have been over 4,000 updates to the code in the SVN repository.

[*] It should be noted that these folks need not leave disappointed. Several projects have arisen to lower the barriers to entry for Asterisk. By far the most popular and well known is trixbox (http://www.trixbox.org). If you have an old PC lying around (or a copy of VMware), trixbox will build a GUI-based PBX for you simply by answering a few questions during the automated install process. This does not make it easier to learn Asterisk, because you are no longer involved in the platform or dialplan configuration, but it will deliver a working PBX to you much faster than the more hands-on approach we employ in this book.

remember when rotary dial mattered, enterprise telecom people who recall when voicemail was the hottest new technology, and data communications geeks and coders who helped build the Internet. These people all share a common belief—that the telecommunications industry needs a *proper* revolution.[†]

Asterisk is the catalyst.

Asterisk: The Hacker's PBX

Telecommunications companies who choose to ignore Asterisk do so at their peril. The flexibility it delivers creates possibilities that the best proprietary systems can scarcely dream of. This is because Asterisk is the ultimate hacker's PBX.

If someone asks you not to use the term *hacker*, refuse. This term does not belong to the mass media. They stole it and corrupted it to mean "malicious cracker." It's time we took it back. Hackers built the networking engine that is the Internet. Hackers built the Apple Macintosh and the Unix operating system. Hackers are also building your next telecom system. Do not fear; these are the good guys, and they'll be able to build a system that's far more secure than anything that exists today. Rather than being constricted by the dubious and easily cracked security of closed systems, the hackers will be able to quickly respond to changing trends in security and fine-tune the telephone system in response to both corporate policy and industry best practices.

Like other open source systems, Asterisk will be able to evolve into a far more secure platform than any proprietary system, not in spite of its hacker roots, but rather because of them.

Asterisk: The Professional's PBX

Never in the history of telecommunications has a system so suited to the needs of business been available, at any price. Asterisk is an enabling technology and, as with Linux, it will become increasingly rare to find an enterprise that is not running some version of Asterisk, in some capacity, somewhere in the network, solving a problem as only Asterisk can.

This acceptance is likely to happen much faster than it did with Linux, though, for several reasons:

- Linux has already blazed the trail that led to open source acceptance. Asterisk is following that lead.
- The telecom industry is crippled, with no leadership being provided by the giant industry players. Asterisk has a compelling, realistic, and exciting vision.

[†] The telecom industry has been predicting a revolution since before the crash; time will tell how well they respond to the *open source* revolution.

- End users are fed up with incompatible, limited functionality, and horrible support. Asterisk solves the first two problems; entepreneurs and the community are addressing the latter.

The Asterisk Community

One of the compelling strengths of Asterisk is the passionate community that developed and supports it. This community, led by Mark Spencer of Digium, is keenly aware of the cultural significance of Asterisk, and is giddy about the future.

One of the more powerful side effects caused by the energy of the Asterisk community is the cooperation it has spawned among the telecommunications professionals, networking professionals, and information technology professionals who share a love for this phenomenon. While these professions have traditionally been at odds with each other, in the Asterisk community they delight in each others' skills. The significance of this cooperation cannot be underestimated.

Still, if the dream of Asterisk is to be realized, the community must grow—yet one of the key challenges that the community currently faces is a rapid influx of new users. The members of the existing community, having birthed this thing called Asterisk, are generally welcoming of new users, but they've grown impatient with being asked the kinds of questions whose answers can often be obtained independently, if one is willing to put forth the time needed to research and experiment.

Obviously, new users do not fit any particular kind of mold. While some will happily spend hours experimenting and reading various blogs describing the trials and tribulations of others, many people who have become enthusiastic about this technology are completely uninterested in such pursuits. They want a simple, straightforward, step-by-step guide that'll get them up and running, followed by some sensible examples describing the best methods of implementing common functionality (such as voicemail, auto attendants, and the like).

To the members of the expert community, who (correctly) perceive that Asterisk is like a web development language, this approach doesn't make any sense. To them, it's clear that you have to immerse yourself in Asterisk to appreciate its subtleties. Would one ask for a step-by-step guide to programming and expect to learn from it all that a language has to offer?

Clearly, there's no one approach that's right for everyone. Asterisk is a different animal altogether, and it requires a totally different mind-set. As you explore the community, though, be aware that there are people with many different skill sets and attitudes here. Some of these folks do not display much patience with new users, but that's often due to their passion for the subject, not because they don't welcome your participation.

The Asterisk Mailing Lists

As with any community, there are places where members of the Asterisk community meet to discuss matters of mutual interest. Of the mailing lists you will find at *http://lists.digium.com*, these four are currently the most important:

Asterisk-Biz

Anything commercial with respect to Asterisk belongs in this list. If you're selling something Asterisk-related, sell it here. If you want to buy an Asterisk service or product, post here.

Asterisk-Dev

The Asterisk developers hang out here. The purpose of this list is the discussion of the development of the software that is Asterisk, and its participants vigorously defend that purpose. Expect a lot of heat if you post anything to this list not relating to programming or development of the Asterisk code base specifically. General coding questions (such as interfacing with AGI or AMI), should be directed to the Asterisk-Users list.

 The Asterisk-Dev list is not second-level support! If you scroll through the mailing list archives, you'll see this is a strict rule. The Asterisk-Dev mailing list is about discussion of core Asterisk development, and questions about interfacing your external programs via AGI or AMI should be posted on the Asterisk-Users list.

Asterisk-Users

This is where most Asterisk users hang out. This list generates several hundred messages per day and has over ten thousand subscribers. While you can go here for help, you are expected to have done some reading on your own before you post a query.

Asterisk-BSD

This is where users who are implementing Asterisk on FreeBSD (and other BSD dialects) hang out.

The Asterisk Wiki

The Asterisk Wiki (which exists in large part due to the tireless efforts of James Thompson—*thanks James!*) is a source of much enlightenment and confusion. A community-maintained repository of VoIP knowledge (*http://www.voip-info.org*) contains a truly inspiring cornucopia of fascinating, informative, and frequently contradictory information about many subjects, just one of which is Asterisk.

Since Asterisk documentation forms by far the bulk of the information on this web site,[‡] and it probably contains more Asterisk knowledge than all other sources put together (with the exception of the mailing-list archives), it is commonly referred to as *the* place to go for Asterisk knowledge.

The IRC Channels

The Asterisk community maintains Internet Relay Chat (IRC) channels on *irc.freenode.net*. The two most active channels are *#asterisk* and *#asterisk-dev*.[§] To cut down on spam-bot intrusions, both of these channels now require registration to join.[||]

Asterisk User Groups

In many cites around the world, lonely Asterisk users began to realize that there were other like-minded people in their towns. Asterisk User Groups (AUGs) began to spring up all over the place. While these groups don't have any official affiliation with each other, they generally link to each others' web sites and welcome members from anywhere. Type "Asterisk User Group" into Google to track down one in your area.

The Asterisk Documentation Project

The Asterisk Documentation Project was started by Leif Madsen and Jared Smith, but several people in the community have contributed.

The goal of the documentation project is to provide a structured repository of written work on Asterisk. In contrast with the flexible and ad hoc nature of the Wiki, the Docs project is passionate about building a more focused approach to various Asterisk-related subjects.

As part of the efforts of the Asterisk Docs project to make documentation available online, this book is available at the *http://www.asteriskdocs.org* web site, under a Creative Commons license.

The Business Case

It is very rare to find businesses these days that do not have to reinvent themselves every few years. It is equally rare to find a business that can afford to replace its communications infrastructure each time it goes in a new direction. Today's businesses need extreme flexibility in all of their technology, including telecom.

In his book *Crossing the Chasm* (HarperBusiness), Geoffrey Moore opines, "The idea that the value of the system will be discovered rather than known at the time of installation implies, in turn, that product flexibility and adaptability, as well as ongoing account service, should be critical components of any buyer's evaluation checklist."

[‡] More than 30 percent, at last count.

[§] The *#asterisk-dev* channel is for the discussion of changes to the underlying code base of Asterisk and is also not second-tier support. Discussions related to programming external applications that interface with Asterisk via AGI or AMI are meant to be in *#asterisk*.

[||] `/msg nickserv help` when you connect to the service via your favorite IRC client.

What this means, in part, is that the true value of a technology is often not known until it has been deployed.

How compelling, then, to have a system that holds at its very heart the concept of openness and the value of continuous innovation.

This Book

So where to begin? Well, when it comes to Asterisk, there is far more to talk about than we can fit into one book. For now, we're not going to take you down all the roads that the über-geeks follow—we're just going to give you the basics.

In Chapter 2, we cover some of the engineering considerations you should keep in mind when designing a telecommunications system. You can skip much of this material if you want to get right to installing, but these are important concepts to understand, should you ever plan on putting an Asterisk system into production.

Chapter 3 covers obtaining, compiling, and installing Asterisk, and Chapter 4 deals with the initial configuration of Asterisk. Here we cover the important configuration files that must exist to define the channels and features available to your system. This will prepare you for Chapter 5, where we introduce the heart of Asterisk—the dialplan. Chapter 6 will introduce some more advanced dialplan concepts.

We will take a break from Asterisk in Chapter 7 and discuss some of the more important technologies in use in the PSTN. Naturally, following the discussion of legacy telephony, Chapter 8 discusses Voice over IP.

Chapter 9 introduces one of the more amazing components, the Asterisk Gateway Interface (AGI). Using Perl, PHP, and Python, we demonstrate how external programs can be used to add nearly limitless functionality to your PBX. In Chapter 14, we briefly cover what is, in fact, a rich and varied cornucopia of incredible features and functions, all of which are part of the Asterisk phenomenon. To conclude, Chapter 15 looks forward, predicting a future where open source telephony completely transforms an industry desperately in need of a revolution. You'll also find a wealth of reference information in the book's five appendixes.

This book can only lay down the basics, but from this foundation you will be able to come to an understanding of the concept of Asterisk—and from that, who knows what you will build?

## CHAPTER 2

Preparing a System for Asterisk

*Very early on, I knew that someday in some "perfect"
future out there over the horizon, it would be commonplace for computers to handle all of the necessary processing functionality internally, making the necessary
external hardware to connect up to telecom interfaces
very inexpensive and, in some cases, trivial.*

—Jim Dixon, "The History of Zapata Telephony and
How It Relates to the Asterisk PBX"

By this point, you must be anxious to get your Asterisk system up and running. If you
are building a hobby system, you can probably jump right to the next chapter and begin
the installation. For a mission-critical deployment, however, some thought must be
given to the environment in which the Asterisk system will run. Make no mistake:
Asterisk, being a very flexible piece of software, will happily and successfully install on
nearly any Linux platform you can conceive of, and several non-Linux platforms as
well.* However, to arm you with an understanding of the type of operating environment
Asterisk will really thrive in, this chapter will discuss issues you need to be aware of in
order to deliver a reliable, well-designed system.

In terms of its resource requirements, Asterisk's needs are similar to those of an embedded, real-time application. This is due in large part to its need to have priority access
to the processor and system buses. It is, therefore, imperative that any functions on the
system not directly related to the call-processing tasks of Asterisk be run at a low priority, if at all. On smaller systems and hobby systems, this might not be as much of an
issue. However, on high-capacity systems, performance shortcomings will manifest as
audio quality problems for users, often experienced as echo, static, and the like. The
symptoms will resemble those experienced on a cell phone when going out of range,

---

* People have successfully compiled and run Asterisk on WRAP boards, Linksys WRT54G routers, Soekris
systems, Pentium 100s, PDAs, Apple Macs, Sun SPARCs, laptops, and more. Of course, whether you would
want to put such a system into production is another matter entirely. (Actually, the AstLinux distribution,
by Kristian Kielhofner, runs very well indeed on the Soekris 4801 board. Once you've grasped the basics of
Asterisk, this is something worth looking into further. Check out *http://www.astlinux.org*.)

although the underlying causes will be different. As loads increase, the system will have increasing difficulty maintaining connections. For a PBX, such a situation is nothing short of disastrous, so careful attention to performance requirements is a critical consideration during the platform selection process.

Table 2-1 lists some very basic guidelines that you'll want to keep in mind when planning your system. The next section takes a close look at the various design and implementation issues that will affect its performance.

 The size of an Asterisk system is actually not dictated by the number of users or sets, but rather by the number of simultaneous calls it will be expected to support. These numbers are very conservative, so feel free to experiment and see what works for you.

Table 2-1. System requirement guidelines

Purpose	Number of channels	Minimum recommended
Hobby system	No more than 5	400 MHz x86, 256 MB RAM
SOHO system (small office/home office—less than three lines and five sets)	5 to 10	1 GHz x86, 512 MB RAM
Small business system	Up to 25	3 GHz x86, 1 GB RAM
Medium to large system	More than 25	Dual CPUs, possibly also multiple servers in a distributed architecture

With large Asterisk installations, it is common to deploy functionality across several servers. One or more central units will be dedicated to call processing; these will be complemented by one or more ancillary servers handling peripherals (such as a database system, a voicemail system, a conferencing system, a management system, a web interface, a firewall, and so on). As is true in most Linux environments, Asterisk is well suited to growing with your needs: a small system that used to be able to handle all your call-processing and peripheral tasks can be distributed among several servers when increased demands exceed its abilities. Flexibility is a key reason why Asterisk is extremely cost-effective for rapidly growing businesses; there is no effective maximum or minimum size to consider when budgeting the initial purchase. While some scalability is possible with most telephone systems, we have yet to hear of one that can scale as flexibly as Asterisk. Having said that, distributed Asterisk systems are not simple to design—this is not a task for someone new to Asterisk.

If you are sure that you need to set up a distributed Asterisk system, you will want to study the DUNDi protocol, Asterisk Realtime Architecture (ARA), func_odbc, and the various other database tools at your disposal. This will help you to abstract the data your system requires from the dialplan logic your Asterisk systems will utilize, allowing a generic set of dialplan logic that can be used across multiple boxes, thereby allowing you to scale more simply by adding additional boxes to the system. However, this is far beyond the scope of this book and will be left as an exercise for the reader. If you want a teaser of some tools you can use for scaling, see Chapter 12.

A Set of Load Test Results

Joshua Colp was able to produce the results in Table 2-2 on an AMD Athlon64 X2 4200 + with 1 GB RAM and 80 GB SATA hard drive, testing with the default scenario in the SIPp application: a simple call setup, Playback() an audio file, and Wait() a short time. Notice the massive savings in CPU utilization while reading data from the RAM disk versus the hard drive. This could be interpreted as the CPU waiting for data to process before delivering it to the requesting channel. However, this is just a simple test and in no way reflects the amount of calls your system will be able to handle. You are encouraged to load test your own system to determine the number of simultaneous calls that can be handled utilizing your dialplan and combination of applications.

Table 2-2. Sample test results for SIPp default scenario using simple Wait() and Playback() application; SIPp echoed media back to Asterisk

Simultaneous calls	330	330	550
CPU utilization	149%	14.8%	57.6%
Load average	49	25	60
Storage	Hard drive	RAM disk	RAM disk

Server Hardware Selection

The selection of a server is both simple and complicated: simple because, really, any x86-based platform will suffice, but complicated because the reliable performance of your system will depend on the care that is put into the platform design. When selecting your hardware, you must carefully consider the overall design of your system and what functionality you need to support. This will help you determine your requirements for the CPU, motherboard, and power supply. If you are simply setting up your first Asterisk system for the purpose of learning, you can safely ignore the information in this section. If, however, you are building a mission-critical system suitable for deployment, these are issues that require some thought.

Performance Issues

Among other considerations, when selecting the hardware for an Asterisk installation you must bear in mind this critical question: how powerful must the system be? This is not an easy question to answer, because the manner in which the system is to be used will play a big role in the resources it will consume. There is no such thing as an Asterisk performance-engineering matrix, so you will need to understand how Asterisk uses the system in order to make intelligent decisions about what kinds of resources will be required. You will need to consider several factors, including:

The maximum number of concurrent connections the system will be expected to support
> Each connection will increase the workload on the system.

The percentage of traffic that will require processor-intensive DSP of compressed codecs (such as G.729 and GSM)
> The Digital Signal Processing (DSP) work that Asterisk performs in software can have a staggering impact on the number of concurrent calls it will support. A system that might happily handle 50 concurrent G.711 calls could be brought to its knees by a request to conference together 10 G.729 compressed channels. We'll talk more about G.729, GSM, G.711, and many other codecs in Chapter 8.

Whether conferencing will be provided, and what level of conferencing activity is expected
> Will the system be used heavily? Conferencing requires the system to transcode and mix each individual incoming audio stream into multiple outgoing streams. Mixing multiple audio streams in near-real-time can place a significant load on the CPU.

Echo cancellation
> Echo cancellation may be required on any call where a Public Switched Telephone Network (PSTN) interface is involved. Since echo cancellation is a mathematical function, the more of it the system has to perform, the higher the load on the CPU will be. [†] Do not fear. Echo cancellation is another topic for Chapter 8.

Dialplan scripting logic
> Whenever Asterisk has to pass call control to an external program, there is a performance penalty. As much logic as possible should be built into the dialplan. If external scripts are used, they should be designed with performance and efficiency as critical considerations.

As for the exact performance impact of these factors, it's difficult to know for sure. The effect of each is known in general terms, but an accurate performance calculator has not yet been successfully defined. This is partly because the effect of each component of the system is dependent on numerous variables, such as CPU power, motherboard chipset and overall quality, total traffic load on the system, Linux kernel optimizations, network traffic, number and type of PSTN interfaces, and PSTN traffic—not to mention

---

[†] Roughly 30 MHz of CPU power per channel.

any non-Asterisk services the system is performing concurrently. Let's take a look at the effects of several key factors:

Codecs and transcoding
>   Simply put, a *codec* (short for coder/decoder, or compression/decompression) is a set of mathematical rules that define how an analog waveform will be digitized. The differences between the various codecs are due in large part to the levels of compression and quality that they offer. Generally speaking, the more compression that's required, the more work the DSP must do to code or decode the signal. Uncompressed codecs, therefore, put far less strain on the CPU (but require more network bandwidth). Codec selection must strike a balance between bandwidth and processor usage.

Central processing unit (and Floating Point Unit)
>   A CPU is comprised of several components, one of which is the floating point unit (FPU). The speed of the CPU, coupled with the efficiency of its FPU, will play a significant role in the number of concurrent connections a system can effectively support. The next section ("Choosing a Processor) offers some general guidelines for choosing a CPU that will meet the needs of your system.

Other processes running concurrently on the system
>   Being Unix-like, Linux is designed to be able to multitask several different processes. A problem arises when one of those processes (such as Asterisk) demands a very high level of responsiveness from the system. By default, Linux will distribute resources fairly among every application that requests them. If you install a system with many different server applications, those applications will each be allowed their fair use of the CPU. Since Asterisk requires frequent high-priority access to the CPU, it does not get along well with other applications, and if Asterisk must coexist with other apps, the system may require special optimizations. This primarily involves the assignment of priorities to various applications in the system and, during installation, careful attention to which applications are installed as services.

Kernel optimizations
>   A kernel optimized for the performance of one specific application is something that very few Linux distributions offer by default and, thus, it requires some thought. At the very minimum—whichever distribution you choose—a fresh copy of the Linux kernel (available from *http://www.kernel.org*) should be downloaded and compiled on your platform. You may also be able to acquire patches that will yield performance improvements, but these are considered hacks to the officially supported kernel.

IRQ latency
>   Interrupt request (IRQ) latency is basically the delay between the moment a peripheral card (such as a telephone interface card) requests the CPU to stop what it's doing and the moment when the CPU actually responds and is ready to handle the task. Asterisk's peripherals (especially the Zaptel cards) are extremely intoler-

ant of IRQ latency. This is not due to any problem with the cards so much as part of the nature of how a software-based TDM engine has to work. If we buffer the TDM data and send it on the bus as a larger packet, that may be more efficient from a system perspective, but it will create a delay between the time the audio is received on the card, and when it is delivered to the CPU. This makes real-time processing of TDM data next to impossible. In the design of Zaptel, it was decided that sending the data every 1 ms would create the best trade-off, but a side effect of this is that any card in the system that uses the Zaptel interface is going to ask the system to process an interrupt every millisecond. This used to be a factor on older motherboards, but it has largely ceased to be a cause for concern.

 Linux has historically had problems with its ability to service IRQs quickly; this problem has caused enough trouble for audio developers that several patches have been created to address this shortcoming. So far, there has been some mild controversy over how to incorporate these patches into the Linux kernel.

Kernel version
Asterisk is officially supported on Linux Version 2.6.

Linux distribution
Linux distributions are many and varied. In the next chapter, we will discuss the challenge of selecting a Linux distribution, and how to obtain and install both Linux and Asterisk.

Choosing a Processor

Since the performance demands of Asterisk will generally involve a large number of math calculations, it is essential that you select a processor with a powerful FPU. The signal processing that Asterisk performs can quickly demand a staggering quantity of complex mathematical computations from the CPU. The efficiency with which these tasks are carried out will be determined by the power of the FPU within the processor.

To actually name a best processor for Asterisk in this book would fly in the face of Moore's law. Even in the time between the authoring and publishing of this book, processor speeds will undergo rapid improvements, as will Asterisk's support for various architectures. Obviously, this is a good thing, but it also makes the giving of advice on the topic a thankless task. Naturally, the more powerful the FPU is, the more concurrent DSP tasks Asterisk will be able to handle, so that is the ultimate consideration. When you are selecting a processor, the raw clock speed is only part of the equation. How well it handles floating-point operations will be a key differentiator, as DSP operations in Asterisk will place a large demand on that process.

Both Intel and AMD CPUs have powerful FPUs. Current-generation chips from either of those manufacturers can be expected to perform well.[‡]

The obvious conclusion is that you should get the most powerful CPU your budget will allow. However, don't be too quick to buy the most expensive CPU out there. You'll need to keep the requirements of your system in mind; after all, a Formula 1 Ferrari is ill-suited to the rigors of rush-hour traffic. Slower CPUs will often run cooler and, thus, you might be able to build a lower-powered, fanless Asterisk system for a small office, which could work well in a dusty environment, perhaps.

In order to attempt to provide a frame of reference from which we can contemplate our platform decision, we have chosen to define three sizes of Asterisk systems: small, medium, and large.

Small systems

Small systems (up to 10 phones) are not immune to the performance requirements of Asterisk, but the typical load that will be placed on a smaller system will generally fall within the capabilities of a modern processor.

If you are building a small system from older components you have lying around, be aware that the resulting system cannot be expected to perform at the same level as a more powerful machine, and will run into performance degradation under a much lighter load. Hobby systems can be run successfully on very low-powered hardware, although this is by no means recommended for anyone who is not a whiz at Linux performance tuning.[§]

If you are setting up an Asterisk system for the purposes of learning, you will be able to build a fully featured platform using a relatively low-powered CPU. The authors of this book run several Asterisk lab systems with 433 MHz to 700 MHz Celeron processors, but the workload of these systems is minimal (never more than two concurrent calls).

AstLinux and Asterisk on OpenWRT

If you are really comfortable working with Linux on embedded platforms, you will want to join the AstLinux mailing list and run Kristian Kielhofner's creation, AstLinux, or get yourself a Linksys WRT54GL and install Brian Capouch's version of Asterisk for that platform.

These projects strip Asterisk down to its essentials, and allow incredibly powerful PBX applications to be deployed on very inexpensive hardware.

[‡] If you want to be completely up to the minute on which CPUs are leading the performance race, surf on over to Tom's Hardware (*http://www.tomshardware.com*) or AnandTech (*http://www.anandtech.com*), where you will find a wealth of information about both current and out-of-date CPUs, motherboards, and chipsets.

[§] Greg Boehnlein once compiled and ran Asterisk on a 133 MHz Pentium system, but that was mostly as an experiment. Performance problems are far more likely, and properly configuring such a system requires an expert knowledge of Linux. We do not recommend running Asterisk on anything less than a 500 MHz system (for a production system, 2 GHz might be a sensible minimum). Still, we think the fact that Asterisk is so flexible is remarkable.

> While both projects require a fair amount of knowlege and effort on your part, they also share a huge coolness factor, are extrememly popular, and are of excellent quality.

Medium systems

Medium-sized systems (from 10 to 50 phones) are where performance considerations will be the most challenging to resolve. Generally, these systems will be deployed on one or two servers only and, thus, each machine will be required to handle more than one specific task. As loads increase, the limits of the platform will become increasingly stressed. Users may begin to perceive quality problems without realizing that the system is not faulty in any way, but simply exceeding its capacity. These problems will get progressively worse as more and more load is placed on the system, with the user experience degrading accordingly. It is critical that performance problems be identified and addressed before they are noticed by users.

Monitoring performance on these systems and quickly acting on any developing trends is key to ensuring that a quality telephony platform is provided.

Large systems

Large systems (more than 120 channels) can be distributed across multiple systems and sites and, thus, performance concerns can be managed through the addition of machines. Very large Asterisk systems have been created in this way.

Building a large system requires an advanced level of knowledge in many different disciplines. We will not discuss it in detail in this book, other than to say that the issues you'll encounter will be similar to those encountered during any deployment of multiple servers handling a single, distributed task.

Choosing a Motherboard

Just to get any anticipation out of the way, we also cannot recommend specific motherboards in this book. With new motherboards coming out on a weekly basis, any recommendations we could make would be rendered moot by obsolescence before the published copy hit the shelves. Not only that, but motherboards are like automobiles: while they are all very similar in principle, the difference is in the details. And as Asterisk is a performance application, the details matter.

What we will do, therefore, is give you some idea of the kinds of motherboards that can be expected to work well with Asterisk, and the features that will make for a good motherboard. The key is to have both stability and high performance. Here are some guidelines to follow:

- The various system buses must provide the minimum possible latency. If you are planning a PSTN connection using analog or PRI interfaces (discussed later in this chapter), having Zaptel cards in the system will generate 1,000 interrupt requests per second. Having devices on the bus that interfere with this process will result in degradation of call quality. Chipsets from Intel (for Intel CPUs) and nVidia nForce (for AMD CPUs) seem to score the best marks in this area. Review the specific chipset of any motherboard you are evaluating to ensure that it does not have known problems with IRQ latency.

- If you are running Zaptel cards in your system, you will want to ensure that your BIOS allows you maximum control over IRQ assignment. As a rule, high-end motherboards will offer far greater flexibility with respect to BIOS tweaking; value-priced boards will generally offer very little control. This may be a moot point, however, as APIC-enabled motherboards turn IRQ control over to the operating system.

- Server-class motherboards generally implement a different PCI standard than workstation-class motherboards. While there are many differences, the most obvious and well known is that the two versions have different voltages. Depending on which cards you purchase, you will need to know if you require 3.3V or 5V PCI slots.[||] Figure 2-1 shows the visual differences between 3.3V and 5V slots. Most server motherboards will have both types, but workstations will typically have only the 5V version.

> There is some evidence that suggests connecting together two completely separate, single-CPU systems may provide far more benefits than simply using two processors in the same machine. You not only double your CPU power, but you also achieve a much better level of redundancy at a similar cost to a single-chassis, dual-CPU machine. Keep in mind, though, that a dual-server Asterisk solution will be more complex to design than a single-machine solution.

- Consider using multiple processors, or processors with multiple cores. This will provide an improvement in the system's ability to handle multiple tasks. For Asterisk, this will be of special benefit in the area of floating-point operations.

- If you need a modem, install an external unit that connects to a serial port. If you must have an internal modem, you will need to ensure that it is not a so-called "Win-modem"—it must be a completely self-sufficient unit (note that these are very difficult, if not impossible, to find).

- Consider that with built-in networking, if you have a network component failure, the entire motherboard will need to be replaced. On the other hand, if you install a peripheral Network Interface Card (NIC), there may be an increased chance of

---

[||] With the advent of PCI-X and PCI-Express, it is becoming harder and harder to select a motherboard with the correct type of slots. Be very certain that the motherboard you select has the correct type and quantity of card slots for your hardware. Keep in mind that most companies that produce hardware cards for Asterisk offer PCI and PCI-Express versions, but it's still up to you to make sure they make sense in whatever motherboard and chassis combination you choose.

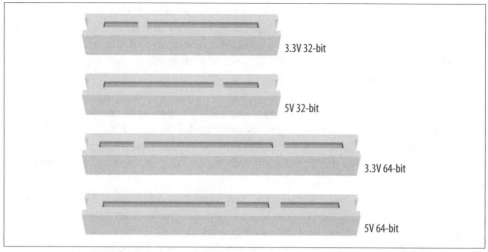

Figure 2-1. Visual identification of PCI slots

failure due to the extra mechanical connections involved. It can also be useful to have separate network cards serving sets and users (the internal network) and VoIP providers and external sites (the external network). NICs are cheap; we suggest always having at least two.

- The stability and quality of your Asterisk system will be dependent on the components you select for its architecture. Asterisk is a beast, and it expects to be fed the best. As with just about anything, high cost is not always synonymous with quality, but you will want to become a connoisseur of computer components.

Having said all that, we need to get back to the original point: Asterisk can and will happily install on pretty much any system that will run Linux. The lab systems used to write this book, for example, included everything from a Linksys WRT to a dual-Xeon locomotive.[#] We have not experienced any performance or stability problems running less than five concurrent telephone connections. For the purposes of learning, do not be afraid to install Asterisk on whatever system you can scrounge up. When you are ready to put your system into production, however, you will need to understand the ramifications of the choices you make with respect to your hardware.

Power Supply Requirements

One often-overlooked component in a PC is the power supply (and the supply of power). For a telecommunications system,[*] these components can play a significant role in the quality of the user experience.

[#] OK, it wasn't *actually* a locomotive, but it sure sounded like one. Does anyone know where to get quiet CPU fans for Xeon processors? It's getting too loud in the lab here.

[*] Or any system that is expected to process audio.

Computer power supplies

The power supply you select for your system will play a vital role in the stability of the entire platform. Asterisk is not a particularly power-hungry application, but anything relating to multimedia (whether it be telephony, professional audio, video, or the like) is generally sensitive to power *quality*.

This oft-neglected component can turn an otherwise top-quality system into a poor performer. By the same token, a top-notch power supply might enable an otherwise cheap PC to perform like a champ.

The power supplied to a system must provide not only the energy the system needs to perform its tasks but also stable, clean signal lines for all of the voltages your system expects from it.

Spend the money and get a top-notch power supply (gamers are pretty passionate about this sort of thing, so there are lots of choices out there).

Redundant power supplies

In a carrier-grade or high-availability environment, it is common to deploy servers that use a redundant power supply. Essentially, this involves two completely independent power supplies, either one of which is capable of meeting the power requirements of the system.

If this is important to you, keep in mind that best practices suggest that to be properly redundant, these power supplies should be connected to completely independent uninterruptible power supplies (UPSes) that are in turn fed by totally separate electrical circuits. In truly mission-critical environments (such as hospitals), even the main electrical feeds into the building are redundant, and diesel-powered generators are on-site to generate electricity during extended power failures (such as the one that hit Northeastern North America on August 15, 2003).

Environment

Your system's environment consists of all of those factors that are not actually part of the server itself but nevertheless play a crucial role in the reliability and quality that can be expected from the system. Electrical supplies, room temperature and humidity, sources of interference, and security are all factors that should be contemplated.

Power Conditioning and Uninterruptible Power Supplies

When selecting the power sources for your system, consideration should be given not only to the amount of power the system will use, but also to the manner in which this power is delivered.

Power is not as simple as voltage coming from the outlet in the wall, and you should never just plug a production system into whatever electrical source is near at hand.[†] Giving some consideration to the supply of power to your system can provide a far more stable power environment, leading to a far more stable system.

One of the benefits of clean power is a reduction in heat, which means less stress on components, leading to a longer life expectancy.

Properly grounded, conditioned power feeding a premium-quality power supply will ensure a clean *logic ground* (a.k.a. 0 volt) reference[‡] for the system and keep electrical noise on the motherboard to a minimum. These are industry-standard best practices for this type of equipment, which should not be neglected. A relatively simple way to achieve this is through the use of a *power-conditioned* UPS.[§]

Power-conditioned UPSes

The UPS is well known for its role as a battery backup, but the power-conditioning benefits that high-end UPS units also provide are less well understood.

Power conditioning can provide a valuable level of protection from the electrical environment by regenerating clean power through an isolation transformer. A quality power conditioner in your UPS will eliminate most electrical noise from the power feed and help to ensure a rock-steady supply of power to your system.

Unfortunately, not all UPS units are created equal; many of the less expensive units do not provide clean power. What's worse, manufacturers of these devices will often promise all kinds of protection from surges, spikes, overvoltages, and transients. While such devices may protect your system from getting fried in an electrical storm, they will not clean up the power being fed to your system and, thus, will do nothing to contribute to stability.

Make sure your UPS is *power conditioned*. If it doesn't say exactly that, it isn't.

Grounding

Voltage is defined as the difference in electrical potential between two points. When considering a *ground* (which is basically nothing more than an electrical path to earth), the common assumption is that it represents 0 volts. But if we do not define that 0V in

---

[†] Okay, look, you *can* plug it in wherever you'd like, and it'll probably work, but if your system has strange stability problems, please give this section another read. Deal?

[‡] In electronic devices, a binary zero (0) is generally related to a 0 volt signal, while a binary one (1) can be represented by many different voltages (commonly between 2.5 and 5 volts). The grounding reference that the system will consider 0 volts is often referred to as the *logic ground*. A poorly grounded system might have electrical potential on the logic ground to such a degree that the electronics mistake a binary zero for a binary one. This can wreak havoc with the system's ability to process instructions.

[§] It is a common misconception belief that all UPSes provide clean power. This is not at all true.

---

relation to something, we are in danger of assuming things that may not be so. If you measure the voltage between two grounding references, you'll often find that there is a voltage potential between them. This voltage potential between grounding points can be significant enough to cause logic errors—or even damage—in a system where more than one path to ground is present.

One of the authors recalls once frying a sound card he was trying to connect to a friend's stereo system. Even though both the computer and the stereo were in the same room, more than 6 volts of difference was measured between the ground conductors of the two electrical outlets they were plugged into! The wire between the stereo and the PC (by way of the sound card) provided a path that the voltage eagerly followed, thus frying a sound card that was not designed to handle that much current on its signal leads. Connecting both the PC and the stereo to the same outlet fixed the problem.

When considering electrical regulations, the purpose of a ground is primarily human safety. In a computer, the ground is used as a 0V logic reference. An electrical system that provides proper safety will not always provide a proper logic reference—in fact, the goals of safety and power quality are sometimes in disagreement. Naturally, when a choice must be made, safety has to take precedence.

Since the difference between a binary zero and a binary one is represented in computers by voltage differences of sometimes less than 3V, it is entirely possible for unstable power conditions caused by poor grounding or electrical noise to cause all kinds of intermittent system problems. Some power and grounding advocates estimate that more than 80 percent of unexplained computer glitches can be traced to power quality. Most of us blame Microsoft.

Modern switching power supplies are somewhat isolated from power quality issues, but any high-performance system will always benefit from a well-designed power environment. In mainframes, proprietary PBXes, and other expensive computing platforms, the grounding of the system is never left to chance. The electronics and frames of these systems are always provided with a dedicated ground that does not depend on the safety grounds supplied with the electrical feed.

Regardless of how much you are willing to invest in grounding, when you specify the electrical supply to any PBX, ensure that the electrical circuit is completely dedicated to your system (as discussed in the next section) and that an insulated, isolated grounding conductor is provided. This can be expensive to provision, but it will contribute greatly to a quality power environment for your system.‖

It is also vital that each and every peripheral you connect to your system be connected to the same electrical receptacle (or, more specifically, the same ground reference). This will cut down on the occurrence of ground loops, which can cause anything from buzzing and humming noises to damaged or destroyed equipment.

Electrical Circuits

If you've ever seen the lights dim when an electrical appliance kicks in, you've seen the effect that a high-energy device can have on an electrical circuit. If you were to look at the effects of a multitude of such devices, each drawing power in its own way, you would see that the harmonically perfect 50 or 60 Hz sine wave you may think you're getting with your power is anything but. Harmonic noise is extremely common on electrical circuits , and it can wreak havoc on sensitive electronic equipment. For a PBX, these problems can manifest as audio problems, logic errors, and system instability.

Ideally, you should never install a server on an electrical circuit that is shared with other devices. There should be only one outlet on the circuit, and you should connect only your telephone system (and associated peripherals) to it. The wire (including the ground) should be run unbroken directly back to the electrical panel. The grounding conductor should be insulated and isolated. There are far too many stories of photocopiers, air conditioners, and vacuum cleaners wreaking havoc with sensitive electronics to ignore this rule of thumb.

 The electrical regulations in your area must always take precedence over any ideas presented here. If in doubt, consult a power quality expert in your area on how to ensure that you adhere to electrical regulations. Remember, electrical regulations take into account the fact that human safety is far more important than the safety of the equipment.

The Equipment Room

Environmental conditions can wreak havoc on systems, and yet it is quite common to see critical systems deployed with little or no attention given to these matters. When the system is installed, everything works well, but after as little as six months, components begin to fail. Talk to anyone with experience in maintaining servers and systems, and it becomes obvious that attention to environmental factors can play a significant role in the stability and reliability of systems.

Humidity

Simply put, humidity is water in the air. Water is a disaster for electronics for two main reasons: 1) water is a catalyst for corrosion, and 2) water is conductive enough that it

---

‖ On a hobby system, this is probably too much to ask, but if you are planning on using Asterisk for anything important, at least be sure to give it a fighting chance; don't put anything like air conditioners, photocopiers, laser printers, or motors on the same circuit. The strain such items place on your power supply will shorten its life expectancy.

can cause short circuits. Do not install any electronic equipment in areas of high humidity without providing a means to remove the moisture.

Temperature

Heat is the enemy of electronics. The cooler you keep your system, the more reliably it will perform, and the longer it will last. If you cannot provide a properly cooled room for your system, at a minimum ensure that it is placed in a location that ensures a steady supply of clean, cool air. Also, keep the temperature steady. Changes in temperature can lead to condensation and other damaging changes.

Dust

An old adage in the computer industry holds that dust bunnies inside of a computer are lucky. Let's consider some of the realities of dust bunnies:

- Significant buildup of dust can restrict airflow inside the system, leading to increased levels of heat.
- Dust can contain metal particles, which, in sufficient quantities, can contribute to signal degradation or shorts on circuit boards.

Put critical servers in a filtered environment, and clean out dust bunnies on a regular schedule.

Security

Server security naturally involves protecting against network-originated intrusions, but the environment also plays a part in the security of a system. Telephone equipment should always be locked away, and only persons who have a need to access the equipment should be allowed near it.

Telephony Hardware

If you are going to connect Asterisk to any traditional telecommunications equipment, you will need the correct hardware. The hardware you require will be determined by what it is you want to achieve.

Connecting to the PSTN

Asterisk allows you to seamlessly bridge circuit-switched telecommunications networks[#] with packet-switched data networks.[*] Because of Asterisk's open architecture (and open source code), it is ultimately possible to connect any standards-compliant

---

[#] Often referred to as *TDM networks*, due to the Time Division Multiplexing used to carry traffic through the PSTN.

interface hardware. The selection of open source telephony interface boards is currently limited, but as interest in Asterisk grows, that will rapidly change.[†] At the moment, one of the most popular and cost-effective ways to connect to the PSTN is to use the interface cards that evolved from the work of the Zapata Telephony Project (*http://www.zapatatelephony.org*).

Analog interface cards

Unless you need a lot of channels (or a have lot of money to spend each month on telecommunications facilities), chances are that your PSTN interface will consist of one or more analog circuits, each of which will require a Foreign eXchange Office (FXO) port.

Digium, the company that sponsors Asterisk development, produces analog interface cards for Asterisk. Check out its web site for its extensive line of analog cards, including the venerable TDM400P, the latest TDM800P, and the high-density TDM2400P. As an example, the TDM800P is an eight-port base card that allows for the insertion of up to two daughter cards, which each deliver either four FXO or four FXS ports.[‡] The TDM800P can be purchased with these modules preinstalled, and a hardware echo-canceller can be added as well. Check out Digium's web site (*http://www.digium.com*) for more information about these cards.

Other companies that produce Asterisk-compatible analog cards include:

- Rhino (*http://www.channelbanks.com*)
- Sangoma (*http://www.sangoma.com*)
- Voicetronix (*http://www.voicetronix.com*)
- Pika Technologies (*http://www.pikatechnologies.com*)

These are all well-established companies that produce excellent products.

Digital interface cards

If you require more than 10 circuits, or require digital connectivity, chances are you're going to be in the market for a T1 or E1 card.[§] Bear in mind, though, that the monthly charges for a digital PSTN circuit vary widely. In some places, as few as five circuits can justify a digital circuit; in others, the technology may never be cost-justifiable. The more

---

[*] Popularly called VoIP networks, although Voice over IP is not the only method of transmitting voice over packet networks (Voice over Frame Relay was very popular in the late 1990s).

[†] The evolution of inexpensive, commodity-based telephony hardware is only slightly behind the telephony software revolution. New companies spring up on a weekly basis, each one bringing new and inexpensive standards-based devices into the market.

[‡] FXS and FXO refer to the opposing ends of an analog circuit. Which one you need will be determined by what you want to connect to. Chapter 7 discusses these in more detail.

[§] T1 and E1 are digital telephony circuits. We'll discuss them further in Chapter 7.

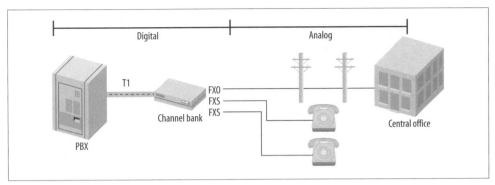

Figure 2-2. One way you might connect a channel bank

competition there is in your area, the better chance you have of finding a good deal. Be sure to shop around.

The Zapata Telephony Project originally produced a T1 card, the Tormenta, that is the ancestor of most Asterisk-compatible T1 cards. The original Tormenta cards are now considered obsolete, but they do still work with Asterisk.

Digium makes several different digital circuit interface cards. The features on the cards are the same; the primary differences are whether they provide T1 or E1 interfaces, and how many spans each card provides. Digium has been producing Zaptel cards for Linux longer than anyone else, as they were deeply involved with the development of Zaptel on Linux, and have been the driving force behind Zaptel development over the years.

Sangoma, which has been producing open source WAN cards for many years, added Asterisk support for its T1/E1 cards a few years ago.[||] Rhino has had T1 hardware for Asterisk for a while now, and there are many other companies that offer digital interface cards for Asterisk as well.

Channel banks

A *channel bank* is loosely defined as a device that allows a digital circuit to be de-multiplexed into several analog circuits (and vice versa). More specifically, a channel bank lets you connect analog telephones and lines into a system across a T1 line. Figure 2-2 shows how a channel bank fits into a typical office phone system.

Although they can be expensive to purchase, many people feel very strongly that the only proper way to integrate analog circuits and devices into Asterisk is through a channel bank. Whether that is true or not depends on a lot of factors, but if you have the budget, they can be very useful.[#] You can often pick up used channel banks on eBay. Look for units from Adtran and Carrier Access Corp. (Rhino makes great channel

---

[||] It should be noted that a Sangoma Frame Relay card played a role in the original development of Asterisk (see *http://linuxdevices.com/articles/AT8678310302.html*); Sangoma has a long history of supporting open source WAN interfaces with Linux.

banks, and they are very competitively priced, but they may be hard to find used on eBay.) Don't forget that you will need a T1 card in order to connect a channel bank to Asterisk.

Other types of PSTN interfaces

Many VoIP gateways exist that can be configured to provide access to PSTN circuits. Generally speaking, these will be of most use in a smaller system (one or two lines). They can also be very complicated to configure, as grasping the interaction between the various networks and devices requires a solid understanding of both telephony and VoIP fundamentals. For that reason, we will not discuss these devices in detail in this book. They are worth looking into, however; popular units are made by Sipura, Grand-stream, Digium, and many other companies.

Another way to connect to the PSTN is through the use of Basic Rate Interface (BRI) ISDN circuits. BRI is a digital telecom standard that specifies a two-channel circuit that can carry up to 144 Kbps of traffic. It is very rarely used in North America, but in Europe it is very widely deployed. Due to the variety of different ways this technology has been implemented, and a lack of testing equipment, we will not be discussing BRI in very much detail in this book. Please note, however, that BRI is very popular in Europe, and Digium has produced the B410P card to address this need.

Connecting Exclusively to a Packet-Based Telephone Network

If you do not need to connect to the PSTN, Asterisk requires no hardware other than a server with a Network Interface Card (NIC).

However, if you are going to be providing music on hold* or conferencing and you have no physical timing source, you will need the *ztdummy* Linux kernel module. *ztdummy* is a clocking mechanism designed to provide a timing source to a system where no hardware timing source exists. Think of it as a kind of metronome to allow the system to mix multiple audio streams in a properly synchronized manner.

Echo Cancellation

One of the issues that can arise if you use analog interfaces on a VoIP system is echo. *Echo* is simply what you say being reflected back to you a short time later. The echo is caused by the far end, but you are the one that hears it. It is a little known fact that echo would be a massive problem in the PSTN were it not for the fact that the carriers employ complex (and expensive) strategies to eliminate it. We will talk about echo a bit more later on, but with respect to hardware we would suggest that you consider adding echo-

---

We use channel banks to simulate a central office. One 24-port channel bank off an Asterisk system can provide up to 24 analog lines—perfect for a classroom or lab.

* Technically, no timing source is needed for music on hold, but it generally works better with one.

cancellation hardware to any card you purchase for use as a PSTN interface. While Asterisk can do some work with echo in software, it does not provide nearly enough power to deal with the problem. Also, echo cancellation in software imposes a load on the processor; hardware echo cancellers built into the PSTN card take this burden away from the CPU.

Hardware echo cancellation can add several hundred dollars to your equipment cost, but if you are serious about having a quality system, invest the extra money now instead of suffering later. Echo problems are not pleasant at all, and your users will hate the system if they experience it.

As of this writing, several software echo cancellers have become available. We have not had a chance to evaluate any of them, but we know that they employ the same algorythems the hardware echo cancellers do. If you have a recently purchased Digium analog card, you can call Digium sales for a keycode to allow its latest software echo canceller to work with your system.[†] There are other software options available for other types of cards, but you will have to look into whether you have to purchase a license to use them.[‡] Keep in mind that there is a performance cost to using software echo cancellers. They will place a measureable load on the CPU that needs to be taken into account when you design a system using these technologies.

Types of Phones

Since the title of this book is *Asterisk: The Future of Telephony*, we would be remiss if we didn't discuss the devices that all of this technology ultimately has to interconnect: telephones!

We all know what a telephone is—but will it be the same five years from now? Part of the revolution that Asterisk is contributing to is the evolution of the telephone, from a simple audio communications device into a multimedia communications terminal providing all kinds of yet-to-be-imagined functions.

As an introduction to this exciting concept, we will briefly discuss the various kinds of devices we currently call "telephones" (any of which can easily be integrated with Asterisk). We will also discuss some ideas about what these devices may evolve into in the future (devices that will also easily integrate with Asterisk).

[†] This software is not part of a normal Asterisk download because Digium has to pay to license it separately. Nevertheless, it has grandfathered it into all of its cards, so it is available for free to anyone who has a Digium analog card that is still under warranty. If you are running a non-Digium analog card, you can purchase a keycode for this software echo canceller from Digium's web site.

[‡] Sangoma also offers free software echo cancellation on their analog cards (up to six channels).

Physical Telephones

Any physical device whose primary purpose is terminating an on-demand audio communications circuit between two points can be classified as a physical telephone. At a minimum, such a device has a handset and a dial pad; it may also have feature keys, a display screen, and various audio interfaces.

This section takes a brief look at the various user (or endpoint) devices you might want to connect to your Asterisk system. We'll delve more deeply into the mechanics of analog and digital telephony in Chapter 7.

Analog telephones

Analog phones have been around since the invention of the telephone. Up until about 20 years ago, all telephones were analog. Although analog phones have some technical differences in different countries, they all operate on similar principles.

 This contiguous connection is referred to as a *circuit*, which the telephone network used to use electromechanical switches to create—hence the term *circuit-switched network*.

When a human being speaks, the vocal cords, tongue, teeth, and lips create a complex variety of sounds. The purpose of the telephone is to capture these sounds and convert them into a format suitable for transmission over wires. In an analog telephone, the transmitted signal is *analogous* to the sound waves produced by the person speaking. If you could see the sound waves passing from the mouth to the microphone, they would be proportional to the electrical signal you could measure on the wire.

Analog telephones are the only kind of phone that are commonly available in any retail electronics store. In the next few years, that can be expected to change dramatically.

Proprietary digital telephones

As digital switching systems developed in the 1980s and 1990s, telecommunications companies developed digital Private Branch eXchanges (PBXes) and Key Telephone Systems (KTSes). The proprietary telephones developed for these systems were completely dependent on the systems to which they were connected and could not be used on any other systems. Even phones produced by the same manufacturer were not cross-compatible (for example, a Nortel Norstar set will not work on a Nortel Meridian 1 PBX). The proprietary nature of digital telephones limits their future. In this emerging era of standards-based communications, they will quickly be relegated to the dustbin of history.

The handset in a digital telephone is generally identical in function to the handset in an analog telephone, and they are often compatible with each other. Where the digital phone is different is that inside the telephone, the analog signal is sampled and converted into a digital signal—that is, a numerical representation of the analog waveform. We'll leave a detailed discussion of digital signals until Chapter 7; for now, suffice it to say that the primary advantage of a digital signal is that it can be transmitted over limitless distances with no loss of signal quality.

The chances of anyone ever making a proprietary digital phone directly compatible with Asterisk are slim, but companies such as Citel (*http://www.citel.com*)[§] have created gateways that convert the proprietary signals to Session Initiation Protocol (SIP).[‖]

ISDN telephones

Prior to VoIP, the closest thing to a standards-based digital telephone was an ISDN-BRI terminal. Developed in the early 1980s, ISDN was expected to revolutionize the telecommunications industry in exactly the same way that VoIP promises to finally achieve today.

 There are two types of ISDN: *Primary Rate Interface* (PRI) and *Basic Rate Interface* (BRI). PRI is commonly used to provide trunking facilities between PBXes and the PSTN, and is widely deployed all over the world. BRI is not at all popular in North America, but is common in Europe.

While ISDN was widely deployed by the telephone companies, many consider the standard to have been a flop, as it generally failed to live up to its promises. The high costs of implementation, recurring charges, and lack of cooperation among the major industry players contributed to an environment that caused more problems than it solved.

BRI was intended to service terminal devices and smaller sites (a BRI loop provides two digital circuits). A wealth of BRI devices have been developed, but BRI has largely been deprecated in favor of faster, less expensive technologies such as ADSL, cable modems, and VoIP.

BRI is still very popular for use in video-conferencing equipment, as it provides a fixed bandwidth link. Also, BRI does not have the type of quality of service issues a VoIP connection might, as it is circuit-switched.

[§] Citel has produced a fantastic product that is limited by the fact that it is too expensive. If you have old proprietary PBX telephones, and you want to use them with your Asterisk system, Citel's technology can do the job, but make sure you understand how the per-port cost of these units stacks up against replacing the old sets with pure VoIP telephones.

[‖] The SIP is currently the most well-known and popular protocol for VoIP. We will discuss it further in Chapter 8.

BRI is still sometimes used in place of analog circuits to provide trunking to a PBX. Whether or not this is a good idea depends mostly on how your local phone company prices the service, and what features it is willing to provide.[#]

IP telephones

IP telephones are heralds of the most exciting change in the telecommunications industry. Already now, standards-based IP telephones are available in retail stores. The wealth of possibilities inherent in these devices will cause an explosion of interesting applications, from video phones to high-fidelity broadcasting devices, to wireless mobility solutions, to purpose-built sets for particular industries, to flexible all-in-one multimedia systems.

The revolution that IP telephones will spawn has nothing to do with a new type of wire to connect your phone to, and everything to do with giving you the power to communicate the way you want.

The early-model IP phones that have been available for several years now do not represent the future of these exciting appliances. They are merely a stepping-stone, a familiar package in which to wrap a fantastic new way of thinking.

The future is far more promising.

Softphones

A *softphone* is a software program that provides telephone functionality on a non-telephone device, such as a PC or PDA. So how do we recognize such a beast? What might at first glance seem a simple question actually raises many. A softphone should probably have some sort of dial pad, and it should provide an interface that reminds users of a telephone. But will this always be the case?

The term *softphone* can be expected to evolve rapidly, as our concept of what exactly a telephone is undergoes a revolutionary metamorphosis.[*] As an example of this evolution, consider the following: would we correctly define popular communication programs such as Instant Messenger as softphones? IM provides the ability to initiate and receive standards-based VoIP connections. Does this not qualify it as a softphone? Answering that question requires knowledge of the future that we do not yet possess. Suffice it to say that while at this point in time, softphones are expected to look and sound like traditional phones, that conception is likely to change in the very near future.

As standards evolve and we move away from the traditional telephone and toward a multimedia communications culture, the line between softphones and physical telephones will become blurred indeed. For example, we might purchase a communica-

---

[#] If you are in North America, give up on this idea, unless you have a lot of patience and money, and are a bit of a masochist.

[*] Ever heard of Skype?

tions terminal to serve as a telephone and install a softphone program onto it to provide the functions we desire.

Having thus muddied the waters, the best we can do at this point is to define what the term *softphone* will refer to in relation to this book, with the understanding that the meaning of the term can be expected to undergo a massive change over the next few years. For our purposes, we will define a softphone as any device that runs on a personal computer, presents the look and feel of a telephone, and provides as its primary function the ability to make and receive full-duplex audio communications (formerly known as "phone calls")[†] through E.164 addressing.[‡]

Telephony Adaptors

A *telephony adaptor* (usually referred to as an ATA, or Analog Terminal Adaptor) can loosely be described as an end-user device that converts communications circuits from one protocol to another. Most commonly, these devices are used to convert from some digital (IP or proprietary) signal to an analog connection that you can plug a standard telephone or fax machine into.

These adaptors could be described as gateways, for that is their function. However, popular usage of the term *telephony gateway* would probably best describe a multiport telephony adaptor, generally with more complicated routing functions.

Telephony adaptors will be with us for as long as there is a need to connect incompatible standards and old devices to new networks. Eventually, our reliance on these devices will disappear, as did our reliance on the modem—obsolescence through irrelevance.

Communications Terminals

Communications terminal is an old term that disappeared for a decade or two and is being reintroduced here, very possibly for no other reason than that it needs to be discussed so that it can eventually disappear again—once it becomes ubiquitous.

First, a little history. When digital PBX systems were first released, manufacturers of these machines realized that they could not refer to their endpoints as telephones— their proprietary nature prevented them from connecting to the PSTN. They were therefore called *terminals*, or *stations*. Users, of course, weren't having any of it. It looked like a telephone and acted like a telephone, and therefore it *was* a telephone. You will still occasionally find PBX sets referred to as terminals, but for the most part they are called telephones.

---

† OK, so you think you know what a phone call is? So did we. Let's just wait a few years, shall we?

‡ E.164 is the ITU standard that defines how phone numbers are assigned. If you've used a telephone, you've used E.164 addressing.

The renewed relevance of the term *communications terminal* has nothing to do with anything proprietary—rather, it's the opposite. As we develop more creative ways of communicating with each other, we gain access to many different devices that will allow us to connect. Consider the following scenarios:

- If I use my PDA to connect to my voicemail and retrieve my voice messages (converted to text), does my PDA become a phone?
- If I attach a video camera to my PC, connect to a company's web site, and request a live chat with a customer service rep, is my PC now a telephone?
- If I use the IP phone in my kitchen to surf for recipes, is that a phone call?

The point is simply this: we'll probably always be "phoning" each other, but will we always be using "telephones" to do so?

Linux Considerations

If you ask anyone at the Free Software Foundation, they will tell you that what we know as Linux is in fact GNU/Linux. All etymological arguments aside, there is some valuable truth to this statement. While the kernel of the operating system is indeed Linux, the vast majority of the utilities installed on a Linux system and used regularly are in fact GNU utilities. "Linux" is probably only 5 percent Linux, possibly 75 percent GNU, and perhaps 20 percent everything else.

Why does this matter? Well, the flexibility of Linux is both a blessing and a curse. It is a blessing because with Linux you can truly craft your very own operating system from scratch. Since very few people ever do this, the curse is in large part due to the responsibility you must bear in determining which of the GNU utilities to install, and how to configure the system.

If this seems overwhelming, do not fear. In the next chapter, we will discuss the selection, installation, and configuration of the software environment for your Asterisk system.

Conclusion

In this chapter, we've discussed all manner of issues that can contribute to the stability and quality of an Asterisk installation. Before we scare you off, we should tell you that many people have installed Asterisk on top of a graphical Linux workstation—running a web server, a database, and who knows what else—with no problems whatsoever.§ How much time and effort you should devote to following the best practices and en-

---

§ Just don't ever install the X-windowing environment (which is anything that delivers a desktop, such as GNOME, KDE, and such). You are almost guaranteed to have audio quality problems, as Asterisk and the GUI will fight for control of the CPU.

gineering tips in this chapter all depends on how much work you expect the Asterisk server to perform, and how much quality and reliability your system must provide. If you are experimenting with Asterisk, don't worry too much; just be aware that any problems you have may not be the fault of the Asterisk system.

What we have attempted to do in this chapter is give you a feel for the kinds of best practices that will help to ensure that your Asterisk system will be built on a reliable, stable platform. Asterisk is quite willing to operate under far worse conditions, but the amount of effort and consideration you decide to give these matters will play a part in the stability of your PBX. Your decision should depend on how critical your Asterisk system will be.

## CHAPTER 3

Installing Asterisk

*I long to accomplish great and noble tasks, but it is my
chief duty to accomplish humble tasks as though they
were great and noble. The world is moved along, not
only by the mighty shoves of its heroes, but also by the
aggregate of the tiny pushes of each honest worker.*

—Helen Keller

In the previous chapter, we discussed preparing a system to install Asterisk. Now it's time to get our hands dirty!

Although a large number of Linux* distributions and PC architectures are excellent candidates for Asterisk, we have chosen to focus on a single distribution in order to maintain brevity and clarity throughout the book. The instructions that follow have been made as generic as possible, but you will notice a leaning toward CentOS directory structure and system utilities. We have chosen to focus on CentOS (arguably, the most popular distro for Asterisk) because its command set, directory structure, and so forth are likely to be familiar to a larger percentage of readers (we have found that many Linux administrators are familiar with CentOS, even if they don't prefer it). This doesn't mean that CentOS is the only choice, or even the best one for you. A question that often appears on the mailing lists is: "Which distribution of Linux is the best to use with Asterisk?" The multitude of answers generally boils down to "the one you like the best."†

---

* And some non-Linux operating systems as well, such as Solaris, *BSD, and OS X. You should note that while people have managed to successfully run Asterisk on these alternative systems, Asterisk was, and continues to be, actively developed for Linux.

† We will be using CentOS Server 4.4 in this book, which we usually install with nothing except the Editors package selected. If you are not sure what distribution to choose, CentOS is an excellent choice. CentOS can be obtained from *http://www.centos.org*.

What Packages Do I Need?

Most Asterisk configurations are composed of three main packages : the main Asterisk program (*asterisk*), the Zapata telephony drivers (*zaptel*), and the PRI libraries (*libpri*). If you plan on a pure VoIP network, the only real requirement is the *asterisk* package, but we recommend installing all three packages; you can choose what modules to activate later. The *zaptel* drivers are required if you are using analog or digital hardware, or if you're using the *ztdummy* driver (discussed later in this chapter) as a timing source. The *libpri* library is optional unless you're using ISDN PRI interfaces, and you may save a small amount of RAM if you don't load it, but we recommend that it be installed in conjunction with the *zaptel* package for completeness.

In the first edition of this book, we recommended that you install the additional *asterisk-sounds* package. This was a separate compressed archive that you would download, extract, and then install. As of Asterisk version 1.4.0, there are now two sets of sounds packages: the Core Sound package and the Extra Sound package. Since Asterisk supports several different audio formats, these packages can be obtained in a number of different sound formats, such as G.729 and GSM. The reason for all of the different formats is that Asterisk can use the sound format that requires the least amount of CPU transcode. For example, if you have a lot of connections coming in on VoIP channels that are running GSM, you would want to have the GSM version of the sound files. You can select one or more sound prompt types in the menuselect screen (discussed later in this chapter). We recommend that you install at least one type of sounds file from both the Core Sound package and Extra Sound package menu items. Since we may make use of some of the Extra Sound files throughout this book, we will assume you have at least one of the formats installed.

Linux Package Requirements

To compile Asterisk, you must have the *GCC compiler* (version 3.x or later) and its dependencies on your system. Asterisk also requires *bison*, a parser generator program that replaces *yacc*, and *ncurses* for CLI functionality. The cryptographic library in Asterisk requires *OpenSSL* and its development packages.

Zaptel requires *libnewt* and its development packages for the *zttool* program (see "Using ztcfg and zttool later in this chapter). If you're using PRI interfaces, Zaptel also requires the *libpri* package (again, even if you aren't using PRI circuits, we recommend that you install *libpri* along with *zaptel*).

If you install the Software Development packages in CentOS, you will have all of these tools. If you are looking to keep things trim, and wish to install the bare minimum to compile Asterisk and its related packages, Table 3-1 will prove useful.

In the following table, the -y switch to the *yum* application means to answer yes to all prompts, and using it will install the application and all dependencies without prompting you. If this is not what you want, omit the -y switch.

If you just want to install all of the above packages in one go, you can specify more than one package on the command line, e.g.:

```
# yum install -y gcc ncurses-devel libtermcap-devel [...]
```

Table 3-1. List of packages required to compile libpri, zaptel, and asterisk

Package name	Installation command	Note	Used by
GCC 3.x	yum install -y gcc	Required to compile zaptel, libpri, and asterisk	libpri, zaptel, asterisk
ncurses-devel	yum install -y ncurses-devel	Required by menuselect	menuselect
libtermcap-devel	yum install -y libtermcap-devel	Required by asterisk	asterisk
Kernel Development Headers	yum install -y kernel-devel	Required to compile zaptel	zaptel
Kernel Development Headers (SMP)	yum install -y kernel-smp-devel	Required to compile zaptel	zaptel
GCC C++ 3.x	yum install -y gcc-c++	Required by asterisk	asterisk
OpenSSL (optional)	yum install -y openssl-devel	Dependency of OSP, IAX2 encryption, res_crypto (RSA key support)	asterisk
newt-devel (optional)	yum install -y newt-devel	Dependency of zttool	zaptel
zlib-devel (optional)	yum install -y zlib-devel	Dependency of DUNDi	asterisk
unixODBC; unixODBC-devel (optional)	yum install -y unixODBC-devel	Dependency of func_odbc, cdr_odbc, res_config_odbc, res_odbc, ODBC_STORAGE	asterisk
libtool (optional; recommended)	yum install -y libtool	Dependency of ODBC-related modules	asterisk
GNU make (version 3.80 or higher) [a]	yum install -y make	Required to compile zaptel and asterisk	asterisk

[a] It is a common problem among new installs on some Linux distriebutons to see *GNU make* versions of 3.79 or lower. Note that Asterisk will no longer build correctly unless you have at least version 3.80 of *GNU make*.

Obtaining the Source Code

The best place to get source code for Asterisk and it's packages is directly from the *http://www.asterisk.org* web site or FTP server.

Release Versus Trunk

The Asterisk code base is under a constant state of change. Developers use a sofware revision tool called Subversion (SVN)[‡] to manage the code base. Subversion allows a communty of developers to collaborate with each other on complex programming projects.

There are two main areas where Asterisk is developed, and these are referred to as the Branch and the Trunk. In the Trunk, new features, architectural changes, and any of the brand-new stuff that is going on is performed. This place in the code base contains all the new toys, but at any time can be in a nonworking state, and is absolutely forbidden from production use (see figure).

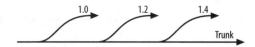

Just like a tree, a Trunk will have Branches. These Branches have the major revision numbers such as 1.0, 1.2, and 1.4 (in the future we will likely see 1.6, 1.8, 1.8.2, 1.8.4. 1.8.6, 1.8.8. 1.8.8.2...um...etc...).[§] Within the Branch there are no major architectural changes or new features—simply bug and security fixes. In a production environment, stability is far more important than feature evolution.

Roughly every 14 months (although Asterisk does not follow a formal release timeline like many commercial software packages), a version of Asterisk is released intended for use in production environments. The first version of Asterisk was 1.0, which was released at the very first AstriCon in Atlanta in September of 2004. Asterisk 1.2 was released at IP4IT in November 2005, and Asterisk 1.4 was released in December of 2006.

Obtaining Asterisk Source Code

The easiest way to obtain the most recent *release* is through the use of the program *wget*.

[‡] Subversion is an excellent code management system, available at *http://subversion.tigris.org/*. It also has an equally excellent Creative Commons released book, *Version Control with Subversion*, by Ben Collins Sussman et al. (O'Reilly), available online at *http://svnbook.red-bean.com/*.

[§] As of the release date of this book, there has been no determination that the next Asterisk release will be 1.6. It could just as easily be 2.0. Therefore, when discussing new features, you'll see us talk about what's in Trunk or what will be in the next release—without mentioning the specific version.

Note that we will be making use of the */usr/src/* directory to extract and compile the Asterisk source, although some system administrators may prefer to use */usr/local/src*. Also be aware that you will need *root* access to write files to the */usr/src/* directory and to install Asterisk and its associated packages.

> See Chapter 13 for information on running Asterisk as non-*root*. All security professionals will recommend that you run your daemons as a non-*root* user in case there are security vulnerabilities in the software. This helps to lower (but obviously does not eliminate) the risk of someone compromising the *root* user.

To obtain the latest release source code via *wget*, enter the following commands on the command line:

```
# cd /usr/src/
# wget http://downloads.digium.com/pub/asterisk/asterisk-1.4-current.tar.gz
# wget http://downloads.digium.com/pub/libpri/libpri-1.4-current.tar.gz
# wget http://downloads.digium.com/pub/zaptel/zaptel-1.4-current.tar.gz
```

> The latest versions of the *asterisk*, *libpri*, and *zaptel* packages may not necessarily be the same version number.

Alternatively, during development and testing you will probably want to work with the latest branch. To check it out from SVN, run:

```
# svn co http://svn.digium.com/svn/asterisk/branches/1.4 asterisk-1.4
```

If you retrieved the described source code via the release files on the Digium FTP server, then extract the files as described in the next section before continuing on with compiling.

Extracting the Source Code

The packages you downloaded from the FTP server are compressed archives containing the source code; thus, you will need to extract them before compiling. If you didn't download the packages to */usr/src/*, either move them there now or specify the full path to their location. We will be using the GNU *tar* application to extract the source code from the compressed archive. This is a simple process that can be achieved through the use of the following commands:

```
# cd /usr/src/
# tar zxvf zaptel-1.4-current.tar.gz
# tar zxvf libpri-1.4-current.tar.gz
# tar zxvf asterisk-1.4-current.tar.gz
```

In bash (and other shell systems which support it), you can use an extremely handy feature called Tab completion. This will allow you to type part of a filename and have the rest of it completed automatically. For example, if you type `tar zxvf zap<tab>` that will complete the full *zaptel* filename for you. If more than one filename matches the pattern and you hit Tab twice, it will list the files matching that pattern.

These commands will extract the packages and source code to their respective directories. When you extract the *asterisk-1.4-current.tar.gz* file, you will find that the file will extract to the current version of Asterisk, i.e. `asterisk-1.4.4`.

 It's always a good idea to keep the source code of the most recently *working* version of a package in case you have to "roll back" out of a new bug introduced, or some other strange behavior you can't solve immediately.

Menuselect

In the 1.4.0 version of Asterisk and its related packages, a new build system, *autoconf*, was implemented. This has changed the build process slightly, but has given more flexibilty to control what modules are being built at build time. This has an advantage in that we only have to build the modules we want and need instead of building everything.

Along with the new build system, a new menu-based selection system was introduced, courtesy of Russell Bryant. This new system permits a finer-grained selection to which modules are built before compiling the software and no longer requires the user to edit *Makefiles*. So instead of discussing how to use menuselect in every "Compiling ..." section, we will discuss it here, so when you see `make menuselect` you will understand what to do once inside the menuselect configuration screen.

In Figure 3-1, we see the opening menuselect screen for the Asterisk software. Other packages will look extremely similar, but with less options. We can navigate up and down the list using the arrow keys. We can select one of the menu options by pressing Enter or by using the right arrow key. The left arrow key can be used to go back.

Figure 3-2 shows a list of possible dialplan applications that can be built for use in Asterisk. Modules to be built are marked as [*]. A module is marked as not being built by [ ]. Modules that have XXX in front of them are missing a package dependency which must be satisfied before it will be available to be built. In Figure 3-2, we can see that the *app_flash* module cannot be built due to a missing dependency of Zaptel (i.e., the Zaptel module has not been built and installed on the system since the last time ./configure was run). If you have satisfied a dependency since the last time you

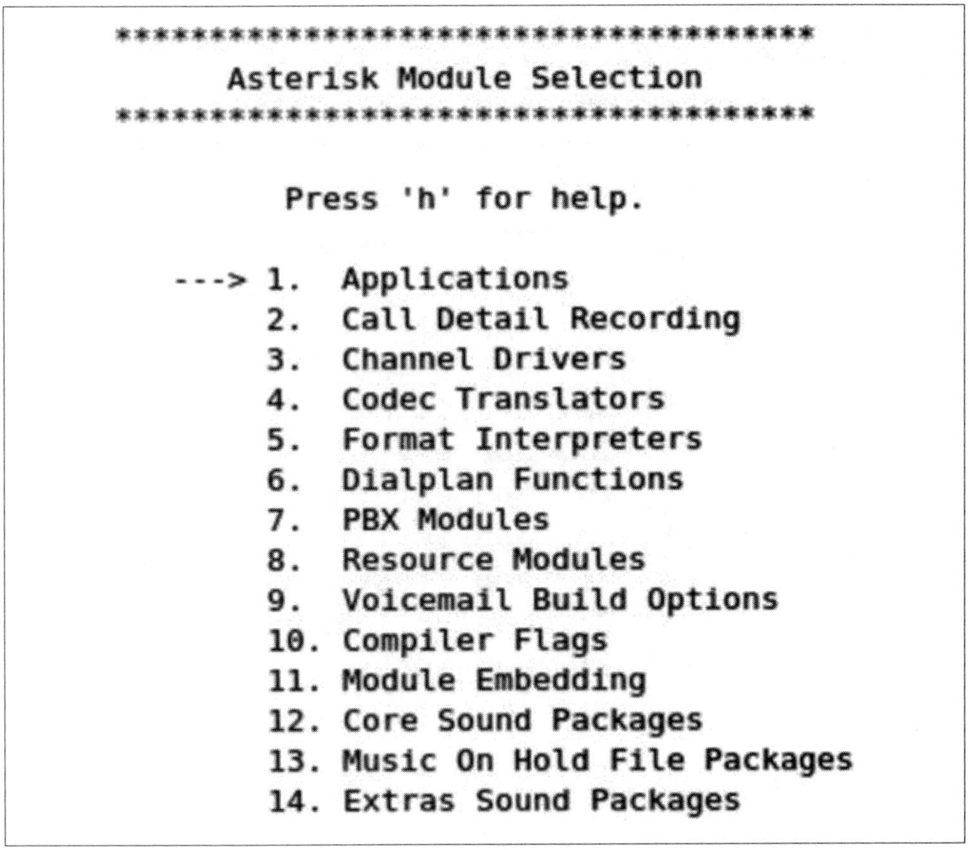

```
****************************************
        Asterisk Module Selection
****************************************

              Press 'h' for help.

   ---> 1.   Applications
        2.   Call Detail Recording
        3.   Channel Drivers
        4.   Codec Translators
        5.   Format Interpreters
        6.   Dialplan Functions
        7.   PBX Modules
        8.   Resource Modules
        9.   Voicemail Build Options
       10.   Compiler Flags
       11.   Module Embedding
       12.   Core Sound Packages
       13.   Music On Hold File Packages
       14.   Extras Sound Packages
```

Figure 3-1. Sample menuselect screen

ran `./configure`, then run it again, and rerun menuselect. Your module should now be available for building.

After you have finished making changes to menuselect, type x to save and quit. q will also quit out of menuselect, but it will not save the changes. If you make changes and type q, your changes may be lost!

Compiling Zaptel

Figure 3-3 shows the layers of interaction between Asterisk and the Linux kernel with respect to hardware control. On the Asterisk side is the Zapata channel module, *chan_zap*. Asterisk uses this interface to communicate with the Linux kernel, where the drivers for the hardware are loaded.

The Zaptel interface is a kernel loadable module that presents an abstraction layer between the hardware drivers and the Zapata module in Asterisk. It is this concept that allows the device drivers to be modified without any changes being made to the Asterisk

```
**********************************
    Asterisk Module Selection
**********************************

        Press 'h' for help.

    [*] 15. app_disa
    [*] 16. app_dumpchan
    [*] 17. app_echo
    [*] 18. app_exec
    [*] 19. app_externalivr
    [*] 20. app_festival
    XⓍX 21. app_flash
    [*] 22. app_followme
    [*] 23. app_forkcdr



    Flash channel application
    Depends on: zaptel
```

Figure 3-2. List of modules to be built

source itself. The device drivers are used to communicate with the hardware directly and to pass the information between Zaptel and the hardware.

> While Asterisk itself compiles on a variety of platforms, the Zaptel drivers are Linux-specific—they are written to interface directly with the Linux kernel. There is a project at *http://www.solarisvoip.com* that provides Zaptel support for Solaris. There is also a project that is working to provide Zapata drivers for BSD, located at *http://www.voip-info.org/tiki-index.php?page=FreeBSD+zaptel*.

First we will discuss the *ztdummy* driver, used on systems that require a timing interface but that do not have hardware. Then we will look at compiling and installing the drivers. (The configuration of Zaptel drivers will be discussed in the next chapter.)

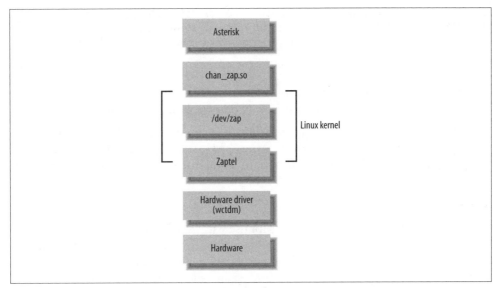

Figure 3-3. *Layers of device interaction with Asterisk*

Before compiling the Zaptel drivers on a system running a Linux 2.4 kernel, you should verify that */usr/src/* contains a symbolic link named *linux-2.4* pointing to your kernel source. If the symbolic link doesn't exist, you can create it with the following command (assuming you've installed the source in */usr/src/*):

```
# ln -s /usr/src/'uname -r' /usr/src/linux-2.4
```

Computers running Linux 2.6 kernel-based distributions do not usually require the use of the symbolic link, as these distributions will search for the kernel build directory automatically. However, if you've placed the build directory in a nonstandard place (i.e., somewhere other than */lib/modules/ <kernel version> /build/*), you will require the use of the symbolic link.

While Asterisk and the other related packages run on Linux 2.4.x kernels, development is done first and foremost on 2.6.x kernels and support for 2.4.x kernels is not guarenteed in the future.

The ztdummy Driver

In Asterisk, certain applications and features require a timing device in order to operate (Asterisk won't even compile them if no timing device is found). All Digium PCI hardware provides a 1 kHz timing interface that satisfies this requirement. If you lack the PCI hardware required to provide timing, the *ztdummy* driver can be used as a timing device. On Linux 2.4 kernel-based distributions, *ztdummy* must use the clocking provided by the UHCI USB controller.

Many older systems (and some newer ones) use an OHCI USB controller chip, which is incompatible with *ztdummy*. However, if you're using a 2.6 kernel there is no need to worry about which USB controller chip your system has.

The driver looks to see that the *usb-uhci* module is loaded and that the kernel version is at least 2.4.5. Older kernel versions are incompatible with *ztdummy*.

On a 2.6 kernel-based distribution, *ztdummy* does not require the use of the USB controller. (As of v2.6.0, the kernel now provides 1 kHz timing[||] with which the driver can interface; thus, the USB controller hardware requirement is no longer necessary.)

The Zapata Telephony Drivers

Compiling the Zapata telephony drivers for use with your Digium hardware is straight-forward; however, the method employed between the 1.2 and 1.4 versions is slightly different due to the new build environment. First we need to run `./configure` in order to determine what applications and libraries are installed on the system. This will ensure that everything Zaptel needs is installed. The following commands will build Zaptel and its modules:

```
# cd /usr/src/zaptel-version

# make clean
# ./configure
# make menuselect
# make
# make install
```

While running `make clean` is not always necessary, it's a good idea to run it before recompiling any of the modules, as it will remove the compiled binary files from within the source code directory. You can also use it to clean up after installing if you don't like to leave the compiled binaries floating around. Note that this removes the binaries only from the source directory, not from the system.

In addition to the executables, `make clean` also removes the intermediary files (i.e., the object files) after compilation. You don't need them occupying space on your hard drive.

If you're using a system that makes use of the */etc/rc.d/init.d/* or */etc/init.d/* directories (such as CentOS and other Red Hat-based distros), you may wish to run the `make con`

[||] Note that this is configurable in the kernel, so it is possible certain distributions may not have this set to 1,000 Hz; CentOS, however, does have this set at the correct frequency.

---

fig command as well. This will install the startup scripts and configure the system, using the chkconfig command to load the *zaptel* module automatically at startup:

```
# make config
```

The Debian equivalent of chkconfig is update-rc.d.

While Digium only officially supports Zaptel on Linux, several projects to port Zaptel to other platforms should be noted:

- Solaris (*http://www.solarisvoip.com*)
- BSD (*http://lists.digium.com/mailman/listinfo/asterisk-bsd*)

Using ztcfg and zttool

Two programs installed along with Zaptel are *ztcfg* and *zttool*. The *ztcfg* program is used to read the configuration in */etc/zaptel.conf* to configure the hardware. The *zttool* program can be used to check the status of your installed hardware. For instance, if you are using a T1 card and there is no communication between the endpoints, you will see a red alarm. If everything is configured correctly and communication is possible, you should see an "OK." The *zttool* application is also useful for analog cards, because it tells you their current state (configured, off-hook, etc.). The use of these programs will be explored further in the next chapter.

The *libnewt* libraries and their development packages (*newt-devel* on Red Hat-based distributions) must be installed for *zttool* to be compiled.

The *ztcfg* and *zttool* applications, along with other useful utilities, are located under the *Utilities* section of the Zaptel menuselect screen.

Compiling libpri

The *libpri* libraries do not make use of the autoconf build environment or the menu-select feature as they are unnecessary; thus, the installation is simplified. *libpri* is used by various makers of Time Division Multiplexing (TDM) hardware, but even if you don't have the hardware installed, it is safe to compile and install this library. You must compile and install *libpri* before Asterisk, as it will be detected and used when Asterisk is compiled. Here are the commands (replace *version* with your version of *libpri*):

```
# cd /usr/src/libpri-version
```

```
# make clean
# make
# make install
```

Compiling Asterisk

Once you've compiled and installed the *zaptel* and *libpri* packages (if you need them), you can move on to Asterisk. This section walks you through a standard installation and introduces some of the alternative `make` arguments that you may find useful.

Standard Installation

Asterisk is compiled with *gcc* through the use of the GNU *make* program. To get started compiling Asterisk, simply run the following commands (replace **version** with your version of Asterisk):

```
# cd /usr/src/asterisk-version

# make clean
# ./configure
# make menuselect
# make install
# make samples
```

Be aware that compile times will vary between systems. On a current-generation processor, you shouldn't need to wait more than five minutes. At AstriCon (*http://www.astricon.net*), someone reported successfully compiling Asterisk on a 133 MHz Pentium, but it took approximately five hours. You do the math.

Run the `make samples` command to install the default configuration files. Installing these files (instead of configuring each file manually) will allow you to get your Asterisk system up and running much faster. Many of the default values are fine for Asterisk. Files that require editing will be explained in future chapters.

> If you already have configuration files installed in */etc/asterisk/* when you run the `make samples` command, *.old* will be appended to the end of each of your current configuration files, for example, *extensions.conf* will be renamed *extensions.conf.old*. Be careful, though, because if you run `make samples` more than once you will overwrite your original configuration files!
>
> The sample configuration files can also be found in the *configs/* subdirectory within your Asterisk *sources* directory.

If you're using a system that makes use of the */etc/rc.d/init.d/* or */etc/init.d/* directories, you may wish to run the `make config` command as well. This will install the startup scripts and configure the system (through the use of the `chkconfig` command) to execute Asterisk automatically at startup:

```
# make config
```

Alternative make Arguments

There are several other make arguments that you can pass at compile time. While some of these will be discussed here, the remainder are used internally within the file and really have no bearing or use for the end user. (Of course, new functions may have been added, so be sure to check the *Makefile* for other options.)

Let's take a look at some useful make arguments.

make clean

The make clean command is used to remove the compiled binaries from within the source directory. This command should be run before you attempt to recompile or, if space is an issue, if you would like to clean up the files.

make distclean

The make distclean command is used to remove the compiled binaries and to clean the source directory back to its original state after being extracted from the compressed archive.

make update

The make update command is used to update the existing code from the Digium SVN server. If you downloaded the source code from the FTP server, you will receive a notice stating so.

make webvmail

The Asterisk Web Voicemail script is used to give a graphical interface to your voicemail account, allowing you to manage and interact with your voicemail remotely from a web browser.

When you run the make webvmail command, the Asterisk Web Voicemail script will be placed into the *cgi-bin/* directory of your HTTP daemon. If you have specific policies with respect to security, be aware that it uses a setuid root Perl script. This command will install only on a CentOS or Fedora box, as other distributions may have different paths to their *cgi-bin/* directories. (This, of course, can be changed by editing the HTTP_CFGDIR variable in the *Makefile* at line 133 at the time of this writing.)

make progdocs

The make progdocs command will create documentation using the *doxygen* software from comments placed within the source code by the developers. You must have the appropriate *doxygen* software installed on your system in order for this to work. Note that *doxygen* assumes that the source code is well documented, which, sadly, is not always the case, although much work was published since the first edition of this book! The information contained within the *doxygen* system will be useful only to developers.

make config

The `make config` command will install Red Hat-style initialization scripts, if the */etc/rc.d/init.d* or */etc/init.d* directories are found to exist. If they do exist, the scripts are installed with file permissions equal to `755`. If the script detects that */etc/rc.d/init.d/* exists, the `chkconfig --add asterisk` command will also be run to cause Asterisk to be started automatically at boot time. This is not the case, however, with distributions that only use the */etc/init.d/* directory. Running `make config` will not do anything to an already running Asterisk process, or start one if it's not running.

This script currently is really only useful on a Red Hat-based system, although initialization scripts are available for other distributions (such as Gentoo, Mandrake, and Slackware) in the *./contrib./init.d/* directory of your Asterisk source directory.

Using Precompiled Binaries

While the documented process of installing Asterisk expects you to compile the source code yourself, there are Linux distributions (such as Debian) that include precompiled Asterisk binaries. Failing that, you may be able to install Asterisk with the package managers that those distributions of Linux provide (such as *apt-get* for Debian and *portage* for Gentoo).[#] However, you may also find that many of these prebuilt binaries are quite out of date and do not follow the same furious development cycle as Asterisk.

Finally, there do exist basic, precompiled Asterisk binaries that can be downloaded and installed in whatever Linux distribution you have chosen. However, the use of precompiled binaries doesn't really save much time, and we have found that compiling Asterisk with each install is not a very cumbersome task. We believe that the best way to install Asterisk is to compile from the source code, so we won't discuss prebuilt binaries very much in this book—and besides, don't you want to be l33t?[*] In the next chapter, we'll look at how to initially configure Asterisk and several kinds of channels.

Installing Additional Prompts

Additional prompts are installed via the *menuselect* application in your Asterisk source directory. There are three sets of audio packages: Core Sound, Extra Sound, and Music On Hold File. Each set of packages is broken down into different formats (and the Core Sound packages are available in multiple languages). Using the *menuselect* application,

---

[#] Gentoo doesn't actually use a precompiled binary, but rather pulls the source from a repository, and builds and installs the software using its own package management system. But the version you get is still dependant upon the maintainers packaging it for you, when you could simply build it yourself!

[*] l33t is a funny way of saying "elite," known as *leetspeak* (computer slang). Even more funny is a well-written, serious article by Microsoft about leetspeak at *http://www.microsoft.com/athome/security/children/ leetspeak.mspx*.

you can select combinations of audio packages for use in your environment. Some of the formats available include:

- WAV
- μlaw
- alaw
- GSM
- G.729
- G.722 (wideband, 16-bit)

As of this writing, the Core Sound packages are available in the following languages:

- English
- Spanish
- French

 Selecting any sounds in menuselect will cause the system to download the files from the Digium FTP server upon install. The size of these files ranges anywhere from 2 MB to 27 MB, so be aware of this when installing offline, or on slow and expensive links.

Other Useful Add-ons

The *asterisk-addons* package contains code to allow the storage of Call Detail Records (CDRs) to a MySQL database. There is also code that allows Asterisk to natively play MP3s (which we don't recommend unless you have a powerful system with very few phones on it). Some folks may also be interested in the interpreter that allows you to load Perl code into memory for the life of an Asterisk process (which can be very helpful if you have a large number of AGI calls to the Perl interpreter). Programs are placed into *asterisk-addons* when there are licensing issues preventing them from being implemented directly into the Asterisk source code, or when they are not considred mature enough to be integrated with Asterisk.

The *http://ftp.digium.com/pub/asterisk/g729/* directory contains the code and registration program for the proprietary G.729A codec. If you install the g729 sounds packages, Asterisk will be able to communicate with devices that natively support the G.729A codec, but will not be able to transcode between other codecs and G.729A until a license is obtained to use it.

Common Compiling Issues

There are many common compiling issues that users often run into. Here are some of the more common problems, and how to resolve them.

Asterisk

First, let's take a look at some of the errors you may encounter when running the *configure* script.

configure: error: no acceptable C compiler found in $PATH

If you receive the following error while attempting to run the *configure* script, you must install the *gcc* compiler and its dependencies:

```
configure: error: no acceptable C compiler found in $PATH
```

The following packages are required for *gcc*:

- *gcc*
- *cpp*
- *glibc-headers*
- *glibc-devel*
- *glibc-kernheaders*

These can be installed manually, by copying the files off of your distribution disks, or through the *yum* package manager, with the command yum `install gcc`.

configure: error: C++ preprocessor "/lib/cpp" fails sanity check

The following error will be displayed if no C++ preprocessor is found installed on the system. You must install the gcc-c++ package and its dependencies:

```
configure: error: C++ preprocessor "/lib/cpp" fails sanity check
```

The following packages are required for the *gcc-c++* preprocessor; installed by running yum `install gcc-c++`:

- *gcc-c++*
- *libstdc++-devel*

configure: error: *** termcap support not found

The following error may be encountered during initialization of the *configure* script if the *libtermcap-devel* package is not installed:

```
configure: error: *** termcap support not found
```

The following file is required in order to compile Asterisk; it can be installed with the yum `install libtermcap-devel` command:

- *libtermcap-devel*

Zaptel

You may also run into errors when compiling Zaptel. Here are some of the most commonly occurring problems, and what to do about them. If your error is not listed below, see the previous section as your error may be covered there.

make: cc: Command not found

You will receive the following error if you attempt to build Zaptel without the *gcc* compiler installed:

```
make: cc: Command not found
make: *** [gendigits.o] Error 127
```

Be sure to install *gcc* and its dependencies. For more information, see "configure: error: no acceptable C compiler found in $PATH in the previous section.

FATAL: Module wctdm/fxs/fxo not found

The TDM400P cards require the PCI bus to be version 2.2. If you attempt to load the Zapata telephony drivers with an older version, you may get the following errors:

- When attempting to load the *wctdm* driver, you may see this error:

    ```
    FATAL: Module wctdm not found
    ```

- When attempting to load the *wctdm* or *wcfxo* driver, you may see an error such as this:

    ```
    ZT_CHANCONFIG failed on channel 1: No such device or address (6)
    FATAL: Module wctdm not found
    ```

The only way to resolve these errors is to use a newer motherboard that supports PCI version 2.2:

> You may also encounter these errors if the power has not been attached to the Molex connector found on the TDM400P card.

Unresolved symbol link when loading ztdummy

The *ztdummy* driver requires that a UHCI USB controller be available on Linux 2.4 kernels (the USB controller is not a requirement on Linux 2.6 kernels, because they are capable of generating the 1 kHz timing reference). There exists a secondary kind of controller, known as OHCI, which is not compatible with the *ztdummy* driver. If the UHCI USB controller is not accessible on Linux 2.4 kernels, the following error will occur:

```
/lib/modules/2.4.22/misc/ztdummy.o: /lib/modules/2.4.22/misc/ztdummy.o: unresolved
symbol unlink_td
/lib/modules/2.4.22/misc/ztdummy.o: /lib/modules/2.4.22/misc/ztdummy.o: unresolved
```

```
symbol alloc_td
/lib/modules/2.4.22/misc/ztdummy.o: /lib/modules/2.4.22/misc/ztdummy.o: unresolved
symbol delete_desc
/lib/modules/2.4.22/misc/ztdummy.o: /lib/modules/2.4.22/misc/ztdummy.o: unresolved
symbol uhci_devices
/lib/modules/2.4.22/misc/ztdummy.o: /lib/modules/2.4.22/misc/ztdummy.o: unresolved
symbol uhci_interrupt
/lib/modules/2.4.22/misc/ztdummy.o: /lib/modules/2.4.22/misc/ztdummy.o: unresolved
symbol fill_td
/lib/modules/2.4.22/misc/ztdummy.o: /lib/modules/2.4.22/misc/ztdummy.o: unresolved
symbol insert_td_horizontal
/lib/modules/2.4.22/misc/ztdummy.o: insmod /lib/modules/2.4.22/misc/ztdummy.o failed
/lib/modules/2.4.22/misc/ztdummy.o: insmod ztdummy failed
```

You can verify that you have the correct style of USB controller and its associated drivers with the lsmod command:

```
# lsmod
Module                Size  Used by
usb_uhci             26412  0
usbcore             79040  1 [hid usb-uhci]
```

As you can see in the example above, you are looking to make sure that the *usbcore* and *usb_uhci* modules are loaded. If these modules are not loaded, be sure that USB has been activated within your BIOS and that the modules exist.

If the USB drivers are not loaded, you can still check which type of USB controller you have with the dmesg command:

```
# dmesg | grep -i usb
```

To verify that you indeed have a UHCI USB controller, look for the following lines:

```
uhci_hcd 0000:00:04.2: new USB bus registered, assigned bus number 1
hub 1-0:1.0: USB hub found
uhci_hcd 0000:00:04.3: new USB bus registered, assigned bus number 2
hub 2-0:1.0: USB hub found
```

Depmod errors during compilation

If you experience depmod errors during compilation, you more than likely don't have a symbolic link to your Linux kernel sources. If you don't have your Linux kernel sources installed, retrieve the sources for your installed kernel, install them, and create a symbolic link against */usr/src/linux-2.4*. The following is an example of a depmod error:

```
depmod: *** Unresolved symbols in /lib/modules/2.4.22/kernel/drivers/block/
loop.o
```

Loading Asterisk and Zaptel Quickly

If you run make config in the Asterisk or Zaptel source directories, then the initialization scripts used to control Asterisk or Zaptel will be copied to */etc/rc.d/init.d/*. The scripts can be used to easily load and unload Asterisk and Zaptel. They will also run the `chkconfig` command for you so Asterisk and Zaptel will be started automatically upon system boot. The following shows their usage:

```
# service zaptel start
# service asterisk start
```

Each initialization script has several options that can be utilized to control the PBX or the drivers. Tables 3-2 and 3-3 show the commands run by the script as if you had typed them into the command-line interface (CLI) yourself:

Table 3-2. Asterisk initialization script options

service asterisk <option>	Manual equivalent
start	asterisk
stop	killproc asterisk
restart	stop; start
reload	asterisk -rx "reload"
status	ps aux | grep [a]sterisk

Table 3-3. Zaptel initialization script options

service zaptel <option>	Manual equivalent
start	modprobe zaptel;modprobe <module>;/sbin/ztcfg
stop	rmmod ztdummy;rmmod zaptel
restart	stop;start
reload	/sbin/ztcfg

Loading Zaptel Modules Without Scripts

In this section, we'll take a quick look at how to load the *zaptel* and *ztdummy* modules without the CentOS initialization script. The *zaptel* module does not require any configuration if it's being used only for the *ztdummy* module. If you plan on loading the *ztdummy* module as your timing source (and thus, you will not be running any PCI hardware in your system), now is a good time to load both drivers.

Systems Running udevd

In the early days of Linux, the system's */dev/* directory was populated with a list of devices with which the system could potentially interact. At the time, nearly 18,000 devices were listed. That all changed when *devfs* was released, allowing dynamic creation of devices that are active within the system. Some of the recently released distributions have incorporated the *udev* daemon into their systems to dynamically populate */dev/* with device nodes.

To allow Zaptel and other device drivers to access the PCI hardware installed in your system, you must add some rules. Using your favorite text editor, open up your *udevd* rules file. On CentOS, for example, this file is located at */etc/udev/rules.d/50-udev.rules*. Add the following lines to the end of your rules file:

```
# Section for zaptel

device
    KERNEL="zapctl",     NAME="zap/ctl"
    KERNEL="zaptimer",   NAME="zap/timer"
    KERNEL="zapchannel", NAME="zap/channel"
    KERNEL="zappseudo",  NAME="zap/pseudo"
    KERNEL="zap[0-9]*",  NAME="zap/%n"
```

Save the file and reboot your system for the settings to take effect.

 You may not have to actually edit anything in your system, as the Zaptel installation script will try to install the rules for you; however, we have left this here as a reference for those systems that are not automatically configured.

Loading Zaptel

The *zaptel* module must be loaded before any of the other modules are loaded and used. Note that if you will be using the *zaptel* module with PCI hardware, you must config-ure */etc/zaptel.conf* before you load it. (We will discuss how to configure *zaptel.conf* for use with hardware in Chapter 4.) If you are using *zaptel* only to access *ztdummy*, you can load it with the **modprobe** command, as follows:

```
# modprobe zaptel
```

If all goes well, you shouldn't see any output. To verify that the *zaptel* module loaded successfully, use the **lsmod** command. You should be returned a line showing the *zaptel* module and the amount of memory it is using, as in the following:

```
# lsmod | grep zaptel
zaptel               201988  0
```

Loading ztdummy

The *ztdummy* module is an interface to a device that provides timing, which in turn allows Asterisk to provide timing to various applications and functions that require it. Use the **modprobe** command to load the *ztdummy* module after *zaptel* has been loaded:

```
# modprobe ztdummy
```

If *ztdummy* loads successfully, no output will be displayed. To verify that *ztdummy* is loaded and is being used by *zaptel*, use the **lsmod** command. The following output is from a computer running the 2.6 kernel:

```
# lsmod | grep ztdummy
     Module                Size   Used by
     ztdummy               3796   0
     zaptel              201988   1 ztdummy
```

If you happen to be running a 2.4 kernel-based computer, your output from `lsmod` will show that *ztdummy* is using the *usb-uhci* module:

```
# lsmod | grep ztdummy
     Module                Size   Used by
     ztdummy               3796   0
     zaptel              201988   0 ztdummy
     usb-uhci             24524   0 ztdummy
```

Loading libpri Without Script

The *libpri* libraries do not need to be loaded like modules. Asterisk looks for *libpri* at compile time and configures itself to use the libraries if they are found.

Starting Asterisk Without Scripts

Asterisk can be loaded in a variety of ways. The easiest way is to start Asterisk by running the binary file directly from the Linux command-line interface. If you are running a system that uses the *init.d* scripts, you can easily start and restart Asterisk that way as well. However, the preferred way of starting Asterisk is via the *safe_asterisk* script.

Console Commands

The Asterisk binary is, by default, located at *the/usr/sbin/asterisk*. If you run */usr/sbin/asterisk*, it will be loaded as a daemon. There are also a few switches you should be aware of that allow you to (re)connect to the Asterisk CLI, set the verbosity of CLI output, and allow core dumps if Asterisk crashes (for debugging with *gdb*). To explore the full range of options, run Asterisk with the -h switch:

```
# /usr/sbin/asterisk -h
```

Here is a list of the most commonly used options:

-c

> Console. This will start Asterisk as a user process (not as a server), and will connect you to the Asterisk CLI. This option is good when you are debugging your startup parameters, but should not be used for a normal system (if Asterisk is already running, this option will not work and will issue a complaint).

-v

> Verbosity. This is used to set the amount of output for CLI debugging. The more "v"s, the more verbose.

-g

Core dump. If Asterisk were to crash unexpectedly, this would cause a core file to be created for later tracing with *gdb*. You generally do not use this in production, unless you are writing code for Asterisk and want to debug any resulting crashes.

-r

Remote. This is used to reconnect remotely to an already running Asterisk process. (The process is remote from the standpoint of the console connecting to it but is actually a local process on the machine. This has nothing to do with connecting to a remote process over a network using a protocol such as IP, as this is not supported.) This is the most common option and it is what you would use to connect to Asterisk on a system where it is running as a daemon/service that was started by init at boot time.

-x "<CLI command>"

Execute. Using this command in combination with -r allows you to execute a CLI command without having to connect to the CLI and type it manually. An example would be to send a restart, which you would do by typing `asterisk -rx "reload"` from the command line.

Let's look at some examples. If you want to start Asterisk as a user program (because you are tweaking your config and will be starting and stopping it several times), and you want a verbosity level of 3, use the following command:

    # /usr/sbin/asterisk -cvvv

If the Asterisk process is already running (for example, if you have installed Asterisk as part of the init process of the system), use the reconnect switch, like so:

    # /usr/sbin/asterisk -vvvr

If you want Asterisk to dump a core file after a crash, you can use the -g switch when starting Asterisk:

    # /usr/sbin/asterisk -g

To execute a command without connecting to the CLI and typing it (perhaps for use within a script), you can use the -x switch in combination with the -r switch:

    # /usr/sbin/asterisk -rx "restart now"
    # /usr/sbin/asterisk -rx "database show"
    # /usr/sbin/asterisk -rx "sip show peers"

If you are experiencing crashes and would like to output to a debug file, use the following command:

    # /usr/sbin/asterisk -vvvvc | tee /tmp/debug.log

Note that you do not have to use the v switch if you do not want the system to provide detailed output of what is going on. On a busy system, you may not want to get any output, as it can interfere with whatever you are doing on the console.

Directories Used by Asterisk

Asterisk uses several directories on a Linux system to manage the various aspects of the system, such as voicemail recordings, voice prompts, and configuration files. This section discusses the necessary directories, all of which are created during installation and configured in the *asterisk.conf* file.

/etc/asterisk/

The */etc/asterisk/* directory contains the Asterisk configuration files. One file, however —*zaptel.conf*—is located in the */etc/* directory. The Zaptel hardware was originally designed by Jim Dixon of the Zapata Telephony Group as a way of bringing reasonable and affordable computer telephony equipment to the world. Asterisk makes use of this hardware, but any other software can also make use of the Zaptel hardware and drivers. Consequently, the *zaptel.conf* configuration file is not directly located in the */etc/asterisk/* directory.

/usr/lib/asterisk/modules/

The */usr/lib/asterisk/modules/* directory contains all of the Asterisk loadable modules. Within this directory are the various applications, codecs, formats, and channels used by Asterisk. By default, Asterisk loads all of these modules at startup. You can disable any modules you are not using in the *modules.conf* file, but be aware that certain modules are required by Asterisk or are dependencies of other modules. Attempting to load Asterisk without these modules will cause an error at startup.

/var/lib/asterisk

The */var/lib/asterisk/* directory contains the *astdb* file and a number of subdirectories. The *astdb* file contains the local Asterisk database information, which is somewhat like the Microsoft Windows Registry. The Asterisk database is a simple implementation based on v1 of the Berkeley database. The *db.c* file in the Asterisk source states that this version was chosen for the following reason: "DB3 implementation is released under an alternative license incompatible with the GPL. Thus, in order to keep Asterisk licensing simplistic, it was decided to use version 1 as it is released under the BSD license."

The subdirectories within */var/lib/asterisk/* include:

agi-bin/
> The *agi-bin/* directory contains your custom scripts, which can interface with Asterisk via the various built-in AGI applications. For more information about AGI, see Chapter 8.

firmware/

The *firmware/* directory contains firmware for various Asterisk-compatible devices. It currently contains only the *iax/* subdirectory, which holds the binary firmware image for Digium's IAXy.

images/

Applications that communicate with channels supporting graphical images look in the *images/* directory. Most channels do not support the transmission of images, so this directory is rarely used. However, if more devices that support and make use of graphical images are released, this directory will become more relevant.

keys/

Asterisk can use a public/private key system to authenticate peers connecting to your box via an RSA digital signature. If you place a peer's public key in your *keys/* directory, that peer can be authenticated by channels supporting this method (such as the IAX2 channels). The private key is never distributed to the public. The reverse is also true: you can distribute your public key to your peers, allowing you to be authenticated with the use of your private key. Both the public and private keys—ending in the *.pub* and *.key* file extensions, respectively—are stored in the *keys/* directory.

mohmp3/

When you configure Asterisk for Music on Hold, applications utilizing this feature look for their MP3 files in the *mohmp3/* directory. Asterisk is a bit picky about how the MP3 files are formatted, so you should use constant bitrate (CBR) encoding and strip the ID3 tags from your files.

sounds/

All of the available voice prompts for Asterisk reside in the *sounds/* directory. The contents of the basic prompts included with Asterisk are in the *sounds.txt* file located in your Asterisk source code directory. Contents of the additional prompts are located in the *sounds-extra.txt* file in the directory to which you extracted the *asterisk-sounds* package earlier in this chapter.

/var/spool/asterisk/

The Asterisk spool directory contains several subdirectories, including *dictate/, meetme/, monitor/, outgoing/, system/, tmp/,* and *voicemail/* (see Figure 3-4). Asterisk monitors the *outgoing* directory for text files containing call request information. These files allow you to generate a call simply by moving the correctly structured file into the *outgoing/* directory.

Call files being placed into the *outgoing/* directory can contain useful information, such as the Context, Extension, and Priority where the answered call should start, or simply the application and its arguments. You can also set variables and specify an account code for Call Detail Records. More information about the use of call files is presented in Chapter 9.

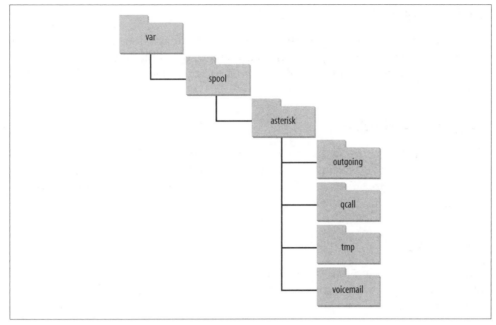

Figure 3-4. /var/spool/asterisk/ directory structure

The *dictate/* directory is the default location where the `Dictate()` application looks for files.

The *meetme/* directory is the location where `MeetMe()` conference recordings are saved.

Recordings from either one-touch recording (the `w` and `W` flags to the `Dial()` application), the `MixMonitor()`, or `Monitor()` applications are stored in the *monitor/* directory.

system/ is used by the `System()` application for temporary storage of data.

The *tmp/* directory is used, funny enough, to hold temporary information. Certain applications may require a place to write files to before copying the complete files to their final destinations. This prevents two processes from trying to write to and read from a file at the same time.

All voicemail and user greetings are contained within the *voicemail/* directory. Extensions configured in *voicemail.conf* that have been logged in to at least once are created as subdirectories of *voicemail/*.

/var/run/

The */var/run/* directory contains the process ID (PID) information for all active processes on the system, including Asterisk (as specified in the *asterisk.conf* file). Note that */var/run/* is OS-dependent and may differ.

/var/log/asterisk/

The */var/log/asterisk/* directory is where Asterisk logs information. You can control the type of information being logged to the various files by editing the *logger.conf* file located in the */etc/asterisk/* directory. Basic configuration of the *logger.conf* file is covered in Appendix D.

/var/log/asterisk/cdr-csv

The */var/log/asterisk/cdr-csv* directory is used to store the CDRs in comma-separated value (CSV) format. By default information is stored in the *Master.csv* file, but individual accounts can store their own CDRs in separate files with the use of the `accountcode` option (see Appendix A for more information).

AsteriskNOW™

In the following sections we will provide a gentle introduction to the AsteriskNOW software, which gives you a complete PBX system with graphical configuration screen all built into one!

What Is AsteriskNOW?

AsteriskNOW is an open source software appliance, a customized Linux distribution that includes Asterisk, the Asterisk GUI, and all other software needed for an Asterisk system. The Asterisk GUI gives you the ability to easily configure your Asterisk system without being a technical expert.

Note: The complete software appliance distribution is provided under the GPL (*http://www.gnu.org/copyleft/gpl.html*) and may legally be used for any purpose, commercial or otherwise.

Before You Begin

AsteriskNOW installation is easy, because the appliance includes only those components necessary to run, debug, and build Asterisk. You no longer have to worry about kernel versions and package dependencies. AsteriskNOW is a custom Linux distribution for Asterisk based on rPath Linux.

What You Will Need

- A system on which you can install AsteriskNOW
- A CD writer and associated software
- Connection to the Internet

- Firefox browser

 The Asterisk GUI currently requires the Firefox browser (available at *http://www.mozilla.com/en-US/* for optimum performance. Wider browser support will be available with future versions.

Installation

You should observe all normal precautions when preparing and installing a new distribution. Any existing operating systems on your hard drive will be removed by the Express Installation. If you are not sure that you are ready to alter your system, try one of the alternate installations (discussed in "Alternate Installations") to give Asterisk-kNOW a try. For more help on Asterisk and rPath see the "For More Information" section at the end of the chapter.

Quick installation

The essential installation of AsteriskNOW is really quite simple and gives you the ability to get up and running in a short amount of time. Use this quick installation procedure if you are comfortable with accepting the defaults. Any help you may need is provided with the installation screens. If you would like more information on the installation procedure, refer to the "Extended procedure" section below:

1. Download the AsteriskNOW ISO file (*http://www.asterisknow.org/downloads*) and create a CD image from the file. This step is required before installation can begin. The process for creating a CD image will vary depending upon the CD authoring software you are using.
2. Insert your newly created AsteriskNOW CD into the CD-ROM drive of the PC.
3. Boot from the CD by restarting the PC. A basic AsteriskNOW boot menu with several options will be provided:
   - To install or upgrade in graphical mode, press Enter.
   - To install or upgrade in Linux text mode, type **linux text** and then press Enter.
   The recommended, and default, installation mode is graphical. If you do not make an entry, the installation will continue in graphical mode.
4. From here, follow the self-explanatory, onscreen prompts to guide you through the installation process.
5. When installation is complete, the system will prompt you to reboot. After rebooting, a URL to access the Asterisk GUI will be displayed.
6. You are now ready to configure and run AsteriskNOW.

Extended procedure

1. Download the AsteriskNOW ISO file (*http://www.asterisknow.org/downloads*) and create a CD image from the file. This step is required before installation can begin. The process for creating a CD image will vary depending upon the CD authoring software you are using.

2. Insert your newly created AsteriskNOW CD into the CD-ROM drive.

3. Boot from the CD by restarting the PC. A basic AsteriskNOW boot menu with several options will be provided:

   • To install or upgrade in graphical mode, press Enter.

   • To install or upgrade in Linux text mode, type `linux text` and then press Enter.

   The recommended, and default, installation mode is graphical. If you do not make an entry, the installation will continue in graphical mode.

   After a bit of processing, the initial installation screen is displayed. The initial screen is similar to the following illustration:

4. From the initial installation screen you can read the release notes or the Help information. When you are ready, click Next to continue the installation.

   The next installation screen lets you choose the type of installation. The two modes of installation available are:

   Express Installation
   > The Express Installation installs all of the software needed to install Asterisk. Debugging and development tools are installed with this installation type.

   Expert
   > Select this installation type if you want to have complete control over all installation options. Among the options you can control are software package selection, partitioning, and language selection.

   The default installation type is Express Installation. This installation type assumes an English language reader and that you aren't concerned with the finer points. Choose Expert if you don't read English, and/or want more control over the installation details. For the purposes of this procedure, Express Installation is discussed.

5. Choose your installation type and then click Next.

   The Automatic Partitioning screen is displayed. The Automatic Partitioning screen gives you several options to choose from before the software partitions your drive. This gives you the opportunity to choose which data (if any) is removed from your system, and how the drive is partitioned. The following options are available:

   Remove All Linux Partitions
   > This option will only remove any Linux partitions created from a previous Linux installation.

Remove All Partitions

> Select this option if you want to remove all partitions on your system, including those created by other operating systems (such as Windows).

Keep All Partitions

> You should choose this option if you want to retain all of your current data and partitions. You will need enough hard drive space for your Asterisk implementation. Twenty GB is a realistic minimum, but the minimum space is dependent on the needs of the system you want to create.

In most cases, you will want to choose Remove All Partitions. A hard drive dedicated to your Asterisk implementation is the best way to ensure maximum performance. Select the Review checkbox on the Automatic Partitioning screen if you want to review or modify your partition selections.

6. A list of the hard drives available for use is listed on the Automatic Partitioning screen. Select the checkbox next to the hard drive(s) you want to use for your system. Click Next to continue with the installation.

   - If you selected Remove All Partitions or Remove All Linux Partitions, a warning dialog will be displayed that asks if you want to proceed. Click Yes to proceed, or No to change your partition selection.

   - If you selected Review on the Automatic Partitioning screen, a screen will be displayed with the partitions created. You can modify your partitions on this screen. To proceed, click Next.

7. The Network Configuration screen is displayed.

   - You can configure the network devices associated with your system on the Network Configuration screen. Any network devices attached to your system are automatically detected by the installation program and displayed in the Network Devices list. You can either accept the device(s) automatically selected by the installation program, or you can edit them by selecting Edit.

   - Set the Hostname by either selecting Automatically via DHCP, or by selecting Manually and enter the hostname for your system. Once you have specified the hostname, click Next to proceed.

8. The Time Zone Selection screen is displayed.

   - The Time Zone Selection screen offers several ways for you to select the time zone appropriate for your installation. You can either use the world map, which displays major cities, select from a list of locations and time zones, or select the System Clock Uses UTC to use the system time. Once you have selected a time zone, click Next.

9. The Administrator Password screen is displayed.

   - You must set a password for the AsteriskNOW administrator account, "admin". This password will be used to log on to the system, as well as the Asterisk GUI. Set and confirm an administrator password, and then click Next to proceed.

- The About to Install screen is displayed, giving you an opportunity to delay or abort the installation process. If you are ready to continue with the installation, click Next.

10. The Installing Packages screen is displayed.
    - While AsteriskNOW is being installed, the Installing Packages screen will be displayed. The installation will continue for a few minutes.
    - Once the installation is complete, the system will prompt you to reboot. Remove the installation disk you created, and click Reboot. After rebooting, a URL to access the Asterisk GUI will be displayed.

Accessing the GUI

Once you have completed your installation and rebooted your machine, you will be able to access the Asterisk GUI. The URL used to access the Asterisk GUI is the IP address or hostname displayed after rebooting your machine. Enter this IP address in your browser URL. You will be able to refine your AsteriskNOW installation by accessing the Asterisk GUI.

Alternate Installations

You can also try out AsteriskNOW using the available VMware Player image (*http://www.vmware.com/download/player/*), Xen universal guest domain image (*http://wiki.rpath.com/wiki/Xen_Solutions_Using_rPath_Technologies*) or the LiveCD (just burn and boot). All alternate installations can be downloaded from the AsteriskNOW download (*http://www.asterisknow.org/downloads*) page.

Note: When using the LiveCD, the default username is "admin" with "password" as the password.

For More Information

An AsteriskNOW Users' Guide is currently under development by the Asterisk community on the Asterisk Forums. For additional information on AsteriskNOW, including step-by-step installation screenshots and configuration screenshots showing the setup wizard, please refer to *http://www.asterisknow.org*, and visit the Asterisk Forums at *http://forums.digium.com*. For more information and help with rPath Linux, please see rPath's wiki, *http://wiki.rpath.com*.

Conclusion

In this chapter, we have reviewed the procedures for obtaining, compiling, and installing Asterisk and the associated packages. In the following chapter, we will touch on the initial configuration of your system with regard to various communications channels, such as analog devices attached to FXS and FXO ports, SIP channels, and IAX2 endpoints.

## CHAPTER 4

Initial Configuration of Asterisk

> *I don't always know what I'm talking about,*
> *but I know I'm right.*
>
> —Muhammad Ali

Completing all the steps in Chapter 3 should have left you with a working Asterisk system. If it did not, please take the time to go back and review the steps, consult the wiki, engage the community, and get your system running.

Unfortunately, we cannot yet make any calls, because we have not yet created any channels. To get this plane to fly, we're going to need some runways. While there are dozens of different channel types, and dozens of different ways to configure each type of channel, we just want to get some calls happening, so let's try and keep things simple. We have decided to guide you through the configuration of four channels: a Foreign eXchange Office (FXO) channel, a Foreign eXchange Station (FXS) channel, a Session Initiation Protocol (SIP) channel, and an Inter-Asterisk eXchange (IAX) channel.[*] We selected these channel types because they are far and away the most popular channel types in use in small Asterisk systems, and one of the goals of this book is to keep things as simple as is reasonable. If we cover the basics of these channels, we will not have done an exhaustive survey of all channel types or topologies, but we will have created a base platform on which to develop your telecommunications system. Further scenarios and channel configuration details can be found in Appendix D.

Our first effort will be to explore the basic configuration of analog interfaces such as FXS and FXO ports with the use of a Digium TDM11B (which is an analog card with one FXS port and one FXO port).[†]

---

[*] Officially, the current version is IAX2, but since all support for IAX1 was dropped many years ago, whether you say "IAX" or "IAX2," you are talking about the same version.

[†] This configuration used to be known as the Digium Dev-lite kit. For more information on FXS versus FXO, keep reading. Put simply, this card will give us one port to connect to a traditional analog line from the phone company (FXO), and one port to connect to an analog telephone (FXS), which is any type of phone that will work with a traditional home telephone circuit.

Next, we'll tackle a few Voice over Internet Protocol (VoIP) interfaces: a local SIP and IAX2 channel connected to a softphone or hardphone, along with connecting two Asterisk boxes via these two popular protocols.

For SIP, we are going to cover Linksys, Polycom, Aastra, Grandstream, and Cisco sets. If we do not cover your phone model, we apologize, but what is important to realize is that while most of these devices have many different parameters that you can define, generally only a few parameters need to be defined in order to get the device to work. That will be our goal, because we figure it's a lot less frustrating to tweak a functioning device than to get it perfectly set up on the first try. We won't discuss all the features you may want your channel to have (such as caller ID or advanced codec and security settings), but you will be able to make and receive calls with your phone, which should put a smile on your face—a good state to be in as we dig deeper into things.

Once you've worked through this chapter, you will have a basic system consisting of many useful interfaces, which will provide the foundation we need to explore the *extensions.conf* file (discussed in detail in Chapter 5), where the dialplan is stored (technically, it contains the instructions Asterisk needs to build the dialplan). If you do not have access to the analog hardware, some of the examples will not be available to you, but you will still have configured a system suitable for a pure-VoIP environment.

What Do I Really Need?

The asterisk character (*) is used as a wildcard in many different applications. It is a good name for this PBX for many reasons, one of which is the enormous number of interface types to which Asterisk can connect. These include:

- Analog interfaces, such as your telephone line and analog telephones
- Digital circuits, such as T1 and E1 lines
- VoIP protocols such as SIP and IAX[‡]

Asterisk doesn't need any specialized hardware—not even a sound card—even though it is common to expect a telephone system to physically connect to a voice network. There are many types of channel cards that allow you to connect your Asterisk to things like analog phones or PSTN circuits, but they are not essential to the functioning of Asterisk. On the user (or station) side of the system, you can choose from all kinds of softphones that are available for Windows, Linux, and other operating systems—or use almost any physical IP phone. That handles the telephone side of the system. On the carrier side, if you don't connect directly to a circuit from your central office, you can still route your calls over the Internet using a VoIP service provider.

[‡] ...and H.323 and SCCP and MGCP and UNISTIM

Working with Interface Configuration Files

In this chapter, we're going to build an Asterisk configuration on the platform we have just installed. For the first few sections on FXO and FXS channels, we'll assume that you have a Digium TDM11B kit (which comes with one FXO and one FXS interface). This will allow you to connect to an analog circuit (FXO) and to an analog telephone (FXS). Note that this hardware interface isn't necessary; if you want to build an IP-only configuration, you can skip to the section on configuring SIP.

The configuration we do in this chapter won't be particularly useful on its own, but it will be a kernel to build on. We're going to touch on the following files:

zaptel.conf
> Here, we'll do low-level configuration for the hardware interface. We'll set up one FXO channel and one FXS channel. This configures the driver for the Linux kernel.

zapata.conf
> In this file, we'll configure Asterisk's interface to the hardware. This file contains a slightly higher-level configuration of the hardware in the Asterisk user-level process.

extensions.conf
> The dialplans we create will be extremely primitive, but they will prove that the system is working.

sip.conf
> This is where we'll configure the SIP protocol.

iax.conf
> This is where we'll configure incoming and outgoing IAX channels.

In the following sections, you will be editing several configuration files. You'll have to reload these files for your changes to take effect. After you edit the *zaptel.conf* file, you will need to reload the configuration for the hardware with `/sbin/ztcfg -vv` (you may omit the -vv if you don't need verbose output). Changes made in *zapata.conf* will require a `module reload` from the Asterisk console; however, changing signaling methods requires a `restart`. You will need to perform an `iax2 reload` and a `sip reload` after editing the *iax.conf* and *sip.conf* files, respectively.

In order to test the new devices we have defined, we must have a dialplan through which we can make connections. Even though we have not discussed the Asterisk dialplan (that's coming up in the next chapter), we want you to create a basic *extensions.conf* file so that we can test our work in this chapter.

Make a backup copy of the sample *extensions.conf* (try the bash command `mv extensions.conf extensions.conf.sample`), and then create a blank *extensions.conf* file (using the bash command `touch extensions.conf`), and insert the following lines:

```
[globals]

[general]
autofallthrough=yes

[default]

[incoming_calls]

[internal]

[phones]
include => internal
```

 In the [general] section, we have set autofallthrough=yes, which tells Asterisk to continue when an extension runs out of things to do. If you set this to no, then Asterisk will sit and wait for input after all priorities have executed. This is most prevalent if the Background() application is the last application executed in an extension. If set to yes (which is now the default in 1.4), Asterisk will drop the call after Background() finishes executing (at the end of the prompt(s) supplied to it). In order to force Asterisk to wait for input after the Background() application finishes playing the voice prompts supplied to it, we use the WaitExten() application.

Do not be afraid if what we've just written doesn't make a whole lot of sense, as we haven't explored the dialplan, applications, priorities, or extensions yet (that is coming up in the next chapter). So for now, just set autofallthrough=yes. It is safest to use the autofallthrough=yes command as we don't want Asterisk hanging around waiting for input unless we explicitly tell it to do so.

There is nothing else for now, but we'll be using this file as we go through this chapter to build a test dialplan so we can ensure that all of our devices are working. Also, be sure to run the dialplan reload command from the Asterisk CLI to update to the latest changes. Verify your changes by running the CLI command dialplan show:

```
*CLI> dialplan show
[ Context 'phones' created by 'pbx_config' ]
  Include =>         'internal'                                [pbx_config]

[ Context 'internal' created by 'pbx_config' ]

[ Context 'incoming_calls' created by 'pbx_config' ]

[ Context 'default' created by 'pbx_config' ]

[ Context 'parkedcalls' created by 'res_features' ]
  '700' =>           1. Park((null))                           [res_features]

-= 1 extension (1 priority) in 5 contexts. =-
```

You will see the `parkedcalls` context because it is an internal context to Asterisk, specified in the *features.conf* file.

Setting Up the Dialplan for Some Test Calls

Now let's expand upon the test dialplan we started in the previous section, allowing us to dial back into the softphone after we have configured it and to use a dialplan application called `Echo()` that will allow us to test bidirectional audio. We'll learn more about dialplans in Chapter 5, so for now, just add the italicized bits to your existing *extensions.conf* file. We'll be making use of this dialplan throughout the chapter, and expanding on it in certain sections. Once you've entered the text into *extensions.conf*, reload the dialplan by running `dialplan reload` from the Asterisk console:

```
[globals]

[general]

[default]
exten => s,1,Verbose(1|Unrouted call handler)
exten => s,n,Answer()
exten => s,n,Wait(1)
exten => s,n,Playback(tt-weasels)
exten => s,n,Hangup()

[incoming_calls]

[internal]
exten => 500,1,Verbose(1|Echo test application)
exten => 500,n,Echo()
exten => 500,n,Hangup()

[phones]
include => internal
```

FXO and FXS Channels

The difference between an FXO channel and an FXS channel is simply which end of the connection provides the dial tone. An FXO port does not generate a dial tone; it accepts one. A common example is the dial tone provided by your phone company. An FXS port provides both the dial tone and ringing voltage to alert the station user of an inbound call. Both interfaces provide bidirectional communication (i.e., communication that is transmitted and received in both directions simultaneously).[§]

[§] Bidirectional communication is also known as *full duplex* in some circles. *Half duplex* means communication is only traveling in one direction at a time.

If your Asterisk server has a compatible FXO port, you can plug a telephone line from your telephone company (or "telco") into this port. Asterisk can then use the telco line to place and receive telephone calls. By the same token, if your Asterisk server has a compatible FXS port, you may plug an analog telephone into your Asterisk server, so that Asterisk may call the phone and you may place calls.

Ports are defined in the configuration by the signaling they use, as opposed to the physical type of port they are. For instance, a physical FXO port will be defined in the configuration with FXS signaling, and an FXS port will be defined with FXO signaling. This can be confusing until you understand the reasons for it. FX_ cards are named not according to what they are, but rather according to what is connected to them. An FXS card, therefore, is a card that connects to a station. Since that is so, you can see that in order to do its job, an FXS card must *behave* like a central office and use FXO signaling. Similarly, an FXO card connects to a central office (CO), which means it will need to behave like a station and use FXS signaling. The modem in your computer is a classic example of an FXO device.

The older Digium X100P card used a Motorola chipset, and the X101P (which Digium sold before completely switching to the TDM400P) is based on the Ambient/Intel MD3200 chipset. These cards are modems with drivers adapted to utilize the card as a single FXO device (the telephone interface cannot be used as an FXS port). Support for the X101P card has been dropped in favor of the TDM series of cards.

These cards (or their clones) *SHOULD NOT* be used in production environments. They are $10 on eBay for a reason.

The X100P/X101P cards are poor cards for production use due to their tendency to introduce echo into your telephone calls, and their lack of remote disconnect supervision. Do yourself a favor and don't waste your time with this hardware. You will find that if you ask the community for support of these cards, many responses will be hostile. You have been warned.

Determining the FXO and FXS Ports on Your TDM400P

Figure 4-1 contains a picture of a TDM400P with an FXS module and an FXO module. You can't see the colors, but module 1 is a green FXS module, and module 2 is an orange-red FXO module. In the bottom-right corner of the picture is the Molex connector, where power is supplied from the computer's power supply.

Plugging an FXS port (the green module) into the PSTN may destroy the module and the card due to voltage being introduced into a system that wants to produce voltage, not receive it!

Figure 4-1. A TDM400P with an FXS module (1 across) and an FXO module (2 across)

 Be sure to connect your computer's power supply to the Molex connector on the TDM400P if you have FXS modules, as it is used to supply the voltage needed to drive the ring generator on the FXS ports. The Molex connector is not required if you have only FXO modules.

Configuring an FXO Channel for a PSTN Connection

We'll start by configuring an FXO channel. First we'll configure the Zaptel hardware, and then the Zapata hardware. We'll set up a very basic dialplan, and we'll show you how to test the channel.

Zaptel Hardware Configuration

The *zaptel.conf* file located in */etc/* is used to configure your hardware. The following minimal configuration defines an FXO port with FXS signaling:

```
fxsks=2
loadzone=us
defaultzone=us
```

In the first line, in addition to indicating whether we are using FXO or FXS signaling, we specify one of the following protocols for channel 2:

- Loop start (`ls`)
- Ground start (`gs`)
- Kewlstart (`ks`)

The difference between *loop start* and *ground start* has to do with how the equipment requests a dial tone: a ground-start circuit signals the far end that it wants a dial tone by momentarily grounding one of the leads; a loop-start circuit uses a short to request a dial tone. Though not common for new installations, analog ground start lines still exist in many areas of the country.[‖] Ground start is really a rather strange thing, because it doesn't exist in its analog form in Asterisk, so technically, there is no ground signal happening, but is rather a signaling bit that is intended for analog circuitry that historically would have been at the end of the T1. If this does not make much sense, don't sweat it; chances are you won't have to worry about ground-start signaling. All home lines (and analog telephones/modems/faxes) in North America use loop-start signaling. *Kewlstart* is in fact the same as loop start, except that it has greater intelligence and is thus better able to detect far-end disconnects.[#] Kewlstart is the preferred signaling protocol for analog circuits in Asterisk.

To configure a signaling method other than kewlstart, replace the `ks` in `fxsks` with either `ls` or `gs` (for loop start or ground start, respectively).

`loadzone` configures the set of indications (as configured in *zonedata.c*) to use for the channel. The *zonedata.c* file contains information about all of the various sounds that a phone system makes in a particular country: dial tone, ringing cycles, busy tone, and so on. When you apply a loaded tone zone to a Zap channel, that channel will mimic the indications for the specified country. Different indication sets can be configured for different channels. The `defaultzone` is used if no zone is specified for a channel.

After configuring *zaptel.conf*, you can load the drivers for the card. `modprobe` is used to load modules for use by the Linux kernel. For example, to load the *wctdm* driver, you would run:

```
# modprobe wctdm
```

If the drivers load without any output, they have loaded successfully.[*] You can verify that the hardware and ports were loaded and configured correctly with the use of the *ztcfg* program:

[‖] Yes, there is such a thing as ground-start signaling on channelized T1s, but that has nothing to do with an actual ground condition on the circuit (which is entirely digital).

[#] A *far-end disconnect* happens when the far end hangs up. In an unsupervised circuit, there is no method of telling the near end that the call has ended. If you are on the phone this is no problem, since you will know the call has ended and will manually hang up your end. If, however, your voicemail system is recording a message, it will have no way of knowing that the far end has terminated and will, thus, keep recording silence, or even the dial tone or reorder tone. Kewlstart can detect these conditions and disconnect the circuit.

```
# /sbin/ztcfg -vv
```

The channels that are configured and the signaling method being used will be displayed. For example, a TDM400P with one FXO module has the following output:

```
Zaptel Configuration
======================
Channel map:
Channel 02: FXS Kewlstart (Default) (Slaves: 02)
1 channels configured.
```

If you receive the following error, you have configured the channel for the wrong signaling method (or there is no hardware present at that address):

```
ZT_CHANCONFIG failed on channel 2: Invalid argument (22)
Did you forget that FXS interfaces are configured with FXO signaling
and that FXO interfaces use FXS signaling?
```

To unload drivers from memory, use the **rmmod** (remove module) command, like so:

```
# rmmod wctdm
```

The *zttool* program is a diagnostic tool used to determine the state of your hardware. After running it, you will be presented with a menu of all installed hardware. You can then select the hardware and view the current state. A state of "OK" means the hardware is successfully loaded:

```
Alarms         Span
OK             Wildcard TDM400P REV E/F Board 1
```

Zapata Hardware Configuration

Asterisk uses the *zapata.conf* file to determine the settings and configuration for telephony hardware installed in the system. The *zapata.conf* file also controls the various features and functionality associated with the hardware channels, such as Caller ID, call waiting, echo cancellation, and a myriad of other options.

When you configure *zaptel.conf* and load the modules, Asterisk is not aware of anything you've configured. The hardware doesn't have to be used by Asterisk; it could very well be used by another piece of software that interfaces with the Zaptel modules. You tell Asterisk about the hardware and control the associated features via *zapata.conf*:*

```
[trunkgroups]
; define any trunk groups

[channels]
; hardware channels
```

---

* It is generally safe to assume that the modules have loaded successfully, but to view the debugging output when loading the module, check the console output (by default this is located on TTY terminal 9, but this is configurable in the *safe_asterisk* script—see the previous chapter for details).

```
; default
usecallerid=yes
hidecallerid=no
callwaiting=no
threewaycalling=yes
transfer=yes
echocancel=yes
echotraining=yes

; define channels
context=incoming          ; Incoming calls go to [incoming] in extensions.conf
signaling=fxs_ks          ; Use FXS signaling for an FXO channel
channel => 2              ; PSTN attached to port 2
```

The [trunkgroups] section is used for connections where multiple physical lines are used as a single logical connection to the telephone network, and won't be discussed further in this book. If you require this type of functionality, see the *zapata.conf.sample* file and your favorite search engine for more information.

The [channels] section determines the signaling method for hardware channels and their options. Once an option is defined, it is inherited down through the rest of the file. A channel is defined using channel =>, and each channel definition inherits all of the options defined above that line. If you wish to configure different options for different channels, remember that the options should be configured *before* the channel => definition.

We've enabled Caller ID with usecallerid=yes and specified that it will not be hidden for outgoing calls with hidecallerid=no. Call waiting is deactivated on an FXO line with callwaiting=no. Enabling three-way calling with threewaycalling=yes allows an active call to be placed on hold with a hook switch flash (discussed in Chapter 7) to suspend the current call. You may then dial a third party and join them to the conversation with another hook switch. The default is to not enable three-way calling.

Allowing call transfer with a hook switch is accomplished by configuring transfer=yes; it requires that three-way calling be enabled. The Asterisk echo canceller is used to remove the echo that can be created on analog lines. You can enable the echo canceller with echocancel=yes. The echo canceller in Asterisk requires some time to learn the echo, but you can speed this up by enabling echo training (echotraining=yes). This tells Asterisk to send a tone down the line at the start of a call to measure the echo, and therefore learn it more quickly.

When a call comes in on an FXO interface, you will want to perform some action. The action to be performed is configured inside a block of instructions called a *context*. Incoming calls on the FXO interface are directed to the incoming context with context=incoming. The instructions to perform inside the context are defined within *extensions.conf*.

Finally, since an FXO channel uses FXS signaling, we define it as such with signaling=fxs_ks.

Dialplan Configuration

The following minimal dialplan makes use of the Echo() application to verify that bi-directional communications for the channel are working:

```
[incoming]
; incoming calls from the FXO port are directed to this context
;from zapata.conf
exten => s,1,Answer()
exten => s,n,Echo()
```

Whatever you say, the Echo() application will relay back to you.

Dialing In

Now that the FXO channel is configured, let's test it. Run the *zttool* application and connect your PSTN line to the FXO port on your TDM400P. Once you have a phone line connected to your FXO port, you can watch the card come out of a RED alarm.

Now dial the PSTN number from another external phone (such as a cell phone). Asterisk will answer the call and execute the Echo() application. If you can hear your voice being reflected back, you have successfully installed and configured your FXO channel.

Configuring an FXS Channel for an Analog Telephone

The configuration of an FXS channel is similar to that of an FXO channel. Let's take a look.

Zaptel Hardware Configuration

The following is a minimal configuration for an FXS channel on a TDM400P. The configuration is identical to the FXO channel configuration above, with the addition of fxoks=1.

Recall from our earlier discussion that the opposite type of signaling is used for FXO and FXS channels, so we will be configuring FXO signaling for our FXS channel. In the example below we are configuring channel 1 to use FXO signaling, with the kewlstart signaling protocol:

```
fxoks=1
fxsks=2
loadzone=us
defaultzone=us
```

After loading the drivers for your hardware, you can verify their state with the use of /sbin/ztcfg -vv:

```
Zaptel Configuration
=======================

Channel map:

Channel 01: FXO Kewlstart (Default) (Slaves: 01)
Channel 02: FXS Kewlstart (Default) (Slaves: 02)

2 channels configured.
```

Zapata Hardware Configuration

The following configuration is identical to that for the FXO channel, with the addition of a section for our FXS port and, of the line immediate=no. The context for our FXS port is phones, the signaling is fxoks (kewlstart), and the channel number is set to 1.

FXS channels can be configured to perform one of two different actions when a phone is taken off the hook. The most common (and often expected) option is for Asterisk to produce a dial tone and wait for input from the user. This action is configured with immediate=no. The alternative action is for Asterisk to automatically perform a set of instructions configured in the dialplan instead of producing a dial tone, which you indicate by configuring immediate=yes.[†] The instructions to be performed are found in the context configured for the channel and will match the s extension (both of these topics will be discussed further in the following chapter).

Here's our new *zapata.conf*:

```
[trunkgroups]
; define any trunk groups

[channels]
; hardware channels
; default
usecallerid=yes
hidecallerid=no
callwaiting=no
threewaycalling=yes
transfer=yes
echocancel=yes
echotraining=yes
immediate=no

; define channels
context=phones        ; Uses the [internal] context in extensions.conf
signalling=fxo_ks      ; Uses FXO signalling for an FXS channel


channel => 1          ; Telephone attached to port 1
```

[†] Also referred to as the Batphone method, and more formally known as an Automatic Ringdown or Private Line Automatic Ringdown (PLAR) circuit. This method is commonly used at rental car counters and airports.

```
context=incoming          ; Incoming calls go to [incoming] in extensions.conf
signalling=fxs_ks          ; Use FXS signalling for an FXO channel
channel => 2              ; PSTN attached to port 2
```

Dialplan Configuration

We will make use of our minimal dialplan we configured earlier in the chapter to test
our FXS port with the use of the Echo() application. The relevant section, which should
already exist in your dialplan, looks like this:

```
[internal]
exten => 500,1,Verbose(1|Echo test application)
exten => 500,n,Echo()
exten => 500,n,Hangup()

[phones]
include => internal
```

Whatever you say, the Echo() application will relay back to you.

Configuring SIP Telephones

The Session Initiation Protocol (SIP),[‡] commonly used in VoIP phones (either hard
phones, or softphones), takes care of the setup and teardown of calls, along with any
changes during a call such as call transfers. The purpose of SIP is to help two endpoints
talk to each other (if possible, *directly* to each other). The SIP protocol is simply a
signaling protocol, which means that its purpose is only to get the two endpoints talking
to each other, and not to deal with the media of the call (your voice). Rather, your voice
is carried using another protocol called the Real-Time Transport Protocol (RTP; RFC
3550) to transfer media directly between the two endpoints.

> We use the term *media* to refer to the data transferred between endpoints
> and used to reconstruct your voice at the other end. It may also refer to
> music or prompts from the PBX.

In the world of SIP, we call our endpoints user agents, of which there are two types:
client and server. The *client* is the endpoint that generates the request, and the *server*
processes the request and generates a response. When an endpoint wishes to place a
call to another endpoint (such as our softphone calling another softphone), we generate
our request and send this to a SIP proxy.[§] A proxy server will take the request, determine

---

‡ RFC 3261 is available at *http://www.ietf.org/rfc/rfc3261.txt*. While the document is fairly large, we strongly
  encourage anyone who wishes to become an Asterisk professional to read at least the first 100 or so pages of
  this document and to understand how calls are set up, as this knowledge will be imperative when you're
  looking at a SIP trace (sip debug from the Asterisk console) trying to determine why your calls are not getting
  set up correctly.

where the request is destined for, and forward it on. Once the two user agents have negotiated a successful call setup, the media is transported via the RTP protocol and sent directly between the two user agents. SIP proxies do not handle media; they simply deal with the SIP packets.

Asterisk, on the other hand, is called a Back-To-Back User Agent (B2BUA). This means that Asterisk acts like a user agent in either the server (receiving) or client (sending) role. So when our softphone dials an extension number, the call is set up between the softphone and Asterisk directly. If the logic we've built into Asterisk determines that you mean to call another user agent, then Asterisk acts as a user agent client and sets up another connection (known as a channel) to the other phone. The media between the two phones then flows directly through Asterisk.[||] From the viewpoint of the phones, they are talking with Asterisk directly.

Basic SIP Telephone Configuration in Asterisk

Configuring a SIP phone to work with Asterisk does not require much. However, because there are so many options possible in both Asterisk and the configuration of the particular telephone set or softphone, things can get confusing. Add to this the fact that similar things can have different names, and you have a recipe for frustration. What we are going to do, therefore, is give you the bare-bones basics. If you follow our advice, you should be able to get the sets we cover working (and be well on your way to getting a phone that we have not covered to work as well). We are not saying that this is the best way, or even the right way, but it is the simplest way, and from a working foundation, it is much easier to take a basic configuration and tweak things until you get the solution you need.

 Just as we did with the *extensions.conf* file; run the following commands in your bash shell:

```
# mv sip.conf sip.conf.sample
# touch sip.conf
```

Defining the SIP device in Asterisk

If you put the following in a *sip.conf* file, you will be able to register a phone to the system.

```
[general]
```

§ An excellent open source SIP proxy is OpenSER, available at *http://www.openser.org*.

|| Yes, there are ways to making the media flow directly between the phones once the call is set up. This is done in the *sip.conf* file using either `directrtpsetup=yes` (an experimental option allowing the media to be redirected in the initial call setup) or `canreinvite=yes` (where media initially goes through Asterisk until a re-INVITE happens, at which point the media can be sent directly between the phones).

```
[1000]
type=friend
context=phones
host=dynamic
```

Not pretty, not secure, not flexible, not fully-featured, but this will work.

Even though we have named this SIP device 1000, and we are probably going to assign it that extension number, you should note that we could have named it whatever we wanted. Names such as mysipset, john, 0004f201ab0c are all valid, popular, and may suit your needs better.[#] What we are doing is assigning a unique identifier to a device, which will form part of the credentials when a call is placed using the SIP channel.

Since we want to be able to both send calls to the softphone and allow the client to place calls, we have defined the type as friend. The other two types are user and peer. From the viewpoint of Asterisk, a user is configured to enter the dialplan, and a peer is created for calls leaving the dialplan (via the Dial() application). A friend is simply a shortcut that defines both a user and a peer. If in doubt, define the type as friend.

The host option is used to define where the client exists on the network when Asterisk needs to send a call to it. This can either be defined statically by defining something like host=192.168.1.100, or if the client has a dynamic IP address, then we set host=dynamic. When the host option is set as dynamic, and the client is configured to register, Asterisk will receive a REGISTER packet from the endpoint (i.e., telephone set or softphone), telling Asterisk which IP address the SIP peer is using.

If you do not trust your network, you should probably also force the use of a password by adding the following to the device definition. This is one of those things that is not technically necessary, but is probably a good idea:

```
secret=guessthis
```

Configuring the Device Itself

In the configuration menus of the phone itself (which could be via a web GUI, through menus on the phone itself, or possibly using configuration files that are stored on a server), the unique identifier (which in this case is 1000) again forms part of the credentials for the authentication process. Naturally, for a connection to be successful, this has to match in both Asterisk and in the set itself. The fun begins when you realize that there is no set rule as to what this identifier is formally called. We have elected simply to call it the Unique Identifier.

[#] The maximum length of a username is 255 characters.

In the SIP RFC (*http://www.faqs.org/rfcs/rfc3261.html*), section 19.1 calls this user token, "the identifier of a particular resource at the host being addressed," verbiage consistent with our usage of [1000] as the set identifier in the *sip.conf* file of Asterisk.

Instead, you will want to look for fields that are labeled user name, auth name, authentication name, and so on. The thing to remember is that since you know that the Asterisk end of the equation is configured simply and correctly, you can experiment with the phone setting until you find a combination that works. This is much better than the usual suffering that new users go through, as they change settings in both places and have no luck getting a phone to register.

We're gonna say it again: configure *sip.conf* in the simplest manner possible, and then don't change your Asterisk configuration. Trust us; what we have written here will work. Get your set working (i.e., where you can make and receive calls), and you will be in a far better position to begin experimenting with different settings. We have seen too much suffering (including our own), and we want it to end.

Simplifying sip.conf

The *sip.conf* file (which was copied to the */etc/asterisk* directory by the make samples command we ran in the previous chapter) contains a large number of options and documentation inside it, but the file is actually very minimal if you remove all the commented parameters. The default file really breaks down to just the following few lines being uncommented by default:

```
[general]
context=default           ; Default context for incoming calls
allowoverlap=no           ; Disable overlap dialing support. (Default is yes)
bindport=5060             ; UDP Port to bind to (SIP standard port is 5060)
                          ; bindport is the local UDP port that Asterisk will
                          ; listen on
bindaddr=0.0.0.0          ; IP address to bind to (0.0.0.0 binds to all)
srvlookup=yes             ; Enable DNS SRV lookups on outbound calls
                          ; Note: Asterisk only uses the first host
                          ; in SRV records
                          ; Disabling DNS SRV lookups disables the
                          ; ability to place SIP calls based on domain
                          ; names to some other SIP users on the Internet

[authentication]
```

The [general] section contains the options that will apply to all SIP clients and trunks. Some settings elsewhere are set only in the [general] section, and others can be set in the [general] section as defaults for all conditionals unless overridden. These options are listed under the two columns labeled [users] and [peers] below the [authentication] header.

Generally, the commented-out options will show you the default setting Asterisk uses, or will tell you the default option in the option's description.

You can also check the current state of the SIP channel in Asterisk with the `sip show settings` CLI command.

 If you are running Asterisk and a softphone on the same system (i.e., running an X-Lite softphone and Asterisk on a laptop or desktop), then you will need to modify the SIP port that client listens on. It will need to be changed from 5060 to 5061 (or some other unused port) so that Asterisk and the softphone do not interfere with each other.

Essential Server Components

Before we get into how to define these files, there are a few things that need to be configured on your server. Running the right services on your network will ensure your Polycom sets can autoconfigure from the moment you plug them in. There's a little work involved here, but we promise that the payoff is worth it. Once you've done this a few times, it only really takes a few minutes on each new system, and going forward, it'll save you a lot of mucking about with web interfaces. When you take your new Polycom phone out of the box, plug it into your network, watch it autoconfigure itself, and then successfully register with your Asterisk machine, you will know the sort of joy that only geeks can experience.[*]

It's not really that complicated. Where we think people get confused is in making sense of the various ways this can be achieved, because there are a lot of choices.

DHCP server

Typically, a DHCP server is used to configure basic IP parameters for a device (IP address, default gateway, and DNS), but the DHCP protocol can actually pass many other parameters. In our case, we want it to pass some information to the sets that will tell them where to download their config files from. Here is a sample config from a Linux DHCP server that will do what is required:

```
ddns-update-style interim;
ignore client-updates;

subnet 192.168.1.0 netmask 255.255.255.0 {
  option routers 192.168.1.1;
  option subnet-mask 255.255.255.0;
  option domain-name-servers 192.168.1.1;
  option ntp-servers pool.ntp.org;
```

---

[*] Typically, it's at 4 A.M. on the morning of a critical 8 A.M. meeting, after having worked all weekend. Red Bull is probably the most popular drink of the Asterisk developers. Dr. Pepper would be a close second. Red Bull, anyone?

```
   option time-offset -18000;

range dynamic-bootp 192.168.1.128 192.168.1.254;
default-lease-time 21600;
max-lease-time 43200;
}
```

Keep in mind that this assumes that the only things on this network are devices that belong to the phone system (this setup will hand out an IP address to any device that requests it). If you have a more complex environment, you will need to configure the DHCP daemon to handle the various devices it is serving. For example, you might want to devise a scope that restricts IP addresses in your voice LAN to Polycom phones. Since all Polycom IP desk phones have 00:04:f2 as their OUI (Organizationally Unique Identifier), you might choose to restrict scope based on that.

 In a Microsoft DHCP environment, the tftp-server-name is referred to as *Boot server host name*. It is defined under *option 66*.

The DHCP protocol is far more flexible than is often realized, because in most environments it is not used for complex provisioning tasks. With a little care and attention, you can devise a DHCP environment that serves both your voice and data devices and greatly simplifies administrative workload when adding new devices.

FTP server

FTP is currently our favorite[†]way to configure Polycom sets. We would recommend it over TFTP for any set that allows for both. To install it on your CentOS system, the following command will install VSFTPD, the Very Secure FTP Daemon:

```
# yum -y install vsftpd
```

Then, in order to lock things down, we need to prevent anonymous logins with a simple change to the *vsftpd* config file, */etc/vsftpd/vsftpd.conf*:

```
# anonymous_enable=NO
```

Restart the server with `service vsftpd restart`. To ensure that the daemon runs after every reboot, run `chkconfig vsftpd on`.

Now, we have to create a user account and group for the sets to use. In this case, we will create an account for the Polycom sets:

[†] FTP is preferred over TFTP due to the ability of a Polycom phone to see timestamps on FTP files. This allows the phone to avoid redownloading configuration files and firmware updates that it already has—thus shortening boot time.

```
# groupadd PlcmSpIp
# useradd PlcmSpIp -g PlcmSpIp -p PlcmSpIp
# passwd PlcmSpIp
```

Set the password to *PlcmSpIp* (the default FTP password for Polycom sets). This can be changed, but will then require manual configuration from each set in order to advise them of their nonstandard credentials.[‡]

For added security, let's make sure the FTP server keeps that account in a chroot jail:

```
# echo PlcmSpIp >> /etc/vsftpd/vsftpd.chroot_list
```

That pretty much does it as far as preparing the operating system to provide the required services to the phones.

In the next few sections we have provided instructions for various popular SIP telephones. Choose the section that applies best to the phone that you are planning to use (whether a hard- or soft-phone). You will note that we have given all of these phones the exact same unique identifier. If you plan on installing more than one of them, you will need to ensure that they have unique names, and be sure to update your *sip.conf* file to include those device definitions.

CounterPath's X-Lite Softphone

CounterPath's X-Lite softphone has become very popular with the Asterisk community. It is simple, functional, easy on the eyes, and—most importantly—free.

In this section we will be configuring the X-Lite softphone to connect to Asterisk. The IP address of the phone is 192.168.1.250, and Asterisk is located at 192.168.1.100. The X-Lite is available for Microsoft Windows, Mac, and Linux. You can obtain a copy of X-Lite from *http://www.counterpath.com/index.php?menu=download*.

Now let's configure our softphone for connecting to our Asterisk box. To configure X-Lite, click on the Settings button, as circled in Figure 4-2.

Select *System Settings → SIP Proxy → [Default]*, which will display the default configuration for the softphone. Configure the screen as shown in Figure 4-3.

If you have not already started Asterisk, then start it now (see Chapter 3 for help installing and starting Asterisk). If Asterisk is running in the background, you can reconnect to the CLI by running the following command:

```
# asterisk -rvvv
```

[‡] You can get into assigning complex and unguessable passwords for the phones to use, but unless you are going to input the passwords into each phone manually, you'll have to pass them their FTP username and password from the DHCP server. Any device that can get on the voice network can get the same information from the DHCP server. We're not telling you to ignore security; just don't assume that creating separate passwords for each phone is going to improve security.

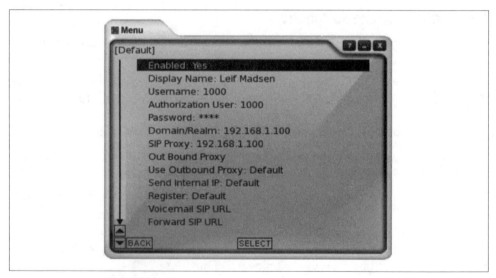

Figure 4-2. X-Lite configuration

Figure 4-3. X-Lite user configuration

You will then be given the Asterisk CLI like so:

```
*CLI>
```

If Asterisk was already running before changing the *sip.conf* as instructed in this chapter, then reload the dialplan and SIP channel with the following two commands:

```
*CLI> dialplan reload
```

```
*CLI> sip reload
```

In your X-Lite softphone client, close the Settings windows by clicking the BACK button until the windows are all closed. You should see X-Lite try to register to Asterisk, and if successful, you will see the following at the Asterisk CLI:

```
-- Registered SIP '1000' at 192.168.1.250 port 5061 expires 3600
```

You can verify the registration status at any time like so:

```
*CLI> sip show peers
```

```
Name/username            Host          Dyn Nat ACL Port    Status
1000/1000                192.168.1.250  D   N       5061    OK (63 ms)
1 sip peers [1 online , 0 offline]
```

More detailed stats of the peer can be shown as follows with sip show peer 1000:

```
*CLI> sip show peer 1000
```

```
  * Name        : 1000
  Secret        : <Not set>
  MD5Secret     : <Not set>
  Context       : phones
  Subscr.Cont.  : <Not set>
  Language      :
  AMA flags     : Unknown
  Transfer mode: open
  CallingPres   : Presentation Allowed, Not Screened
  Callgroup     :
  Pickupgroup   :
  Mailbox       :
  VM Extension  : asterisk
  LastMsgsSent  : 32767/65535
  Call limit    : 0
  Dynamic       : Yes
  Callerid      : "" <>
  MaxCallBR     : 384 kbps
  Expire        : 1032
  Insecure      : no
  Nat           : RFC3581
  ACL           : No
  T38 pt UDPTL  : No
  CanReinvite   : Yes
  PromiscRedir  : No
  User=Phone    : No
  Video Support: No
  Trust RPID    : No
  Send RPID     : No
  Subscriptions: Yes
  Overlap dial  : Yes
  DTMFmode      : rfc2833
  LastMsg       : 0
```

```
ToHost      :
Addr->IP    : 192.168.1.250 Port 5061
Defaddr->IP : 0.0.0.0 Port 5060
Def. Username: 1000
SIP Options : (none)
Codecs      : 0x8000e (gsm|ulaw|alaw|h263)
Codec Order : (none)
Auto-Framing: No
Status      : Unmonitored
Useragent   : X-Lite release 1105d
Reg. Contact : sip:1000@192.168.1.250:5061
```

Polycom's IP 430

A lot of folks say configuring Polycom phones is difficult. From what we can tell, they base this on one of two reasons: 1) The Polycom web-based interface is horrible, or 2) the automatic provisioning process is painful and confusing.

With respect to item 1, we agree. The web interface on the Polycom phones has got to be one of the most annoying web interfaces ever developed for an IP telephone. We don't use it, and we don't recommend it.[§]

So that leaves us with some sort of server-based configuration. Fortunately, in this regard, the Polycom IP phones are superb—so much so that we can pretty much forgive the web interface. Set configurations are stored in files on a server, and each set navigates to the server, downloads the configuration files that are relevant to it, and applies them to itself.

DHCP server

If you cannot control your DHCP server, you may have to manually specify the FTP server information on the phone. This is done by rebooting the set, pressing the *setup* button before the set begins the load process, and specifying the address of the FTP server in the small boot menu that these phones offer.

Protocol to use for downloading

The Polycom phones are able to download their configuration by one of three protocols: TFTP, HTTP, and FTP.

Right off the bat we are going to tell you to avoid TFTP. It is not secure, and the set cannot use date information to determine which versions of various files are the most current. It works, but there are better ways, and we are not going to discuss it further.

Polycom phones can pull their config data using HTTP as well, but it has not proven to be popular, and so we are going to move on.

[§] Actually, it does serve one useful purpose, which is to allow you to log on to a set via a browser and query its configuration.

FTP is currently the preferred method of allowing Polycom phones to obtain their configuration. It works well, is fairly easy to configure, and is well supported by the community.

FTP

FTP is currently our favorite way to configure Polycom sets. To install it on your CentOS system, the following command will install VSFTPD, the Very Secure FTP Daemon:

```
# yum -y install vsftpd
```

Then, in order to lock things down, we need to prevent anonymous logins, with a simple change to the *vsftpd* config file, */etc/vsftpd/vsftpd.conf*:

```
# anonymous_enable=NO
```

Restart the server with **service vsftpd restart**. To ensure that the daemon runs after every reboot, run **chkconfig vsftpd on**.

Now, we have to create a user account and group for the Polycom sets to use:

```
# groupadd PlcmSpIp
# useradd PlcmSpIp -g PlcmSpIp -p PlcmSpIp
# passwd PlcmSpIp
```

Set the password to *PlcmSpIp* (the default FTP password for Polycom sets). This can be changed, but will then require manual configuration from each set in order to advise them of their nonstandard credentials.[‖]

For added security, let's make sure the FTP server keeps that account in a chroot jail:

```
# echo PlcmSpIp >> /etc/vsftpd/vsftpd.chroot_list
```

That pretty much does it as far as preparing the operating system to provide the required services to the phones.

The Polycom configuration files

While there seem to be a lot of different files that are needed to make a Polycom set work, they are each fairly easy to understand.

The bootROM. This can best be described as the BIOS and operating system of the phone. Perhaps there is a more technical explanation, but why make things confusing? The bootROM should not need to be updated regularly, but it is good to keep an eye on the current releases to see if a newer bootROM has features that will be of benefit in your environment. This file will be named *bootrom.ld*.

[‖] You can get into assigning complex and unguessable passwords for the phones to use, but unless you are going to input the passwords into each phone manually, you'll have to pass them their FTP user name and password from the DHCP server. Any device that can get on the voice network can get the same information from the DHCP server. We're not telling you to ignore security, just don't assume that creating separate passwords for each phone is going to improve security.

The application image. Since Polycom sets are capable of supporting other VoIP protocols (MGCP is supported, for example), the protocol that this set will employ forms part of the application image that the phone will download and run. If the image on the set is already correct, this file is not actually needed on the FTP server; however, it is common to have this file available to ensure that the most recent version of the protocol is available for the sets to download. You will sometimes receive phones that are not running the latest version, so having the most current image will ensure that all sets are up-to-date.

The sip.cfg file. There is normally only one version of this file on a system, but it can be named anything you want, and there can be as many different versions of this file as are needed. For example, if you had an office where there were two different languages in use, some users might prefer French on their set, and others English. In that case, you'd create a *french.sip.conf* file and an *english.sip.conf* file to handle each case. Name this file as you see fit, but pick a name that makes sense so that future administrators have a chance to make sense of your design choices.

The master config file for each phone. This file is very simple and small. It is named to match the MAC address of each phone (so each set will need its own copy of this file) and tells the set what other files it needs to download in order to configure itself. This is the first config file each set will read. In this file will be a reference to the application image this set will use (currently named *sip.ld*), as well as the names of the XML files that have the parameters for this phone (the *.cfg* files). A master config file for a set might look something like this:

```
'<?xml version="1.0" standalone="yes"?>'
'<!-- Default Master SIP Configuration File-->'
'<!-- Edit and rename this file to <Ethernet-address>.cfg for each phone.-->'
'<!-- $Revision: 1.14 $  $Date: 2005/07/27 18:43:30 $ -->'
'<APPLICATION APP_FILE_PATH="sip.ld"
  CONFIG_FILES="phone1.cfg, sip.cfg"
  MISC_FILES=""
  LOG_FILE_DIRECTORY=""
  OVERRIDES_DIRECTORY=""
  CONTACTS_DIRECTORY=""
/>'
```

Note the name of the application file that we want this set to use, and the config files that it will be trying to find and apply.

The set-specific config file. We recommend giving the *phone1.cfg* files names that make sense. For example, *SET<xxx>.cfg* (such as *SET201.cfg*) to match the extension number of the phone, or *FLOOR4CUBE23.cfg*, or maybe *BOB_SMITHS_IP430_SET.cfg*, or whatever seems best to you. What's the best way to name them? We're going to answer that question by asking a question. Let's say you have 100 of these phones. When you list the contents of the */home/PlcmSpIp* folder, how do you want the 100 config files for the sets to appear?

Gotchas. Settings that are configured directly on the telephone will be stored on the

---

filesystem of the set, and may take precedence over parameters passed in config files. If you are having any problems applying changes to a set, try reformatting the phone. This will force the set to accept the parameters contained in the config files.

Cisco 7960 Telephone

The venerable old C7960 is now a part of VoIP history. One of the first SIP telephones that could actually be taken seriously, the only real complaint one can have about this phone is the price: they are the Cadillac of SIP phones (meaning that they have all the bells and whistles but are tough to justify at the price, and are a little out of date sometimes).

If you can get one of these, you are getting an excellent SIP telephone. If you buy one new, be prepared to pay.

One of the ways this phone is out of date is the lack of remote provisioning from anything other than TFTP. TFTP has lost favor with networking professionals due to the lack of authentication and encryption, but since it is the only method of remotely provisioning the phone, we are going to have to use the tftp-server daemon. We can install tftp-server with the following command:

```
# yum install -y tftp-server
```

Once installed, we need to enable the server by modifying the */etc/xinetd.d/tftp* file. To enable the TFTP server, change the `disable=yes` line to `disable=no`.

```
service tftp
{
    socket_type          = dgram
    protocol             = udp
    wait                 = yes
    user                 = root
    server               = /usr/sbin/in.tftpd
    server_args          = -s /tftpboot
    disable              = no
    per_source           = 11
    cps                  = 100 2
    flags                = IPv4
}
```

Then start the TFTP server by running:

```
# service xinetd restart
```

We can verify the server is running with the following command:

```
# chkconfig --list | grep tftp
    tftp:   on
```

As long as `tftp: on` was returned, the server is up and running.

Cisco phones by default are loaded with their own communication protocol known as SCCP (or Skinny). We will be showing you how to configure the phone, but due to the proprietary nature of Cisco and its phones, you will need to obtain the SIP firmware from your distributor. Also, there are both *chan_sccp* and *chan_skinny* modules for Asterisk, but they are beyond the scope of this book.

We will be registering our Cisco phone to the SIP friend we configured in "Zaptel Hardware Configuration." The following configuration file should be saved into a file taking the format of *SIP<mac>.cnf*, where *<mac>* represents the MAC address of the telephone device you are configuring. Place this file into the */tftpboot/* directory on your server:

```
# Line 1 Configuration
line1_name: "1000"
line1_authname: "1000"
line1_shortname: "Jimmy Carter"
line1_password: ""
line1_displayname: ""

# The phone label, displayed in the upper-righthand corner of the phone
phone_label: "aristotle" ; Has no effect on SIP messaging

# Phone password used for console or telnet access, limited to 31 characters
phone_password: "cisco"
```

Then configure the address to register in the *SIPDefault.cnf* file, also placed in the */tftpboot/* directory of your server. `proxy1_address` will contain the IP address of your Asterisk server of where the phone should register for line 1. The `image_version` contains the version of the *.loads* and *.sb2* files the phone will load into memory.

```
image_version: P0S3-08-4-00
proxy1_address: 192.168.1.100
```

We need one additional file called *OS79XX.TXT*. This file contains only a single line—the *.bin* and *.sbn* file version to load into memory:

```
P003-08-4-00
```

In order for our Cisco 7960 to use these files, we need to tell the phone where to pull its configuration from. If using the DHCP server from your Linux server, you can modify the */etc/dhcpd.conf* file in order to tell the phone where to pull its configuration from by adding the line:

```
option tftp-server-name "192.168.1.100";
```

which contains the IP address of the server hosting the TFTP server (assuming of course the TFTP server is configured at that address. This is the address we've been using for our Asterisk server, and we again assume you've installed the TFTP server on the same box as Asterisk). See "DHCP server" for more information about configuring the DHCP server:

```
ddns-update-style interim;
ignore client-updates;

subnet 192.168.1.0 netmask 255.255.255.0 {
    option routers 192.168.1.1;
    option subnet-mask 255.255.255.0;
    option domain-name-servers 192.168.1.1;
    option tftp-server-name "192.168.1.100";
    option ntp-servers pool.ntp.org;
    option time-offset -18000;

    range dynamic-bootp 192.168.1.128 192.168.1.254;
    default-lease-time 21600;
    max-lease-time 43200;
}
```

 Alternatively, you can configure from the phone itself to manually use an alternative TFTP server than that given by the DHCP server. To do so, press the *settings* button, (on the G version of the Cisco phones, this looks like a square with a check mark inside of it; G means Global). You will then need to unlock the settings by pressing the 9 key. The default password is *cisco*. Once the phone is unlocked, press the 3 key on the dialpad to enter the Network Configuration. Scroll down to option 32 and set the Alternate TFTP to YES. Then scroll up to option 7 and enter the IP address of the TFTP server you wish to boot from. Accept the settings and back out of the menu until the phone reboots itself. You can also use the *-6-settings* three finger salute to reboot your phone at any time.

You can watch the phone pull its configuration from the TFTP server by using tshark (yum install ethereal). Filter on port 69 using the following command:

```
# tshark port 69
```

You should then be able to watch the network traffic from the phone requesting data from your TFTP server.

If all goes well, then you should see your phone registered to Asterisk!

Linksys SPA-942

Ever since they purchased Sipura Technologies, Linksys has been producing a line of economical VoIP telephones and ATAs (Analog Terminal Adaptors). Linksys has been stealing a lot of business from Cisco. If you have read Clayton M. Christensen's *The Innovator's Dilemma* (HarperCollins), it becomes easier to understand Cisco's strategy with respect to Linksys.

Linksys (and Sipura) products are well regarded for their excellent quality, especially relative to their price, but they are also famous for being painfully difficult to configure.

Figure 4-4. SPA-942 keypad

This is mostly because their configuration GUI offers hundreds of configurable parameters.

We don't care about that. Here's what you need to know to get an SPA-942 working with your Asterisk system (and, we hope, most other Linksys VoIP devices as well).

Logging in to the phone

The first thing you need to do is get the IP address of the phone so you can log in to the GUI interface. From the phone itself, select the icon that looks like a piece of paper with a dog-eared corner (right below the envelope icon). This is the Settings button, and is shown in Figure 4-4.

To get the IP address of the phone, press the Settings button, followed by 9 (or use directional pad and scroll down to Network). Then press the *select* button (there is a row of 4 buttons under the LCD screen—*select* is the leftmost button). The second field should show you the IP address of the phone.

Now open your browser, enter the IP address into the address bar, hit Enter, and you will be at the Info screen of the phone.

Registering your phone to Asterisk

Select the Admin Login link in the upper-right corner of the screen. Once you've done this, you will be given several new tabs, such as Regional, Phone, Ext 1, Ext 2, and User.

Select the Ext 1 tab which will set up our first line. Then make the following menu selections:

1. General → Line Enable → yes
2. NAT Settings → NAT Mapping Enable → no
3. NAT Settings → NAT Keep Alive Enable → no

4. Proxy and Registration → Proxy → enter the IP address of Asterisk (e.g., 192.168.1.100)

5. Proxy and Registration → Register → yes

6. Proxy and Registration → Make Call Without Reg → no

7. Proxy and Registration → Ans Call Without Reg → no

8. Subscriber Information → Display Name → Caller ID information

9. Subscriber Information → User ID → 1000

10. Subscriber Information → Password → (leave blank if you're using the simple configuration from earlier in this chapter)

11. Subscriber Information → Use Auth ID → yes

12. Subscriber Information → Auth ID → 1000

13. Audio Configuration → Preferred Codec → G711u

14. Audio Configuration → Use Pref Codec Only → no

15. Audio Configuration → Silence Supp Enable → no

16. Audio Configuration → DTMF Tx Method → Auto

17. Submit All Changes

And that's it! Your phone should be registered to Asterisk now. You'll know this because the first lighted line button beside the LCD screen will change from orange to green.

Configuring the Dialplan for Testing

In order to allow our phone to call other phones (or, if a multiline phone, to call itself), we need to modify the *extensions.conf* file. Building on what we had in "Setting Up the Dialplan for Some Test Calls," add the following parts to the [internal] context:

```
exten => 1000,1,Verbose(1|Extension 1000)
exten => 1000,n,Dial(SIP/1000,30)
exten => 1000,n,Hangup()
```

If you have two phones, or multiple lines configured, you can duplicate the previous configuration and change the 1000 to the other extension name.

Connecting to a SIP Service Provider

With the advent of Internet telephony, there has been an influx of Internet-based phone companies springing up all over the world! This gives you a large number of choices from which to choose. Many of these service providers allow you to connect your Asterisk-based system to their networks,[#] and some of them are even running Asterisk themselves!

The following configuration should get you connected with an Internet Telephony Service Provider (ITSP),[*] although it is impossible to know the unique configurations each service provider will require from you, and ideally the provider will give you the configuration required to connect your system with its own. However, not all are going to support Asterisk, so we're going to provide you with a generic configuration which should help you get on your way and, ideally, going in a matter of minutes:

```
[my_service_provider]
type=peer
host=10.251.55.100
fromuser=my_unique_id
secret=my_special_secret
context=incoming_calls
dtmfmode=rfc2833
disallow=all
allow=gsm
allow=ulaw
deny=0.0.0.0/0
permit=10.251.55.100/32
insecure=invite
```

Configuring a Local Firewall

If you're running iptables on the same machine as the Asterisk box, then you can run the following commands to open port 5060 for SIP signaling, and ports 10,000 through 20,000 for the RTP traffic. You can also narrow the range of RTP ports in the *rtp.conf* file located in */etc/asterisk*. An excellent book on iptables firewalls is *Linux Firewalls* by Steve Suehring and Robert Ziegler (Novell Press).

```
# iptables -I RH-Firewall-1-INPUT -p udp --dport 5060 -j ACCEPT
# iptables -I RH-Firewall-1-INPUT -p udp --dport 10000:20000 -j ACCEPT
# service iptables save
```

Be aware that this will allow all UDP traffic from any source access to ports 5060 and 10,000 through 20,000.

Most of the previous configuration may be familiar to you by now, but in case it's not, here is a brief rundown.

By defining the type as a peer, we are telling Asterisk not to match on the [my_service_provider] name, but rather to match on the IP address in the INVITE message (when the provider is sending us a call). The host parameter is the IP address that we'll place our calls to, and the IP address we'll be matching on when receiving a call from the provider.

[#] Be sure to check the policy of any service provider you are looking to connect with, as some of them may not allow you to use a PBX system with its service.

[*] Also known as a VoIP Service Provider (VSP).

Matching on Username Instead of IP Address

Some service providers may insteadSession Initiation Protocol be sending their calls to you via multiple IP addresses, requiring you to create a separate peer account for each IP address. If you don't know each of these IP addresses, you may need to match on the username instead. The format for the service provider definition needs to only change slightly, but the biggest change to note is that you will need to set the [*service_provider_header*] as the username your service provider is going to send the call to. We have also changed the type from a peer to a friend, which from the viewpoint of Asterisk creates both a type user and type peer, where the type user will be matched before the peer:

```
[my_unique_id]
type=friend
host=10.251.55.100
fromuser=my_unique_id
secret=my_special_secret
context=incoming_calls
dtmfmode=rfc2833
disallow=all
allow=gsm
allow=ulaw
insecure=invite
```

Note that we've removed the deny and permit parameters since we may not know the IP addresses the calls will be coming from. If you do happen to know them and still wish to match them, you can add back in the deny and permit(s) for the IP addresses.

The fromuser parameter is going to affect the way our INVITE message is structured when sending the call to the provider. By setting our username in the fromuser parameter, we will modify the From: and Contact: fields of the INVITE when sending a call to the provider. This may be required by the provider if it's using these fields as part of its authentication routine. You can see the places Asterisk modifies the header in the next two code blocks.

Without the fromuser:

```
Audio is at 66.135.99.122 port 18154
Adding codec 0x2 (gsm) to SDP
Adding codec 0x4 (ulaw) to SDP
Adding non-codec 0x1 (telephone-event) to SDP
Reliably Transmitting (no NAT) to 10.251.55.100:5060:
INVITE sip:15195915119@10.251.55.100 SIP/2.0
Via: SIP/2.0/UDP 66.135.99.122:5060;branch=z9hG4bK32469d35;rport
From: "asterisk" <sip:asterisk@66.135.99.122>;tag=as4975f3ff
To: <sip:15195915119@10.251.55.100>
Contact: <sip:asterisk@66.135.99.122>
Call-ID: 58e3dfb2584930cd77fe989c00986584@66.135.99.122
CSeq: 102 INVITE
User-Agent: Asterisk PBX
Max-Forwards: 70
Date: Fri, 20 Apr 2007 14:59:24 GMT
```

```
Allow: INVITE, ACK, CANCEL, OPTIONS, BYE, REFER, SUBSCRIBE, NOTIFY
Supported: replaces
Content-Type: application/sdp
Content-Length: 265
```

With the fromuser:

```
Audio is at 66.135.99.122 port 11700
Adding codec 0x2 (gsm) to SDP
Adding codec 0x4 (ulaw) to SDP
Adding non-codec 0x1 (telephone-event) to SDP
Reliably Transmitting (no NAT) to 10.251.55.100:5060:
INVITE sip:15195915119@10.251.55.100 SIP/2.0
Via: SIP/2.0/UDP 66.135.99.122:5060;branch=z9hG4bK635b0b1b;rport
From: "asterisk" <sip:my_unique_id@66.135.99.122>;tag=as3186c1ba
To: <sip:15195915119@10.251.55.100>
Contact: <sip:my_unique_id@66.135.99.122>
Call-ID: 0c7ad6156f92e70b1fecde903550a12f@66.135.99.122
CSeq: 102 INVITE
User-Agent: Asterisk PBX
Max-Forwards: 70
Date: Fri, 20 Apr 2007 15:00:30 GMT
Allow: INVITE, ACK, CANCEL, OPTIONS, BYE, REFER, SUBSCRIBE, NOTIFY
Supported: replaces
Content-Type: application/sdp
Content-Length: 265
```

The deny and permit statements are used to deny all incoming calls to this peer except the IP address defined by the permit parameter. This is simply a security measure used to make sure nothing else matches on this peer except traffic coming from the IP address we expect.

At the end is insecure=invite, which may be required for your provider. This is because the source of the INVITE may originate from its backend platform, but could be directed through its SIP proxy server. Basically what this means is that the IP address that the peer is coming from, and which you are matching on, may not be the IP address that is in the Contact line: field of the INVITE message when you are accepting a call from your provider. This tells Asterisk to ignore this inconsistency and to accept the INVITE anyway.

> You may need to set invite=invite,port if the port address is also inconsistent with what Asterisk is expecting.

Now we need one additional parameter set in the [general] section of our *sip.conf* file: register. register is going to tell the service provider where to send calls when it has a call to deliver to us. This is Asterisk's way of saying to the service provider, "Hey! If you've got a call for me, send it to me at IP address 10.251.55.100." The register parameter takes the following form:

```
register => username:secret@my.service_provider.tld
```

Now we just need to configure a simple dialplan to handle our incoming calls and to send calls via the service provider. We're going to modify the simple dialplan we started building in the "Setting Up the Dialplan for Some Test Calls" section of this chapter. The italicized sections are the new parts that we're adding to the dialplan, with everything else existing previously.[†]

```
[globals]

[general]

[default]
exten => s,1,Verbose(1|Unrouted call handler)
exten => s,n,Answer()
exten => s,n,Wait(1)
exten => s,n,Playback(tt-weasels)
exten => s,n,Hangup()

[incoming_calls]
exten => _X.,1.NoOp()
exten => _X.,n,Dial(SIP/1000)

[outgoing_calls]
exten => _X.,1,NoOp()
exten => _X.,n,Dial(SIP/my_service_provider/${EXTEN})

[internal]
exten => 1000,1,Verbose(1|Extension 1000)
exten => 1000,n,Dial(SIP/1000,30)
exten => 1000,n,Hangup()

exten => 500,1,Verbose(1|Echo test application)
exten => 500,n,Echo()
exten => 500,n,Hangup()

[phones]
include => internal
include => outgoing_calls
```

Connecting Two Asterisk Boxes Together via SIP

There may come a time when you have a pair of Asterisk boxes, and you'd like to pass calls between them. Luckily this isn't very difficult, although it does have some oddities that we need to deal with, but from the configuration viewpoint it isn't really all that difficult.

[†] We also assume you have configured at least one SIP extension from the previous section.

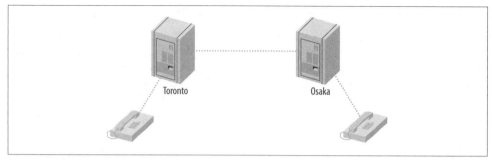

Figure 4-5. SIP trunking topology

Configuring a Local Firewall

If you're running iptables on the same machine as the Asterisk box, then you can run the following commands to open port 5060 for SIP signaling, and ports 10,000 through 20,000 for the RTP traffic. You can also narrow the range of RTP ports in the *rtp.conf* file located in */etc/asterisk*. An excellent book on iptables firewalls is *Linux Firewalls* by Steve Suehring and Robert Ziegler (Novell Press):

```
# iptables -I RH-Firewall-1-INPUT -p udp --dport 5060 -j ACCEPT
# iptables -I RH-Firewall-1-INPUT -p udp --dport 10000:20000 -j ACCEPT
# service iptables save
```

Be aware that this will allow all UDP traffic from any source access to ports 5060 and 10,000 through 20,000.

Our topology will consist of a SIP phone (Alice) registered to Asterisk A (Toronto), and a separate SIP phone (Bob) registered to Asterisk B (Osaka). At the end of this section, you will be able to set up a call from Alice to Bob (and vice versa) through your pair of Asterisk boxes (see Figure 4-5). This is a common scenario when you have two physical locations, such as a company with multiple offices that wants a single logical extension topology.

First, let's configure our Asterisk boxes.

Configuring Our Asterisk Boxes

We have a pair of Asterisk boxes that we're going to call Toronto and Osaka and that we're going to have register to each other. We're going to use the most basic *sip.conf* file that will work in this scenario. Just like the SIP phone configuration earlier in this chapter, it's not necessarily the best way to do it, but it'll work.

Here is the configuration for the Toronto box:

```
[general]
register => toronto:welcome@192.168.1.101/osaka

[osaka]
```

```
type=friend
secret=welcome
context=osaka_incoming
host=dynamic
disallow=all
allow=ulaw
```

And the configuration for the Osaka box:

```
[general]
register => osaka:welcome@192.168.2.202/toronto

[toronto]
type=friend
secret=welcome
context=toronto_incoming
host=dynamic
disallow=all
allow=ulaw
```

Many of the previous options may be familiar to you by now, but let's take a look at them further just in case they are not.

The second line of the file tells our Asterisk box to register to the other box, with the purpose of telling the remote Asterisk box where to send calls when it wishes to send a call to our local Asterisk box. Remember how we mentioned a little oddity in the configuration? Notice that at the end of the registration line we tag on a forward slash and the username of the remote Asterisk box? What this does is tell the remote Asterisk box what digest name to use when it wants to set up a call. If you forget to add this, then when the far end tries to send you a call, you'll see the following at your Asterisk CLI:

```
[Apr 22 18:52:32] WARNING[23631]: chan_sip.c:8117 check_auth: username mismatch,
                       have <toronto>, digest has <s>
```

So by adding the forward slash and username, we tell the other end what to place in the Digest username of the Proxy Authorization field in the SIP INVITE message.

The rest of the file is the authorization block we use to control the incoming and out-going calls from the other Asterisk box. On the Toronto box, we have the [osaka] authorization block, and on the Osaka box, we have the [toronto] block. We define the type as a friend, which allows us to both receive and place calls from the other Asterisk box. The secret is the password the other system should use when authenti-cating. The context is where incoming calls are processed in the dialplan (*extensions.conf*). We set the host parameter to dynamic, which tells our Asterisk box that the other endpoint will register to us, thereby telling us what IP address to set up calls when we want to send a call to the other end. Finally, the disallow and allow parameters control the codecs we wish to use with the other end.

If you save the file and reload the SIP channel on both Asterisk boxes (**sip reload** from the Asterisk console), you should see something like the following, which will tell you the remote box successfully registered:

```
*CLI>      -- Saved useragent "Asterisk PBX" for peer toronto
```

You should see the status of the Host change from (Unspecified) to the IP address of the remote box when you run sip show peers:

```
*CLI> sip show peers
Name/username       Host          Dyn Nat ACL Port     Status
toronto/osaka       192.168.2.202  D          5060      Unmonitored
```

You can verify that your own registration was successful by running sip show registry from the Asterisk console:

```
*CLI> sip show registry
Host                    Username    Refresh State       Reg.Time
192.168.1.101:5060      osaka       105 Registered      Sun, 22 Apr 2007 19:13:20
```

Now that our Asterisk boxes are happy with each other, let's configure a couple of SIP phones so we can call between the boxes.

SIP Phone Configuration

See the "Configuring an FXS Channel for an Analog Telephone" section of this chapter for more information about configuring SIP phones with Asterisk. Below is the configuration for two SIP phones in the *sip.conf* file for each server, which we'll be referencing from the dialplan in the next section, thereby giving us two endpoints to call between. Append this configuration to the end of the *sip.conf* file on each respective server.

Toronto *sip.conf*:

```
[1000]
type=friend
host=dynamic
context=phones
```

Osaka *sip.conf*:

```
[1001]
type=friend
host=dynamic
context=phones
```

You should now have extension 1000 registered to Toronto, and extension 1001 registered to Osaka. You can verify this with the sip show peers command from the Asterisk console. Next, we're going to configure the dialplan logic that will allow us to call between the extensions.

Configuring the Dialplan

Now we can configure a simple dialplan for each server allowing us to call between the two phones we have registered: one to Toronto, the other to Osaka. In the "Working with Interface Configuration Files" section of this chapter, we asked you to create a simple *extensions.conf* file. We are going to build up a dialplan based on this simple configuration. The dialplan for each server will be very similar to the other one, but for clarity we will show both. The new lines we're adding to the file will be italicized.

Toronto *extensions.conf*:

```
[globals]

[general]
autofallthrough=yes

[default]

[incoming_calls]

[phones]
include => internal
include => remote

[internal]
exten => _2XXX,1,NoOp()
exten => _2XXX,n,Dial(SIP/${EXTEN},30)
exten => _2XXX,n,Playback(the-party-you-are-calling&is-curntly-unavail)
exten => _2XXX,n,Hangup()

[remote]
exten => _1XXX,1,NoOp()
exten => _1XXX,n,Dial(SIP/osaka/${EXTEN})
exten => _1XXX,n,Hangup()

[osaka_incoming]
include => internal
```

Osaka *extensions.conf*:

```
[globals]

[general]
autofallthrough=yes

[default]

[incoming_calls]

[phones]
include => internal
include => remote

[internal]
exten => _1XXX,1,NoOp()
exten => _1XXX,n,Dial(SIP/${EXTEN},30)
exten => _1XXX,n,Playback(the-party-you-are-calling&is-curntly-unavail)
exten => _1XXX,n,Hangup()

[remote]
exten => _2XXX,1,NoOp()
```

```
exten => _2XXX,n,Dial(SIP/toronto/${EXTEN})
exten => _2XXX,n,Hangup()

[toronto_incoming]
include => internal
```

Once you've configured your *extensions.conf* file, you can reload it from the Asterisk console with the `dialplan reload` command. Verify your dialplan loaded with the `dialplan show` command.

And that's it! You should be able to place calls between your two Asterisk servers now.

Configuring an IAX Softphone

A major advantage of using the IAX2 protocol is that it is designed to be more friendly to working within odd network configurations, especially working behind NAT. This makes it a fantastic protocol for softphone clients since they are often utilized on laptops that roam into many different networks, often with no control of the network itself (such as when traveling between hotel networks).

The Inter-Asterisk eXchange (IAX) protocol is usually used for server-to-server communication; more hard phones are available that talk SIP. However, there are several softphones that support the IAX protocol, and work is progressing on several fronts for hard phone support in firmware. The primary difference between the IAX and SIP protocols is the way media (your voice) is passed between endpoints.

With SIP, the RTP (media) traffic is passed using different ports than those used by the signaling methods. For example, Asterisk receives the signaling of SIP on port 5060 and the RTP (media) traffic on ports 10,000 through 20,000 by default. The IAX protocol differs in that both the signaling and media traffic are passed via a single port: 4569. An advantage to this approach is that the IAX protocol tends to be better suited to topologies involving NAT.

There exist many IAX-based softphones, but not so many hardware based phones. The most pronounced reason is because IAX2 is not yet an IETF standard, yet many people have become early adopters and have reaped the benefits.

An excellent IAX2 softphone is idefisk, available at *http://www.asteriskguru.com*[‡] for free download. The authors have had excellent results with this softphone, and since it runs on Microsoft Windows, Mac OS X, and Linux, it is an excellent cross-platform softphone to write about. We will be demonstrating version 1.31 in this book, although version 2.0 was recently released (April 2007) but is not yet available for Linux.

[‡] The Asterisk Guru site is also an excellent source of documentation!

Configuring the Channel Configuration File (iax.conf)

Like the rest of this chapter, we're attempting to get you up and running quickly with the smallest configuration file set possible in order to minimize the problems you may have in configuring your devices. Just like the *sip.conf* file, *iax.conf* requires only a few simple lines to get our IAX phone registered to Asterisk. Let's take a look:

```
[general]
autokill=yes

[idefisk]
type=friend
host=dynamic
context=phones
```

Yes, really, that's all you need to get your softphone up and running. It's not the most secure or feature-rich configuration (we're not even using a password), but this will work.

In the [general] section of our *iax.conf* file, we have a single option: autokill= yes. We use this option to avoid things from stalling when a peer does not ACK (reply) to our NEW packet (new call setup request) within 2000 milliseconds. Instead of the reply to value yes, you can set this to the number of milliseconds to wait for the ACK to our NEW packet. You can control the autokill option for each individual peer by defining qualify for those peers that you know may be on poor network connections.

The rest of the file contains the definition for our softphone. We define the type as friend, which tells Asterisk we will send calls to this device and also accept calls from this device. A friend is a shortcut for defining a separate peer (send calls to the softphone), and user (accept calls from the softphone). We could also have defined individual definitions for the peer and user like so:

```
[idefisk]
type=user
context=phones

[idefisk]
type=peer
host=dynamic
```

Once you've configured your *iax.conf* file, save the file and reload the IAX2 channel module from your Asterisk console with module reload chan_iax2.so. Confirm your new peer exists by running iax2 show peers.

```
localhost*CLI> iax2 show peers
Name/Username   Host                  Mask            Port       Status
idefisk         (Unspecified)    (D)  255.255.255.255 0          Unmonitored
1 iax2 peers [0 online, 0 offline, 1 unmonitored]
```

Figure 4-6. idefisk

Figure 4-7. idefisk Account Options screen

Configure the Softphone

Once you've installed the idefisk softphone, open up the client and you'll see the main screen shown in Figure 4-6.

After we've started the softphone, we need to configure our softphone so we can place calls. We also need to register to Asterisk so we can receive calls. To do this, Right-click on the icon in the top-left corner of the screen, which will open a drop-down menu. Select Account Options from the drop-down, which will bring up the screen shown in Figure 4-7.

Start by creating a new account on the softphone by clicking the New button and filling out the relevant information. The Host should point to the IP address or domain name of your Asterisk system, with the username matching that of the value located between the square brackets [ ] in your *iax.conf* file. Leave the password field blank, as we did not configure a secret in *iax.conf*, and the Caller ID and Number can be set to whatever you wish. If you want idefisk to register this account on startup, select the "Register on startup" checkbox. When done, click the OK button to save the new account.

If you clicked the "Register on startup checkbox," then the phone will attempt to register to Asterisk. On the Asterisk console you will see output telling you that the phone has registered:

```
-- Registered IAX2 'idefisk' (UNAUTHENTICATED) at 127.0.0.1:32771
```

You can verify your registration with the **iax2 show peers** command at the Asterisk console:

```
localhost*CLI> iax2 show peers
Name/Username    Host             Mask             Port    Status
idefisk          127.0.0.1    (D) 255.255.255.255 32771   Unmonitored
1 iax2 peers [0 online, 0 offline, 1 unmonitored]
```

Configuring the Dialplan for Testing

One final thing to do is confirm dialing through our phone by configuring a simple dialplan in *extensions.conf*. You can simply test that you have audio in both directions by calling extension 500, or you can modify the dialplan we created in the "Setting Up the Dialplan for Some Test Calls" section of this chapter to place some test calls. If you also configured a SIP phone at extension 1000 in the previous sections, then the following will not overlap with that, as we'll be using extension 1001 (unless you configured multiple SIP extensions, in which case just configure a unique extension number for your IAX2 softphone):

```
[globals]

[general]

[default]
exten => s,1,Verbose(1|Unrouted call handler)
exten => s,n,Answer()
exten => s,n,Wait(1)
exten => s,n,Playback(tt-weasels)
exten => s,n,Hangup()

[incoming_calls]

[internal]
exten => 500,1,Verbose(1|Echo test application)
exten => 500,n,Echo()
exten => 500,n,Hangup()
```

```
exten => 1001,1,Verbose(1|Extension 1000)
exten => 1001,n,Dial(IAX2/idefisk,30)
exten => 1001,n,Hangup()

[phones]
include => phones
```

Connecting to an IAX Service Provider

Some Internet Telephony Service Providers (ITSPs) provide the ability to originate and terminate calls via the IAX2 protocol. Beyond minimizing the number of ports required to be open on the firewall (IAX2 only requires a single port for both signaling and media), the protocol's trunking feature is attractive to both ITSPs and their customers due to the savings in bandwidth that can be obtained when running many simultaneous calls between endpoints.

If your ITSP is offering IAX2 termination, there is a strong chance it is running Asterisk; thus the configuration for connecting to these service providers is more than likely going to be similar to what we are providing here.

The following configuration is a template for connecting to an IAX2 service provider:

```
[general]
autokill=yes

register => username:password@my.service-provider.tld

[my_unique_id]
type=user
secret=my_unique_password
context=incoming_calls
trunking=yes
disallow=all
allow=gsm
allow=ulaw
deny=0.0.0.0/0.0.0.0
permit=10.251.100.1/255.255.255.255

[my_unique_id]
type=peer
host=10.251.100.1
trunking=yes
disallow=all
allow=gsm
allow=ulaw
```

To accept incoming calls from the Direct Inward Dialing (DID) number that your service provider assigned to you, we need to modify our *extensions.conf* file. Perhaps you want to send the call to an auto-attendant, or maybe simply to your desk phone. In either case, you can accept calls from your service provider and match on the incoming DID with the following bit of dialplan logic:

```
[globals]

[general]
autofallthrough=yes

[default]

[incoming_calls]
exten => 14165551212,1,NoOp()
exten => 14165551212,n,Dial(SIP/1000,30)
exten => 14165551212,n,Playback(the-party-you-are-calling&is-curntly-unavail)
exten => 14165551212,n,Hangup()

exten => 4165551212,1,Goto(1${EXTEN})

[internal]

[phones]
include => internal
```

Connecting Two Asterisk Boxes Together via IAX

Often it is desirable to connect two physical Asterisk boxes together via IAX in order to send calls between two physical locations (the distance between these locations may be centimeters or kilometers). One of the advantages to using the IAX protocol to do this is a feature called *trunking*, which utilizes a method of sending the voice data for multiple calls at once with a single header. This has little effect on only one or two simultaneous calls, but if you are sending tens or hundreds of calls between two locations, the bandwidth savings by utilizing trunking can be tremendous.

Configuring a Local Firewall

If you're running iptables on the same machine as the Asterisk box, then you can run the following commands to open port 4569 for the IAX2 protocol. An excellent book on iptables firewalls is *Linux Firewalls* by Steve Suehring and Robert Ziegler (Novell Press).

```
# iptables -I RH-Firewall-1-INPUT -p udp --dport 4569 -j ACCEPT
# service iptables save
```

Be aware that this will allow all UDP traffic from any source access to port 4569.

 You will need a timing interface installed on your system, whether it be hardware from Digium or via the kernel by using the *ztdummy* driver. This will require you to have Zaptel installed on your system and running. See Chapter 3 for more information about installing Zaptel.

Configuring Our Asterisk Boxes

We'll be utilizing a simple topology where we have two Asterisk boxes registered to each other directly, and separate phones registered to each Asterisk box. We'll call the two Asterisk boxes Toronto and Osakafi (see "Connecting Two Asterisk Boxes Together via SIP"). Bob's phone will be registered and connected to Toronto, while Alice's phone will be registered and connected to Osaka.

The first thing we want to do is create a new channel file (*iax.conf*) by renaming the current sample file to *iax.conf.sample* and creating a new blank *iax.conf*:

```
# cd /etc/asterisk
# mv iax.conf iax.conf.sample
# touch iax.conf
```

Next, open up the *iax.conf* file and enter the following configuration on the Toronto Asterisk box:

```
[general]
autokill=yes

register => toronto:welcome@192.168.1.107

[osaka]
type=friend
host=dynamic
trunk=yes
secret=welcome
context=incoming_osaka
deny=0.0.0.0/0.0.0.0
permit=192.168.1.107/255.255.255.255
```

autokill=yes was explained in the previous section, but its purpose is to make sure new calls being set up to a remote system that are not acknowledged within a reasonable amount of time (two seconds by default) are torn down correctly. This saves us from having a lot of hung channels simply waiting for an acknowledgement that probably isn't coming.

The register line is used to tell the remote Asterisk box where we are so that when the box at 192.168.1.107 is ready to send us a call, it sends it to our IP address (in this case our IP address is 192.168.1.104, which you'll see in the *iax.conf* configuration of the Osaka box). We send the username Toronto and the password welcome to Osaka, which authenticates our registration, and if accepted, writes the location of our Asterisk box into its memory for when it needs to send us a call.

The [Osaka] definition is used to control the authentication of the remote box and delivery into our dialplan. *Osaka* is the username used in the incoming authentication. We set the type to friend because we want to have both the ability to send calls to Osaka and to receive calls from Osaka. The host option is set to dynamic which tells Asterisk to send calls to the IP address obtained when the opposite endpoint registers with us.

In the introduction to this section, we mentioned the possible bandwidth savings when utilizing IAX2 trunking. It's simple to enable this functionality in Asterisk, as we just need to add trunk=yes to our friend definition. As long as a timing interface is installed and running (i.e., *dummy*), then we can take advantage of IAX2 trunking.

The secret is straightforward: it's our authentication password. We're defining the [incoming_osaka] context as the place we will process incoming calls for this friend in the *extensions.conf* file. Finally, we block all IP addresses with the deny option from being allowed to authenticate, and explicitly permit 192.168.1.107.

The *iax.conf* configuration for Osaka is nearly identical, except for the changes in IP address and names:

```
[general]
autokill=yes

register => osaka:welcome@192.168.1.104

[toronto]
type=friend
host=dynamic
trunk=yes
secret=welcome
context=incoming_toronto
deny=0.0.0.0/0.0.0.0
permit=192.168.1.104/255.255.255.255
```

IAX Phone Configuration

In the "Configure the Softphone" section, we configured our first IAX2 softphone using idefisk. The configuration we'll be using here is nearly identical except for minor changes in order to cause the peers to be unique. If you've already configured a SIP softphone, then you can also utilize that on one (or both) of the peers. Remember that Asterisk is a multiprotocol application, and you can send a call from a SIP phone to Asterisk, across an IAX2 trunk, and then down to another SIP phone (or H.323, MGCP, etc.).

On Osaka:

```
[1001]
type=friend
host=dynamic
context=phones
```

On Toronto:

```
[2001]
type=friend
host=dynamic
context=phones
```

Next, configure your IAX2 softphone to register to Asterisk. If the phone successfully registers, you'll see something like:

```
*CLI>      -- Registered IAX2 '1001' (UNAUTHENTICATED) at 192.168.1.104:4569
```

Configuring the Dialplan

In order to allow calling between our two Asterisk boxes over the IAX2 trunk, we need to configure a simple dialplan. The following dialplan will send all extensions in the 1000 range (1000–1999) to Osaka, and all extensions in the 2000 range (2000–2999) to Toronto. Our example is going to assume that you have configured a pair of IAX2 softphones, but feel free to utilize a SIP phone if you've already configured one (or two). Just be aware that you'll need to change the Dial() application to send the call to the SIP phone via the SIP protocol instead of IAX2 (i.e. Dial(SIP/${EXTEN},30) instead of Dial(IAX2/${EXTEN},30)).

The *extensions.conf* file on Toronto:

```
[globals]

[general]
autofallthrough=yes

[default]

[incoming_calls]

[phones]
include => internal
include => remote

[internal]
exten => _1XXX,1,NoOp()
exten => _1XXX,n,Dial(IAX2/${EXTEN},30)
exten => _1XXX,n,Playback(the-party-you-are-calling&is-curntly-unavail)
exten => _1XXX,n,Hangup()

[remote]
exten => _2XXX,1,NoOp()
exten => _2XXX,n,Dial(IAX2/toronto/${EXTEN})
exten => _2XXX,n,Hangup()

[toronto_incoming]
include => internal
```

The *extensions.conf* file on Osaka:

```
[globals]

[general]
autofallthrough=yes

[default]
```

```
[incoming_calls]

[phones]
include => internal
include => remote

[internal]
exten => _2XXX,1,NoOp()
exten => _2XXX,n,Dial(IAX2/${EXTEN},30)
exten => _2XXX,n,Playback(the-party-you-are-calling&is-curntly-unavail)
exten => _2XXX,n,Hangup()

[remote]
exten => _1XXX,1,NoOp()
exten => _1XXX,n,Dial(IAX2/osaka/${EXTEN})
exten => _1XXX,n,Hangup()

[osaka_incoming]
include => internal
```

Using Templates in Your Configuration Files

There is a little-known secret in Asterisk config files that is so brilliant that we had to devote a little section to it.

Let us say that you have 20 SIP phones that are all pretty much identical in terms of how they are configured. The documented way to create them is to specify the parameters for each. Part of such a *sip.conf* file might look like this:

```
[1000]
type=friend
context=internal
host=dynamic
disallow=all
allow=ulaw
dtmfmode=rfc2833
maibox=1000
secret=AllYourSetsAreBelongToUs

[1001]
type=friend
context=internal
host=dynamic
disallow=all
allow=ulaw
dtmfmode=rfc2833
maibox=1001
secret=AllYourSetsAreBelongToUs

[1002]
type=friend
context=internal
host=dynamic
```

```
disallow=all
allow=ulaw
dtmfmode=rfc2833
maibox=1002
secret=AllYourSetsAreBelongToUs
```

Seems like a lot of extra typing, cutting, and pasting, yes? And what if you decide that you are going to change the context for your sets to another name. Not looking good, is it?

Enter the template. Let's create the same SIP friends as we did above, only this time using the template construct:

```
[sets](!) ; <== note the exclamation point in parenthesis. That makes this a template.
type=friend
context=internal
host=dynamic
disallow=all
allow=ulaw
dtmfmode=rfc2833
secret=AllYourSetsAreBelongToUs

[1000](sets) ; <== note the template name in parenthesis. All of that templates
             ; settings will be inhereted.
maibox=1000

[1001](sets)
maibox=1001

[1002](sets)
maibox=1002
```

This is one of the best kept secrets of conf file creation. In our experience, very few people use this, but for no other reason than that they don't know about it. Well, that's about to change. Our goal is to see everyone using these from now on; and yes, we will be checking.

Debugging

Several methods of debugging are available in Asterisk. Once you've connected to the console, you can enable different levels of verbosity and debugging output, as well as protocol packet tracing. We'll take a look at the various options in this section.

Connecting to the Console

To connect to the Asterisk console, you can either start the server in the console directly (in which case you will not be able to exit out of the console without killing the Asterisk process), or start Asterisk as a daemon and then connect to a remote console.

To start the Asterisk process directly in the console, use the console flag:

```
# /usr/sbin/asterisk -c
```

To connect to a remote Asterisk console, start the daemon first and then connect with the -r flag:

```
# /usr/sbin/asterisk
# /usr/sbin/asterisk -r
```

If you are having a problem with a specific module not loading, or a module causing Asterisk to not load, start the Asterisk process with the -c flag to monitor the status of modules loading. For example, if you attempt to load the OSS channel driver (which allows the use of the CONSOLE channel), and Asterisk is unable to open /dev/dsp, you will receive the following error on startup:

```
WARNING[32174]: chan_oss.c:470 soundcard_init: Unable to open /dev/dsp:
No such file or directory
   == No sound card detected -- console channel will be unavailable
   == Turn off OSS support by adding 'noload=chan_oss.so' in /etc/
      asterisk/modules.conf
```

Enabling Verbosity and Debugging

Asterisk can output debugging information in the form of WARNING, NOTICE, and ERROR messages. These messages will give you information about your system, such as registrations, status, and progression of calls, and various other useful bits of information. Note that WARNING and NOTICE messages are not errors; however, ERROR messages should be investigated. To enable various levels of verbosity, use set verbose followed by a numerical value. Useful values range from 3 to 10. For example, to set the highest level of verbosity, use:

```
# set verbose 10
```

You can also enable core debugging messages with set debug followed by a numerical value. To enable DEBUG output on the console, you may need to enable it in the *logger.conf* file by adding debug to the console => statement, as follows:

```
console => warning,notice,error,event,debug
```

Useful values for set debug range from 3 to 10. For example:

```
# set debug 10
```

Conclusion

If you've worked through all of the sections in this chapter, you will have configured a pair of analog interfaces, a local SIP and IAX2 channel connected to a softphone and/ or a hardphone, and placed calls across servers using the SIP and IAX2 protocols. These configurations are quite basic, but they give us functional channels to work with. We will make use of them in the following chapters, while we learn to build more useful dialplans.

## CHAPTER 5

Dialplan Basics

*Everything should be made as simple as possible, but not
simpler.*

—Albert Einstein (1879–1955)

The dialplan is truly the heart of any Asterisk system, as it defines how Asterisk handles inbound and outbound calls. In a nutshell, it consists of a list of instructions or steps that Asterisk will follow. Unlike traditional phone systems, Asterisk's dialplan is fully customizable. To successfully set up your own Asterisk system, you will need to understand the dialplan.

If you have attempted to read some sample dialplans and found them overwhelming, or if you've tried to write an Asterisk dialplan and had no success, help is at hand. This chapter explains how dialplans work in a step-by-step manner and teaches the skills necessary to create your own. The examples have been designed to build upon one another, so feel free to go back and reread a section if something doesn't quite make sense. Please also note that this chapter is by no means an exhaustive survey of all the possible things dialplans can do; our aim is to cover just the fundamentals. We'll cover more advanced dialplan topics in later chapters.

Dialplan Syntax

The Asterisk dialplan is specified in the configuration file named *extensions.conf*.

 The *extensions.conf* file usually resides in the */etc/asterisk/* directory, but its location may vary depending on how you installed Asterisk. Other common locations for this file include */usr/local/asterisk/etc/* and */opt/asterisk/etc/*.

The dialplan is made up of four main concepts: contexts, extensions, priorities, and applications. In the next few sections, we'll cover each of these parts and explain how they work together. After explaining the role each of these elements plays in the dialplan, we will step you though the process of creating a basic, functioning dialplan.

Sample Configuration Files

If you installed the sample configuration files when you installed Asterisk, you will most likely have an existing *extensions.conf* file. Instead of starting with the sample file, we suggest that you build your *extensions.conf* file from scratch. This will be very beneficial, as it will give you a better understanding of dialplan concepts and fundamentals.

That being said, the sample *extensions.conf* file remains a fantastic resource, full of examples and ideas that you can use after you've learned the basic concepts. We suggest you rename the sample file to something like *extensions.conf.sample*. That way, you can refer to it in the future. You can also find the sample configuration files in the */configs/* directory of the Asterisk source.

Contexts

Dialplans are broken into sections called *contexts*. Contexts are named groups of extensions, which serve several purposes.

Contexts keep different parts of the dialplan from interacting with one another. An extension that is defined in one context is completely isolated from extensions in any other context, unless interaction is specifically allowed. (We'll cover how to allow interaction between contexts near the end of the chapter.)

As a simple example, let's imagine we have two companies sharing an Asterisk server. If we place each company's voice menu in its own context, they are effectively separated from each other. This allows us to independently define what happens when, say, extension 0 is dialed: people pressing 0 at Company A's voice menu will get Company A's receptionist, and callers pressing 0 at Company B's voice menu will get Company B's receptionist. (This example assumes, of course, that we've told Asterisk to transfer the calls to the receptionists when callers press 0.)

Contexts are denoted by placing the name of the context inside square brackets ([ ]). The name can be made up of the letters A through Z (upper- and lowercase), the numbers 0 through 9, and the hyphen and underscore.* For example, a context for incoming calls looks like this:

```
[incoming]
```

---

* Please note that the space is conspicuously absent from the list of allowed characters. Don't use spaces in your context names—you won't like the result!

Context names have a maximum length of 79 characters (80 characters −1 terminating null)

All of the instructions placed after a context definition are part of that context, until the next context is defined. At the beginning of the dialplan, there are two special contexts named [general] and [globals]. The [general] section contains a list of general dialplan settings (which you'll probably never have to worry about), and we will discuss the [globals] context the "Global variables" section; for now it's just important to know that these two contexts are special. As long as you avoid the names [general] and [globals], you may name your contexts anything you like.

When you define a channel (which is how you connect things to the system), one of the parameters that is defined in the channel definition is the context. In other words, the context is the point in the dialplan where connections from that channel will begin.

Another important use of contexts (perhaps the most important) is to provide security. By using contexts correctly, you can give certain callers access to features (such as long-distance calling) that aren't made available to others. If you don't design your dialplan carefully, you may inadvertently allow others to fraudulently use your system. Please keep this in mind as you build your Asterisk system.

 The *doc/* subdirectory of the Asterisk source code contains a very important file named *security.txt*, which outlines several steps you should take to keep your Asterisk system secure. It is vitally important that you read and understand this file. If you ignore the security precautions outlined there, you may end up allowing anyone and everyone to make long-distance or toll calls at your expense!

If you don't take the security of your Asterisk system seriously, you may end up paying—literally! *Please* take the time and effort to secure your system from toll fraud.

Extensions

In the world of telecommunications, the word *extension* usually refers to a numeric identifier given to a line that rings a particular phone. In Asterisk, however, an extension is far more powerful, as it defines a unique series of steps (each step containing an application) that Asterisk will take that call through. Within each context, we can define as many (or few) extensions as required. When a particular extension is triggered (by an incoming call or by digits being dialed on a channel), Asterisk will follow the steps defined for that extension. It is the extensions, therefore, that specify what happens to calls as they make their way through the dialplan. Although extensions can certainly be used to specify phone extensions in the traditional sense (i.e., extension 153 will cause the SIP telephone set on John's desk to ring), in an Asterisk dialplan, they can be used for much more.

The syntax for an extension is the word **exten**, followed by an arrow formed by the equals sign and the greater-than sign, like this:

```
exten =>
```

This is followed by the name (or number) of the extension. When dealing with traditional telephone systems, we tend to think of extensions as the numbers you would dial to make another phone ring. In Asterisk, you get a whole lot more; for example, extension names can be any combination of numbers and letters. Over the course of this chapter and the next, we'll use both numeric and alphanumeric extensions.

 Assigning names to extensions may seem like a revolutionary concept, but when you realize that many VoIP transports support (or even actively encourage) dialing by name or email address instead of only dialing by number, it makes perfect sense. This is one of the features that makes Asterisk so flexible and powerful.

A complete extension is composed of three components:

- The name (or number) of the extension
- The priority (each extension can include multiple steps; the step number is called the "priority")
- The application (or command) that performs some action on the call

These three components are separated by commas, like this:

```
exten => name,priority,application()
```

Here's a simple example of what a real extension might look like:

```
exten => 123,1,Answer()
```

In this example, the extension name is **123**, the priority is **1**, and the application is **Answer()**. Now, let's move ahead and explain priorities and applications.

Priorities

Each extension can have multiple steps, called *priorities*. Each priority is numbered sequentially, starting with 1, and executes one specific application. As an example, the following extension would answer the phone (in priority number 1), and then hang it up (in priority number 2):

```
exten => 123,1,Answer()
exten => 123,2,Hangup()
```

Don't worry if you don't understand what `Answer()` and `Hangup()` are—we'll cover them shortly. The key point to remember here is that for a particular extension, Asterisk follows the priorities in order.

Unnumbered priorities

In older releases of Asterisk, the numbering of priorities caused a lot of problems. Imagine having an extension that had 15 priorities, and then needing to add something at step 2. All of the subsequent priorities would have to be manually renumbered. Asterisk does not handle missing steps or misnumbered priorities, and debugging these types of errors was pointless and frustrating.

Beginning with version 1.2, Asterisk addressed this problem. It introduced the use of the n priority, which stands for "next." Each time Asterisk encounters a priority named n, it takes the number of the previous priority and adds 1. This makes it easier to make changes to your dialplan, as you don't have to keep renumbering all your steps. For example, your dialplan might look something like this:

```
exten => 123,1,Answer()
exten => 123,n,do something
exten => 123,n,do something else
exten => 123,n,do one last thing
exten => 123,n,Hangup()
```

Internally, Asterisk will calculate the next priority number every time it encounters an n.[†] You should note, however, that you *must always* specify priority number 1. If you accidentally put an n instead of **1** for the first priority, you'll find that the extension will not be available.

Priority labels

Starting with Asterisk version 1.2 and higher, common practice is to assign text labels to priorities. This is to ensure that you can refer to a priority by something other than its number, which probably isn't known, given that dialplans now generally use unnumbered priorities. To assign a text label to a priority, simply add the label inside parentheses after the priority, like this:

```
exten => 123,n(label),application()
```

A very common mistake when writing labels is to insert a comma between the n and the (, like this:

```
exten => 123,n,(label),application() ;<-- THIS IS NOT GOING TO WORK
```

This mistake will break that part of your dialplan, and you will get an error that the application cannot be found.

[†] Asterisk permits simple arithmetic within the priority, such as **n+200** or the priority **s** (for same), but their usage is considered to be an advanced topic. Please note that *extension* **s** and *priority* **s** are two distinct concepts.

In the next chapter, we'll cover how to jump between different priorities based on dialplan logic. You'll be seeing a lot more of priority labels, and you will be using them often in your dialplans.

Applications

Applications are the workhorses of the dialplan. Each application performs a specific action on the current channel, such as playing a sound, accepting touch-tone input, dialing a channel, hanging up the call, and so forth. In the previous example, you were introduced to two simple applications: Answer() and Hangup(). You'll learn more about how these work momentarily.

Some applications, such as Answer()and Hangup(), need no other instructions to do their jobs. Other applications require additional information. These pieces of information, called *arguments*, can be passed on to the applications to affect how they perform their actions. To pass arguments to an application, place them between the parentheses that follow the application name, separated by commas.

> Occasionally, you may also see the pipe character (|) being used as a separator between arguments, instead of a comma. Feel free to use whichever you prefer. For the examples in this book, we will be using the comma to separate arguments to an application, as the authors prefer the look of this syntax. You should be aware, however, that when Asterisk parses the dialplan, it converts any commas in the application arguments to pipes.

As we build our first dialplan in the next section, you'll learn to use applications (and their associated arguments) to your advantage.

A Simple Dialplan

Now we're ready to create our first dialplan. We'll start with a very simple example. We are going to instruct Asterisk to answer a call, play a sound file, and hang up. We'll use this simple example to point out the most important dialplan fundamentals.

For the examples in this chapter to work correctly, we're assuming that at least one channel (either Zap, SIP, or IAX2) has been created and configured (as described in the previous chapter), and that all calls coming into that channel enter the dialplan at the [incoming] context. If you have been creative with any previous examples, you may need to make adjustments to fit your particular channel names.

The s Extension

Because of the technology we are using in our channels, we need to cover one more thing before we get started with our dialplan. We need to explain extension **s**. When calls enter a context without a specific destination extension (for example, a ringing FXO line), they are passed to the **s** extension. (The **s** stands for "start," as this is where a call will start if no extension information was passed with the call.)

Since this is exactly what we need for our dialplan, let's begin to fill in the pieces. We will be performing three actions on the call (answer it, play a sound file, and hang it up), so our extension called **s** will need three priorities. We'll place the three priorities below [incoming], because we have decided that all incoming calls should start in this context.[‡]

```
[incoming]
exten => s,1,application()
exten => s,n,application()
exten => s,n,application()
```

Now all we need to do is fill in the applications, and we've created our first dialplan.

 Note that we could have numbered each priority as shown below, but this is no longer the preferred method, as it makes it harder to make changes to the dialplan at a later time:

```
[incoming]
exten => s,1,application()
exten => s,2,application()
exten => s,3,application()
```

The Answer(), Playback(), and Hangup() Applications

If we're going to answer the call, play a sound file, and then hang up, we'd better learn how to do just that. The Answer() application is used to answer a channel that is ringing. This does the initial setup for the channel that receives the incoming call. (A few applications don't require that you answer the channel first, but properly answering the channel before performing any other actions is a very good habit.) As we mentioned earlier, Answer() takes no arguments.

The Playback() application is used for playing a previously recorded sound file over a channel. When using the Playback() application, input from the user is simply ignored.

[‡] There is nothing special about any context name. We could have named this context [stuff_that_comes_in], and as long as that was the context assigned in the channel definition in *sip.conf*, *iax.conf*, *zaptel.conf*, et al., the channel would enter the dialplan in that context. Having said that, it is strongly recommended that you give your contexts names that help you to understand their purpose. Some good context names might include [incoming], [local_calls], [long_distance], [sip_telephones], [user_services], [experimental], [remote_locations], and so forth. Always remember that a context determines how a channel enters the dialplan, so name accordingly.

Asterisk comes with many professionally recorded sound files, which should be found in the default sounds directory (usually */var/lib/asterisk/sounds/*). When you compile Asterisk, you can choose to install various sets of sample sounds that have been recorded in a variety of languages and file formats. We'll be using these files in many of our examples. Several of the files in our examples come from the Extra Sound Package, so please take the time to install it (see Chapter 3). You can also have your own sound prompts recorded in the same voices as the stock prompts by visiting *http://thevoice.digium.com/*.

To use `Playback()`, specify a filename (without a file extension) as the argument. For example, `Playback(filename)` would play the sound file called *filename.gsm*, assuming it was located in the default sounds directory. Note that you can include the full path to the file if you want, like this:

```
Playback(/home/john/sounds/filename)
```

The previous example would play *filename.gsm* from the */home/john/sounds/* directory. You can also use relative paths from the Asterisk sounds directory as follows:

```
Playback(custom/filename)
```

This example would play *filename.gsm* from the *custom/* subdirectory of the default sounds directory (probably */var/lib/asterisk/sounds/custom/filename.gsm*). Note that if the specified directory contains more than one file with that filename but with different file extensions, Asterisk automatically plays the best file.[§]

The `Hangup()` application does exactly as its name implies: it hangs up the active channel. You should use this application at the end of a context when you want to end the current call to ensure that callers don't continue on in the dialplan in a way you might not have anticipated. The `Hangup()` application takes no arguments.

Our First Dialplan

Now that we have designed our extension, let's put together all the pieces to create our first dialplan. As is typical in many technology books (especially computer programming books), our first example will be called "Hello World!"

In the first priority of our extension, we'll answer the call. In the second, we'll play a sound file named *hello-world.gsm*, and in the third we'll hang up the call. Here's what the dialplan looks like:

[§] Asterisk selects the best file based on translation cost—that is, it selects the file that is the least CPU-intensive to convert to its native audio format. When you start Asterisk, it calculates the translation costs between the different audio formats (they often vary from system to system). You can see these translation costs by typing `show translation` at the Asterisk command-line interface. The numbers shown represent how many milliseconds it takes Asterisk to transcode one second of audio. We'll cover more about the different audio formats (known as *codecs*) in Chapter 8.

```
[incoming]
exten => s,1,Answer()
exten => s,n,Playback(hello-world)
exten => s,n,Hangup()
```

If you have a channel or two configured, go ahead and try it out![||] Simply create a file called *extensions.conf*, (probably in */etc/asterisk*) and insert the four lines of dialplan code we just designed. If it doesn't work, check the Asterisk console for error messages, and make sure your channels are assigned to the [incoming] context.

Even though this example is very short and simple, it emphasizes the core concepts of contexts, extensions, priorities, and applications. If you can get this to work, you have the fundamental knowledge on which all dialplans are built.

Let's build upon our example. After all, a phone system that simply plays a sound file and then hangs up the channel isn't that useful!

Building an Interactive Dialplan

The dialplan we just built was static; it will always perform the same actions on every call. We are going to start adding some logic to our dialplan so that it will perform different actions based on input from the user. To do this, we're going to need to introduce a few more applications.

The Background(), WaitExten(), and Goto() Applications

One of the most important keys to building interactive Asterisk dialplans is the Background()[#] application. Like Playback(), it plays a recorded sound file. Unlike Playback(), however, when the caller presses a key (or series of keys) on her telephone keypad, it interrupts the playback and goes to the extension that corresponds with the pressed digit(s). If a caller presses 5, for example, Asterisk will stop playing the sound prompt and send control of the call to the first priority of extension 5.

The most common use of the Background() application is to create voice menus (often called *auto-attendants* or *phone trees*). Many companies use voice menus to direct callers to the proper extensions, thus relieving their receptionists from having to answer every single call.

[||] In fact, if you don't have any channels configured, now is the time to do so. There is a real satisfaction that comes from passing your first call into an Asterisk system that you built from scratch. People get this funny grin on their face as they realize that they have just created a telephone system. This pleasure can be yours as well, so please, don't go any further until you have made this little dialplan work.

[#] It should be noted that some people expect that Background(), due to its name, would continue in the dialplan while the sound is being played, but its name refers to the fact that it is playing a sound in the background, while waiting for DTMF in the foreground.

Background() has the same syntax as Playback():

```
exten => 123,1,Answer()
exten => 123,n,Background(main-menu)
```

In earlier versions of Asterisk, if the Background() application finished playing the sound prompt and there were no more priorities in the current extension, Asterisk would sit and wait for input from the caller. Asterisk no longer does this by default. If you want Asterisk to wait for input from the caller after the sound prompt has finished playing, you can call the WaitExten() application. The WaitExten() application waits for the caller to enter DTMF digits, and is frequently called directly after the Background() application, like this:

```
exten => 123,1,Answer()
exten => 123,n,Background(main-menu)
exten => 123,n,WaitExten()
```

If you'd like the WaitExten() application to wait a specific number of seconds for a response (instead of using the default timeout), simply pass the number of seconds as the first argument to WaitExten(), like this:

```
exten => 123,n,WaitExten(5)
```

Both Background() and WaitExten() allow the caller to enter DTMF digits. Asterisk then attempts to find an extension in the current context that matches the digits that the caller entered. If Asterisk finds an unambiguous match, it will send the call to that extension. Let's demonstrate by adding a few lines to our example:

```
exten => 123,1,Answer()
exten => 123,n,Background(main-menu)
exten => 123,n,WaitExten()

exten => 2,1,Playback(digits/2)

exten => 3,1,Playback(digits/3)

exten => 4,1,Playback(digits/4)
```

If you call into extension 123 in the example above, it will play a sound prompt that says "main menu." It will then wait for you to enter either 2, 3, or 4. If you press one of those digits, Asterisk will read that digit back to you. You'll also find that if you enter a different digit (such as 5), it won't give you what you expected.

It is also possible that Asterisk will find an ambiguous match. This can be easily explained if we add an extension named 1 to the previous example:

```
exten => 123,1,Answer()
exten => 123,n,Background(main-menu)
exten => 123,n,WaitExten()

exten => 1,1,Playback(digits/1)

exten => 2,1,Playback(digits/2)
```

```
exten => 3,1,Playback(digits/3)

exten => 4,1,Playback(digits/4)
```

Dial extension 123, and then at the main menu prompt dial **1**. Why doesn't Asterisk immediately read back the number one to you? It's because the digit **1** is ambiguous; Asterisk doesn't know whether you're trying to go to extension 1 or extension 123. It waits a few seconds to see if you're going to dial another digit (such as the **2** in extension 123). If you don't dial any more digits, Asterisk will eventually time out and send the call to extension 1. (We'll learn how to choose our own timeout values in Chapter 6.)

Before going on, let's review what we've done so far. When users call into our dialplan, they will hear a greeting. If they press **1**, they will hear the number one, and if they press **2**, they will hear the number two, and so on. While that's a good start, let's embellish it a little. We'll use the `Goto()` application to make the dialplan repeat the greeting after playing back the number.

As its name implies, the `Goto()` application is used to send the call to another part of the dialplan. The syntax for the `Goto()` application requires us to pass the destination context, extension, and priority on as arguments to the application, like this:

```
exten => 123,n,Goto(context,extension,priority)
```

Now, let's use the `Goto()` application in our dialplan:

```
[incoming]
exten => 123,1,Answer()
exten => 123,n,Background(main-menu)

exten => 1,1,Playback(digits/1)
exten => 1,n,Goto(incoming,123,1)

exten => 2,1,Playback(digits/2)
exten => 2,n,Goto(incoming,123,1)
```

These two new lines (highlighted in bold) will send control of the call back to the **123** extension after playing back the selected number.

If you look up the details of the `Goto()` application, you'll find that you can actually pass either one, two, or three arguments to the application. If you pass a single argument, Asterisk will assume it's the destination priority in the current extension. If you pass two arguments, Asterisk will treat them as the extension and priority to go to in the current context.

In this example, we've passed all three arguments for the sake of clarity, but passing just the extension and priority would have had the same effect.

Handling Invalid Entries and Timeouts

Now that our first voice menu is starting to come together, let's add some additional special extensions. First, we need an extension for invalid entries; when a caller presses an invalid entry (e.g., pressing 9 in the above example), the call is sent to the i extension. Second, we need an extension to handle situations when the caller doesn't give input in time (the default timeout is 10 seconds). Calls will be sent to the t extension if the caller takes too long to press a digit after WaitExten() has been called. Here is what our dialplan will look like after we've added these two extensions:

```
[incoming]
exten => 123,1,Answer()
exten => 123,n,Background(enter-ext-of-person)
exten => 123,n,WaitExten()

exten => 1,1,Playback(digits/1)
exten => 1,n,Goto(incoming,123,1)

exten => 2,1,Playback(digits/2)
exten => 2,n,Goto(incoming,123,1)

exten => 3,1,Playback(digits/3)
exten => 3,n,Goto(incoming,123,1)

exten => i,1,Playback(pbx-invalid)
exten => i,n,Goto(incoming,123,1)

exten => t,1,Playback(vm-goodbye)
exten => t,n,Hangup()
```

Using the i and t extensions makes our dialplan a little more robust and user-friendly. That being said, it is still quite limited, because outside callers have no way of connecting to a live person. To do that, we'll need to learn about another application, called Dial().

Using the Dial() Application

One of Asterisk's most valuable features is its ability to connect different callers to each other. This is especially useful when callers are using different methods of communication. For example, caller A might be communicating over the traditional analog telephone network, while user B might be sitting in a café halfway around the world and speaking on an IP telephone. Luckily, Asterisk takes most of the hard work out of connecting and translating between disparate networks. All you have to do is learn how to use the Dial() application.

The syntax of the Dial() application is a little more complex than that of the other applications we've used so far, but don't let that scare you off. Dial() takes up to four arguments. The first is the destination you're attempting to call, which (in its simplest form) is made up of a technology (or transport) across which to make the call, a forward slash, and the remote endpoint or resource. Common technology types include Zap (for analog and T1/E1/J1 channels), SIP, and IAX2. For example, let's assume that we want to call a Zap endpoint identified by `Zap/1`, which is an FXS channel with an analog phone plugged into it. The technology is `Zap`, and the resource is `1`. Similarly, a call to a SIP device (as defined in *sip.conf*) might have a destination of `SIP/Jane`, and a call to an IAX device (defined in *iax.conf*) might have a destination of `IAX2/Fred`. If we wanted Asterisk to ring the `Zap/1` channel when extension 123 is reached in the dialplan, we'd add the following extension:

```
exten => 123,1,Dial(Zap/1)
```

We can also dial multiple channels at the same time, by concatenating the destinations with an ampersand (&), like this:

```
exten => 123,1,Dial(Zap/1&Zap/2&SIP/Jane)
```

The `Dial()` application will ring the specified destinations simultaneously, and bridge the inbound call with whichever destination channel answers the call first. If the `Dial()` application can't contact any of the destinations, Asterisk will set a variable called `DIALSTATUS` with the reason that it couldn't dial the destinations, and continue on with the next priority in the extension.[*]

The `Dial()` application also allows you to connect to a remote VoIP endpoint not previously defined in one of the channel configuration files. The full syntax for this type of connection is:

```
Dial(technology/user[:password]@remote_host[:port][/remote_extension])
```

As an example, you can dial into a demonstration server at Digium using the IAX2 protocol by using the following extension:

```
exten => 500,1,Dial(IAX2/guest@misery.digium.com/s)
```

The full syntax for the `Dial()` application is slightly different when dealing with Zap channels, as shown:

```
Dial(Zap/[gGrR]channel_or_group[/remote_extension])
```

For example, here is how you would dial 1-800-555-1212 on Zap channel number 4.

```
exten => 501,1,Dial(Zap/4/18005551212)
```

The second argument to the `Dial()` application is a timeout, specified in seconds. If a timeout is given, `Dial()` will attempt to call the destination(s) for that number of seconds before giving up and moving on to the next priority in the extension. If no timeout is specified, `Dial()` will continue to dial the called channel(s) until someone answers or the caller hangs up. Let's add a timeout of 10 seconds to our extension:

```
exten => 123,1,Dial(Zap/1,10)
```

---

[*] Don't worry, we'll cover variables (in "Using Variables") and show you how to have your dialplan make decisions based on the value of this `DIALSTATUS` variable.

If the call is answered before the timeout, the channels are bridged and the dialplan is done. If the destination simply does not answer, is busy, or is otherwise unavailable, Asterisk will set a variable called DIALSTATUS and then continue on with the next priority in the extension.

Let's put what we've learned so far into another example:

```
exten => 123,1,Dial(Zap/1,10)
exten => 123,n,Playback(vm-nobodyavail)
exten => 123,n,Hangup()
```

As you can see, this example will play the *vm-nobodyavail.gsm* sound file if the call goes unanswered.

The third argument to Dial() is an option string. It may contain one or more characters that modify the behavior of the Dial() application. While the list of possible options is too long to cover here, one of the most popular options is the m option. If you place the letter m as the third argument, the calling party will hear hold music instead of ringing while the destination channel is being called (assuming, of course, that music on hold has been configured correctly). To add the m option to our last example, we simply change the first line:

```
exten => 123,1,Dial(Zap/1,10,m)
exten => 123,n,Playback(vm-nobodyavail)
exten => 123,n,Hangup()
```

Since the extensions numbered 1 and 2 in our dialplan are somewhat useless now that we know how to use the Dial() application, let's replace them with new extensions that will allow outside callers to reach John and Jane:

```
[incoming]
exten => 123,1,Answer()
exten => 123,n,Background(enter-ext-of-person)
exten => 123,n,WaitExten()

exten => 1,1,Dial(Zap/1,10)
exten => 1,n,Playback(vm-nobodyavail)
exten => 1,n,Hangup()

exten => 2,1,Dial(SIP/Jane,10)
exten => 2,n,Playback(vm-nobodyavail)
exten => 2,n,Hangup()

exten => i,1,Playback(pbx-invalid)
exten => i,n,Goto(incoming,123,1)

exten => t,1,Playback(vm-goodbye)
exten => t,n,Hangup()
```

The fourth and final argument to the Dial() application is a URL. If the destination channel supports receiving a URL at the time of the call, the specified URL will be sent (for example, if you have an IP telephone that supports receiving a URL, it will appear on the phone's display; likewise, if you're using a soft phone, the URL might pop up on your computer screen). This argument is very rarely used.

Note that the second, third, and fourth arguments may be left blank. For example, if you want to specify an option but not a timeout, simply leave the timeout argument blank, like this:

```
exten => 1,1,Dial(Zap/1,,m)
```

Adding a Context for Internal Calls

In our examples thus far, we have limited ourselves to a single context, but it is probably fair to assume that almost all Asterisk installations will have more than one context in their dialplans. As we mentioned at the beginning of this chapter, one important function of contexts is to separate privileges (such as making long-distance calls or calling certain extensions) for different classes of callers. In our next example, we'll add to our dialplan by creating two internal phone extensions, and we'll set up the ability for these two extensions to call each other. To accomplish this, we'll create a new context called [employees].

> As in previous examples, we've assumed that an FXS analog channel (Zap/1, in this case) has already been configured, and that your *zapata.conf* file is configured so that any calls originated by Zap/1 begin in the [employees] context. For a few examples at the end of the chapter, we'll also assume that an FXO Zap channel has been configured as Zap/4, with calls coming in on this channel being sent to the [incoming] context.
>
> We've also assumed you have at least one SIP channel (named SIP/Jane) that is configured to originate in the [employees] context. We've done this to introduce you to using other types of channels.
>
> If you don't have hardware for the channels listed above (such as Zap/4), or if you're using hardware with different channel names (e.g., not SIP/Jane), just change the examples to match your particular system configuration.

Our dialplan now looks like this:

```
[incoming]
exten => 123,1,Answer()
exten => 123,n,Background(enter-ext-of-person)
exten => 123,n,WaitExten()

exten => 1,1,Dial(Zap/1,10)
exten => 1,n,Playback(vm-nobodyavail)
exten => 1,n,Hangup()

exten => 2,1,Dial(SIP/Jane,10)
exten => 2,n,Playback(vm-nobodyavail)
```

```
exten => 2,n,Hangup()

exten => i,1,Playback(pbx-invalid)
exten => i,n,Goto(incoming,123,1)

exten => t,1,Playback(vm-goodbye)
exten => t,n,Hangup()

[employees]
exten => 101,1,Dial(Zap/1)

exten => 102,1,Dial(SIP/Jane)
```

In this example, we have added two new extensions to the [employees] context. This way, the person using channel Zap/1 can pick up the phone and dial the person at channel SIP/Jane by dialing 102. By that same token, the phone registered as SIP/Jane can dial Zap/1 by dialing 101.

We've arbitrarily decided to use extensions 101 and 102 for our examples, but feel free to use whatever numbering convention you wish for your extensions. You should also be aware that you're not limited to three-digit extensions; you can use as few or as many digits as you like. (Well, almost. Extensions must be shorter than 80 characters long, and you shouldn't use single-character extensions for your own use, as they're reserved.) Don't forget that you can use names as well, like so:

```
[incoming]
exten => 123,1,Answer()
exten => 123,n,Background(enter-ext-of-person)
exten => 123,n,WaitExten()

exten => 1,1,Dial(Zap/1,10)
exten => 1,n,Playback(vm-nobodyavail)
exten => 1,n,Hangup()

exten => 2,1,Dial(SIP/Jane,10)
exten => 2,n,Playback(vm-nobodyavail)
exten => 2,n,Hangup()

exten => i,1,Playback(pbx-invalid)
exten => i,n,Goto(incoming,123,1)

exten => t,1,Playback(vm-goodbye)
exten => t,n,Hangup()

[employees]
exten => 101,1,Dial(Zap/1)
exten => john,1,Dial(Zap/1)

exten => 102,1,Dial(SIP/Jane)
exten => jane,1,Dial(SIP/Jane)
```

It certainly wouldn't hurt to add named extensions if you think your users might be dialed via a VoIP protocol such as SIP that supports dialing by name. You can also see that it is possible to have different extensions in the dialplan ring the same endpoint. For example, you could have extension 200 ring SIP/George, and then have extension 201 play a prompt of some kind and *then* ring SIP/George.

Now that our internal callers can call each other, we're well on our way toward having a complete dialplan. Next, we'll see how we can make our dialplan more scalable and easier to modify in the future.

Using Variables

Variables can be used in an Asterisk dialplan to help reduce typing, add clarity, or add additional logic to a dialplan. If you have some computer programming experience, you probably already understand what a variable is. If not, don't worry; we'll explain what variables are and how they are used.

You can think of a variable as a container that can hold one value at a time. So, for example, we might create a variable called JOHN and assign it the value of Zap/1. This way, when we're writing our dialplan, we can refer to John's channel by name, instead of remembering that John is using the channel named Zap/1.

There are two ways to reference a variable. To reference the variable's name, simply type the name of the variable, such as JOHN. If, on the other hand, you want to reference its value, you must type a dollar sign, an opening curly brace, the name of the variable, and a closing curly brace. Here's how we'd reference the variable inside the Dial() application:

```
exten => 555,1,Dial(${JOHN})
```

In our dialplan, whenever we write ${JOHN}, Asterisk will automatically replace it with whatever value has been assigned to the variable named JOHN.

> Note that variable names are case-sensitive. A variable named JOHN is different than a variable named John. For readability's sake, all the variable names in the examples will be written in uppercase. You should also be aware that any variables set by Asterisk will be uppercase as well. Some variables, such as CHANNEL or EXTEN are reserved by Asterisk. You should not attempt to set these variables.

There are three types of variables we can use in our dialplan: global variables, channel variables, and environment variables. Let's take a moment to look at each type.

Global variables

As their name implies, *global* variables apply to all extensions in all contexts. Global variables are useful in that they can be used anywhere within a dialplan to increase readability and manageability. Suppose for a moment that you had a large dialplan and several hundred references to the Zap/1 channel. Now imagine you had to go through your dialplan and change all of those references to Zap/2. It would be a long and error-prone process, to say the least.

On the other hand, if you had defined a global variable with the value Zap/1 at the beginning of your dialplan and then referenced that instead, you would have to change only one line.

Global variables should be declared in the [globals] context at the beginning of the *extensions.conf* file. They can also be defined programmatically, using the GLOBAL() dialplan function.[†] Here is an example of how both methods look inside of a dialplan. The first shows the setting of a global variable named JOHN with a value of Zap/1. This variable is set at the time Asterisk parses the dialplan. The second example shows how a global variable can be set in the dialplan. In this case, the variable named George is being assigned the value of SIP/George when extension 124 is dialed in the [employees] context:

```
[globals]
JOHN=Zap/1

[employees]
exten => 124,1,Set(GLOBAL(GEORGE)=SIP/George)
```

Channel variables

A *channel* variable is a variable that is associated only with a particular call. Unlike global variables, channel variables are defined only for the duration of the current call and are available only to the channels participating in that call.

There are many predefined channel variables available for use within the dialplan, which are explained in the *channelvariables.txt* file in the *doc* subdirectory of the Asterisk source. Channel variables are set via the Set() application:

```
exten => 125,1,Set(MAGICNUMBER=42)
```

We'll cover many uses for channel variables in Chapter 6.

Environment variables

Environment variables are a way of accessing Unix environment variables from within Asterisk. These are referenced using the ENV() dialplan function. The syntax looks like ${ENV(*var*)}, where *var* is the Unix environment variable you wish to reference. Environment variables aren't commonly used in Asterisk dialplans, but they are available should you need them.

Adding variables to our dialplan

Now that we've learned about variables, let's put them to work in our dialplan. We'll add global variables for two people, John and Jane:

[†] Don't worry! We'll cover dialplan functions in the "Dialplan Functions" section.

```
[globals]
JOHN=Zap/1
JANE=SIP/Jane

[incoming]
exten => 123,1,Answer()
exten => 123,n,Background(enter-ext-of-person)
exten => 123,n,WaitExten()

exten => 1,1,Dial(${JOHN},10)
exten => 1,n,Playback(vm-nobodyavail)
exten => 1,n,Hangup()

exten => 2,1,Dial(${JANE},10)
exten => 2,n,Playback(vm-nobodyavail)
exten => 2,n,Hangup()

exten => i,1,Playback(pbx-invalid)
exten => i,n,Goto(incoming,123,1)

exten => t,1,Playback(vm-goodbye)
exten => t,n,Hangup()

[employees]

exten => 101,1,Dial(${JOHN})
exten => john,1,Dial(${JOHN})

exten => 102,1,Dial(${JANE})
exten => jane,1,Dial(${JANE})
```

Pattern Matching

If we want to be able to allow people to dial *through* Asterisk and have Asterisk connect the caller to an outside resource, we need a way to match on any possible phone number that the caller might dial. Can you imagine how tedious it would be to manually write a dialplan with an extension for every possible number you could dial? Luckily, Asterisk has just the thing for situations like this: *pattern matching*. Pattern matching allows you to create one extension in your dialplan that matches many different numbers.

Pattern-matching syntax

When using pattern matching, certain letters and symbols represent what we are trying to match. Patterns always start with an underscore (_). This tells Asterisk that we're matching on a pattern, and not on an explicit extension name. (This means, of course, that you should never start your extension names with an underscore.)

 If you forget the underscore on the front of your pattern, Asterisk will think it's just a named extension and won't do any pattern matching. This is one of the most common mistakes people make when starting to learn Asterisk.

After the underscore, you can use one or more of the following characters.

X

Matches any single digit from 0 to 9.

Z

Matches any single digit from 1 to 9.

N

Matches any single digit from 2 to 9.

[15-7]

Matches a single digit from the range of digits specified. In this case, the pattern matches a single 1, 5, 6, or 7.

. *(period)*

Wildcard match; matches *one or more* characters, no matter what they are.

> If you're not careful, wildcard matches can make your dialplans do things you're not expecting (like matching built-in extensions such as i or h). You should use the wildcard match in a pattern only after you've matched as many other digits as possible. For example, the following pattern match should probably never be used:
>
> ```
> _.
> ```
>
> In fact, Asterisk will warn you if you try to use it. Instead, use this one, if at all possible:
>
> ```
> _X.
> ```

! *(bang)*

Wildcard match; matches *zero or more* characters, no matter what they are.

To use pattern matching in your dialplan, simply put the pattern in the place of the extension name (or number):

```
exten => _NXX,1,Playback(auth-thankyou)
```

In this example, the pattern matches any three-digit extension from 200 through 999 (the N matches any digit between 2 and 9, and each X matches a digit between 0 and 9). That is to say, if a caller dialed any three-digit extension between 200 and 999 in this context, he would hear the sound file *auth-thankyou.gsm*.

One other important thing to know about pattern matching is that if Asterisk finds more than one pattern that matches the dialed extension, it will use the *most specific* one (going from left to right). Say you had defined the following two patterns, and a caller dialed 555-1212:

```
exten => _555XXXX,1,Playback(digits/1)
exten => _55512XX,1,Playback(digits/2)
```

In this case the second extension would be selected, because it is more specific.

Pattern-matching examples

Before we go on, let's look at a few more pattern-matching examples. In each one, see if you can tell what the pattern would match before reading the explanation. We'll start with an easy one:

    _NXXXXXX

This pattern would match any seven-digit number, as long as the first digit was two or higher. This pattern would be compatible with any North American Numbering Plan local seven-digit number. In areas with 10-digit dialing, that pattern would look like this:

    _NXXNXXXXXX

Note that neither of these two patterns would handle long distance calls. We'll cover those shortly.

The NANP and Toll Fraud

The North American Number Plan (NANP) is a shared telephone numbering scheme used by 19 countries in North America and the Caribbean. Countries within NANP share country code 1.

In the United States and Canada, telecom regulations are similar (and sensible) enough that you can place a long-distance call to most numbers in country code 1 and expect to pay a reasonable toll. What many people don't realize, however, is that 19 countries, many of which have very different telecom regulations, share the NANP. (More information can be found at *http://www.nanpa.com*.)

One popular scam using the NANP tries to trick naive North Americans into calling expensive per-minute toll numbers in a Caribbean country; the callers believe that since they dialed 1-NPA-NXX-XXXX to reach the number, they'll be paying their standard national long-distance rate for the call. Since the country in question may have regulations that allow for this form of extortion, the caller is ultimately held responsible for the call charges.

The only way to prevent this sort of activity is to block calls to certain area codes (809, for example) and remove the restrictions only on an as-needed basis.

Let's try another:

    _1NXXNXXXXXX

This one is slightly more difficult. This would match the number 1, followed by an area code between 200 and 999, then any 7-digit number. In the NANP calling area, you would use this pattern to match any long-distance number.[‡]

Now for an even trickier example:

    _011.

If that one left you scratching your head, look at it again. Did you notice the period on the end? This pattern matches any number that starts with 011 and has at least one more digit. In the NANP, this indicates an international phone number. (We'll be using these patterns in the next section to add outbound dialing capabilities to our dialplan.)

Using the ${EXTEN} channel variable

We know what you're thinking... You're sitting there asking yourself, "So what happens if I want to use pattern matching, but I need to know which digits were actually dialed?" Luckily, Asterisk has just the answer. Whenever you dial an extension, Asterisk sets the ${EXTEN} channel variable to the digits that were dialed. We can use an application called SayDigits() to test it out:

```
exten => _XXX,1,SayDigits(${EXTEN})
```

In this example, the SayDigits() application will read back to you the three-digit extension you dialed.

Often, it's useful to manipulate the ${EXTEN} by stripping a certain number of digits off the front of the extension. This is accomplished by using the syntax ${EXTEN:*x*}, where *x* is where you want the returned string to start, from left to right. For example, if the value of EXTEN is 95551212, ${EXTEN:1} equals 5551212. Let's take a look at another example:

```
exten => _XXX,1,SayDigits(${EXTEN:1})
```

In this example, the SayDigits() application would start at the second digit, and thus read back only the last two digits of the dialed extension.

More Advanced Digit Manipulation

The ${EXTEN} variable properly has the syntax ${EXTEN:*x*:*y*}, where *x* is the starting position, and *y* is the number of digits to return. Given the following dial string:

```
94169671111
```

we can extract the following digit strings using the ${EXTEN:x:y} construct:

${EXTEN:1:3} would contain 416.

${EXTEN:4:7} would contain 9671111.

${EXTEN:-4:4} would start four digits from the end, and return four digits, giving us 1111.

---

‡ If you grew up in North America, you may believe that the 1 you dial before a long distance call is "the long distance code." This is incorrect. The number 1 is in fact the international country code for all countries in NANP. Keep this in mind if you ever send your phone number to someone in another country. They may not know what your country code is, and thus be unable to call you with just your area code and phone number. Your full phone number with country code should be printed as +1 NPA NXX XXXX (where NPA is your area code)–e.g., +1 416 555 1212.

${EXTEN:1} would give us everything after the first digit, 4169671111 (if the number of digits to return is left blank, it will return the entire remaining string).

This is a very powerful construct, but most of these variations are not very common in normal use. For the most part, you will be using ${EXTEN:1} to strip off your external access code.

Enabling Outbound Dialing

Now that we've introduced pattern matching, we can go about the process of allowing users to make outbound calls. The first thing we'll do is add a variable to the [globals] context to define which channel will be used for outbound calls:

```
[globals]
JOHN=Zap/1
JANE=SIP/Jane
OUTBOUNDTRUNK=Zap/4
```

Next, we will add contexts to our dialplan for outbound dialing.

You may be asking yourself at this point, "Why do we need separate contexts for outbound calls?" This is so that we can regulate and control which callers have permission to make outbound calls, and which types of outbound calls they are allowed to make.

To begin, let's create a context for local calls. To be consistent with most traditional phone switches, we'll put a 9 on the front of our patterns, so that users have to dial 9 before calling an outside number:

```
[outbound-local]
exten => _9NXXXXXX,1,Dial(${OUTBOUNDTRUNK}/${EXTEN:1})
exten => _9NXXXXXX,n,Congestion()
exten => _9NXXXXXX,n,Hangup()
```

> Note that dialing 9 doesn't actually give you an outside line, unlike with many traditional PBX systems. Once you dial 9 on an analog line, the dial tone will stop. If you'd like the dial tone to continue even after dialing 9, add the following line (right after your context definition):
>
> ```
> ignorepat => 9
> ```
>
> This directive tells Asterisk to continue to provide a dial tone on an analog line, even after the caller has dialed the indicated pattern. This will not work with VoIP telephones, as they usually don't send digits to the system as they are input; they are sent to Asterisk all at once. Luckily, most of the popular VoIP telephones can be configured to emulate the same functionality.

Let's review what we've just done. We've added a global variable called OUTBOUND TRUNK, which simply defines the channel we are using for outbound calls.[§] We've also added a context for local outbound calls. In priority 1, we take the dialed extension, strip off the 9 with the ${EXTEN:1} syntax, and then attempt to dial that number on the channel signified by the variable OUTBOUNDTRUNK. If the call is successful, the caller is bridged with the outbound channel. If the call is unsuccessful (because either the channel is busy or the number can't be dialed for some reason), the Congestion() application is called, which plays a "fast busy signal" (congestion tone) to let the caller know that the call was unsuccessful.

Before we go any further, let's make sure our dialplan allows outbound emergency numbers:

```
[outbound-local]
exten => _9NXXXXXX,1,Dial(${OUTBOUNDTRUNK}/${EXTEN:1})
exten => _9NXXXXXX,n,Congestion()
exten => _9NXXXXXX,n,Hangup()

exten => 911,1,Dial(${OUTBOUNDTRUNK}/911)
exten => 9911,1,Dial(${OUTBOUNDTRUNK}/911) ; So that folks who dial "9"
                                           ; first will also get through
```

Again, we're assuming for the sake of these examples that we're inside the United States or Canada. If you're outside of this area, please replace 911 with the emergency services number in your particular location. This is something you never want to forget to put in your dialplan!

Next, let's add a context for long-distance calls:

```
[outbound-long-distance]
exten => _91NXXNXXXXXX,1,Dial(${OUTBOUNDTRUNK}/${EXTEN:1})
exten => _91NXXNXXXXXX,n,Playtones(congestion)
exten => _91NXXNXXXXXX,n,Hangup()
```

Now that we have these two new contexts, how do we allow internal users to take advantage of them? We need a way for contexts to be able to use the functionality contained in other contexts.

Includes

Asterisk has a feature that enables us to use the extensions from one context within another context via the include directive. This is used to control access to different sections of the dialplan. We'll use the include functionality to allow users in our [employees] context the ability to make outbound phone calls. But first, let's cover the syntax.

The include statement takes the following form, where *context* is the name of the remote context we want to include in the current context:

```
include => context
```

---

§ The advantage of this is that if one day we decide to send all of our calls through some other channel, we have to edit the channel name assigned to the variable OUTBOUNDTRUNK only in the [globals] context, instead of having to manually edit every reference to the channel in our dialplan.

When we include other contexts within our current context, we have to be mindful of the order in which we are including them. Asterisk will first try to match the dialed extension in the current context. If unsuccessful, it will then try the first included context (including any contexts included in that context), and then continue to the other included contexts in the order in which they were included.

As it sits, our current dialplan has two contexts for outbound calls, but there's no way for people in the [employees] context to use them. Let's remedy that by including the two outbound contexts in the [employees] context, like this:

```
[globals]
JOHN=Zap/1
JANE=SIP/Jane
OUTBOUNDTRUNK=Zap/4

[incoming]
exten => 123,1,Answer()
exten => 123,n,Background(enter-ext-of-person)
exten => 123,n,WaitExten()

exten => 1,1,Dial(${JOHN},10)
exten => 1,n,Playback(vm-nobodyavail)
exten => 1,n,Hangup()

exten => 2,1,Dial(${JANE},10)
exten => 2,n,Playback(vm-nobodyavail)
exten => 2,n,Hangup()

exten => i,1,Playback(pbx-invalid)
exten => i,n,Goto(incoming,123,1)

exten => t,1,Playback(vm-goodbye)
exten => t,n,Hangup()

[employees]
include => outbound-local
include => outbound-long-distance

exten => 101,1,Dial(${JOHN})
exten => john,1,Dial(${JOHN})
exten => 102,1,Dial(${JANE})
exten => jane,1,Dial(${JANE})

[outbound-local]
exten => _9NXXXXXX,1,Dial(${OUTBOUNDTRUNK}/${EXTEN:1})
exten => _9NXXXXXX,n,Congestion()
exten => _9NXXXXXX,n,Hangup()

exten => 911,1,Dial(${OUTBOUNDTRUNK}/911)
exten => 9911,1,Dial(${OUTBOUNDTRUNK}/911)

[outbound-long-distance]
exten => _91NXXNXXXXXX,1,Dial(${OUTBOUNDTRUNK}/${EXTEN:1})
```

```
exten => _91NXXNXXXXXX,n,Playtones(congestion)
exten => _91NXXNXXXXXX,n,Hangup()
```

These two `include` statements make it possible for callers in the `[employees]` context to make outbound calls. We should also note that for security's sake you should always make sure that your `[inbound]` context never allows outbound dialing. (If by chance it did, people could dial into your system and then make outbound toll calls that would be charged to you!)

Conclusion

And there you have it—a basic but functional dialplan. It's not exactly fully featured, but we've covered all of the fundamentals. In the following chapters, we'll continue to add features to this foundation.

If parts of this dialplan don't make sense, you may want to go back and re-read a section or two before continuing on to the next chapter. It's imperative that you understand these principles and how to apply them, as the next chapters build on this information.

## CHAPTER 6

More Dialplan Concepts

For a list of all the ways technology has failed to improve
the quality of life, please press three.

—Alice Kahn

Alrighty. You've got the basics of dialplans down, but you know there's more to come. If you don't have the last chapter sorted out yet, please go back and give it another read. We're about to get into more advanced topics.

Expressions and Variable Manipulation

As we begin our dive into the deeper aspects of dialplans, it is time to introduce you to a few tools that will greatly add to the power you can exercise in your dialplan. These constructs add incredible intelligence to your dialplan by enabling it to make decisions based on different criteria you want to define. Put on your thinking cap, and let's get started.

Basic Expressions

Expressions are combinations of variables, operators, and values that you string together to produce a result. An expression can test values, alter strings, or perform mathematical calculations. Let's say we have a variable called COUNT. In plain English, two expressions using that variable might be "COUNT plus 1" and "COUNT divided by 2." Each of these expressions has a particular result or value, depending on the value of the given variable.

In Asterisk, expressions always begin with a dollar sign and an opening square bracket and end with a closing square bracket, as shown here:

```
$[expression]
```

Thus, we would write the above two examples like this:

```
$[${COUNT} + 1]
$[${COUNT} / 2]
```

When Asterisk encounters an expression in a dialplan, it replaces the entire expression with the resulting value. It is important to note that this takes place *after* variable substitution. To demonstrate, let's look at the following code:[*]

```
exten => 321,1,Set(COUNT=3)
exten => 321,n,Set(NEWCOUNT=$[${COUNT} + 1])
exten => 321,n,SayNumber(${NEWCOUNT})
```

In the first priority, we assign the value of 3 to the variable named COUNT.

In the second priority, only one application—Set()—is involved, but three things actually happen:

1. Asterisk substitutes ${COUNT} with the number 3 in the expression. The expression effectively becomes this:

    ```
    exten => 321,n,Set(NEWCOUNT=$[3 + 1])
    ```

2. Asterisk evaluates the expression, adding 1 to 3, and replaces it with its computed value of 4:

    ```
    exten => 321,n,Set(NEWCOUNT=4)
    ```

3. The value 4 is assigned to the NEWCOUNT variable by the Set() application.

The third priority simply invokes the SayNumber() application, which speaks the current value of the variable ${NEWCOUNT} (set to the value 4 in priority two).

Try it out in your own dialplan.

Operators

When you create an Asterisk dialplan, you're really writing code in a specialized scripting language. This means that the Asterisk dialplan—like any programming language—recognizes symbols called *operators* that allow you to manipulate variables. Let's look at the types of operators that are available in Asterisk:

Boolean operators

These operators evaluate the "truth" of a statement. In computing terms, that essentially refers to whether the statement is something or nothing (nonzero or zero, true or false, on or off, and so on). The Boolean operators are:

expr1 | expr2

This operator (called the "or" operator, or "pipe") returns the evaluation of *expr1* if it is true (neither an empty string nor zero). Otherwise, it returns the evaluation of *expr2*.

---

[*] Remember that when you *reference* a variable you can call it by its name, but when you refer to a variable's *value*, you have to use the dollar sign and brackets around the variable name.

expr1 & expr2

> This operator (called "and") returns the evaluation of *expr1* if both expressions are true (i.e., neither expression evaluates to an empty string or zero). Otherwise, it returns zero.

expr1 {=, >, >=, <, <=, !=} expr2

> These operators return the results of an integer comparison if both arguments are integers; otherwise, they return the results of a string comparison. The result of each comparison is 1 if the specified relation is true, or 0 if the relation is false. (If you are doing string comparisons, they will be done in a manner that's consistent with the current local settings of your operating system.)

Mathematical operators

> Want to perform a calculation? You'll want one of these:

expr1 {+, -} expr2

> These operators return the results of the addition or subtraction of integer-valued arguments.

expr1 {, /, %} expr2*

> These return, respectively, the results of the multiplication, integer division, or remainder of integer-valued arguments.

Regular expression operator

> You can also use the regular expression operator in Asterisk:

expr1 : expr2

> This operator matches *expr1* against *expr2*, where *expr2* must be a regular expression.[†] The regular expression is anchored to the beginning of the string with an implicit ^.[‡]
>
> If the match succeeds and the pattern contains at least one regular expression subexpression—\(... \)—the string corresponding to \1 is returned; otherwise, the matching operator returns the number of characters matched. If the match fails and the pattern contains a regular expression subexpression, the null string is returned; otherwise, 0 is returned.

In Asterisk version 1.0 the parser was quite simple, so it required that you put at least one space between the operator and any other values. Consequently, the following might not have worked as expected:

```
exten => 123,1,Set(TEST=$[2+1])
```

[†] For more on regular expressions, grab a copy of the ultimate reference, Jeffrey E.F. Friedl's *Mastering Regular Expressions* (O'Reilly) or visit *http://www.regular-expressions.info*.

[‡] If you don't know what a ^ has to do with regular expressions, you simply must obtain a copy of *Mastering Regular Expressions*. It will change your life!

This would have assigned the variable TEST to the string "2+1", instead of the value 3. In order to remedy that, we would put spaces around the operator like so:

```
exten => 234,1,Set(TEST=$[2 + 1])
```

This is no longer necessary in Asterisk 1.2 or 1.4 as the expression parser has been made more forgiving in these types of scenarios, however, for readability's sake, we still recommend the spaces around your operators.

To concatenate text onto the beginning or end of a variable, simply place them together in an expression, like this:

```
exten => 234,1,Set(NEWTEST=$[blah${TEST}])
```

Dialplan Functions

Dialplan functions allow you to add more power to your expressions; you can think of them as intelligent variables. Dialplan functions allow you to calculate string lengths, dates and times, MD5 checksums, and so on, all from within a dialplan expression.

Syntax

Dialplan functions have the following basic syntax:

```
FUNCTION_NAME(argument)
```

Much like variables, you reference a function's *name* as above, but you reference a function's *value* with the addition of a dollar sign, an opening curly brace, and a closing curly brace:

```
${FUNCTION_NAME(argument)}
```

Functions can also encapsulate other functions, like so:

```
${FUNCTION_NAME(${FUNCTION_NAME(argument)})}
 ^              ^ ^                 ^      ^^^^
 1              2 3                 4      4321
```

As you've probably already figured out, you must be very careful about making sure you have matching parentheses and braces. In the above example, we have labeled the opening parentheses and curly braces with numbers and their corresponding closing counterparts with the same numbers.

Examples of Dialplan Functions

Functions are often used in conjunction with the Set() application to either get or set the value of a variable. As a simple example, let's look at the LEN() function. This function calculates the string length of its argument. Let's calculate the string length of a variable and read back the length to the caller:

```
exten => 123,1,Set(TEST=example)
exten => 123,n,SayNumber(${LEN(${TEST})})
```

The above example would evaluate the string example as having seven characters, assign the number of characters to the variable length, and then speak the number to the user with the SayNumber() application.

Let's look at another simple example. If we wanted to set one of the various channel timeouts, we could use the TIMEOUT() function. The TIMEOUT() function accepts one of three arguments: absolute, digit, and response. To set the digit timeout with the TIMEOUT() function, we could use the Set() application, like so:

```
exten => s,1,Set(TIMEOUT(digit)=30)
```

Notice the lack of ${ } surrounding the function. Just as if we were assigning a value to a variable, we assign a value to a function without the use of the ${ } encapsulation.

A complete list of available functions can be found by typing core show functions at the Asterisk command-line interface. You can also look them up in Appendix F.

Conditional Branching

Now that you've learned a bit about expressions and functions, it's time to put them to use. By using expressions and functions, you can add even more advanced logic to your dialplan. To allow your dialplan to make decisions, you'll use *conditional branching*. Let's take a closer look.

The GotoIf() Application

The key to conditional branching is the GotoIf() application. GotoIf() evaluates an expression and sends the caller to a specific destination based on whether the expression evaluates to true or false.

GotoIf() uses a special syntax, often called the *conditional syntax*:

```
GotoIf(expression?destination1:destination2)
```

If the expression evaluates to true, the caller is sent to *destination1*. If the expression evaluates to false, the caller is sent to the second destination. So, what is true and what is false? An empty string and the number 0 evaluate as false. Anything else evaluates as true.

The destinations can each be one of the following:

- A priority label within the same extension, such as weasels
- An extension and a priority label within the same context, such as 123,weasels
- A context, extension, and priority label, such as incoming,123,weasels

Either of the destinations may be omitted, but not both. If the omitted destination is to be followed, Asterisk simply goes on to the next priority in the current extension.

Let's use GotoIf() in an example:

```
exten => 345,1,Set(TEST=1)
exten => 345,n,GotoIf($[${TEST} = 1]?weasels:iguanas)
exten => 345,n(weasels),Playback(weasels-eaten-phonesys)
exten => 345,n,Hangup()
exten => 345,n(iguanas),Playback(office-iguanas)
exten => 345,n,Hangup()
```

 You will notice that we have used the Hangup() application following each Playback() application. This is done so that when we jump to the weasels label, the call stops before execution gets to the office-iguanas sound file. It is becoming increasingly common to see extensions broken up in to multiple components (protected from each other by the Hangup() command), each one acting as steps executed following a GotoIf().

Providing Only a False Conditional Path

If we wanted to, we could have crafted the preceding example like this:

```
exten => 345,1,Set(TEST=1)
exten => 345,n,GotoIf($[${TEST} = 1]?:iguanas) ; we don't have the weasels label anymore,
; but this will still work
exten => 345,n,Playback(weasels-eaten-phonesys)
exten => 345,n,Hangup()
exten => 345,n(iguanas),Playback(office-iguanas)
exten => 345,n,Hangup()
```

There is nothing between the ? and the :, so if the statement evaluates to true, execution of the dialplan will continue at the next step. Since that is what we want, a label is not needed.

We don't really recommend doing this, because this is hard to read, but you will see dialplans like this, so it's good to be aware that this syntax is totally correct.

Typically when you have this type of layout where you end up wanting to limit Asterisk from falling through to the next priority after you've performed that jump, it's probably better to jump to separate extensions instead of priority labels. If anything, it makes it a bit more clear when reading the dialplan. We could rewrite the previous bit of dialplan like this:

```
exten => 345,1,Set(TEST=1)
exten => 345,n,GotoIf($[${TEST} = 1]?weasels,1:iguanas,1); now we're going to
; extension,priority
exten => weasels,1,Playback(weasels-eaten-phonesys); this is NOT a label.
; It is a different extension
exten => weasels,n,Hangup()
```

```
exten => iguanas,1,Playback(office-iguanas)
exten => iguanas,n,Hangup()
```

By changing the value assigned to TEST in the first line, you should be able to have your Asterisk server play a different greeting.

Let's look at another example of conditional branching. This time, we'll use both Goto() and GotoIf() to count down from 10 and then hang up:

```
exten => 123,1,Set(COUNT=10)
exten => 123,n(start),GotoIf($[${COUNT} > 0]?:goodbye)
exten => 123,n,SayNumber(${COUNT})
exten => 123,n,Set(COUNT=$[${COUNT} - 1])
exten => 123,n,Goto(start)
exten => 123,n(goodbye),Hangup()
```

Let's analyze this example. In the first priority, we set the variable COUNT to 10. Next, we check to see if COUNT is greater than 0. If it is, we move on to the next priority. (Don't forget that if we omit a destination in the GotoIf() application, control goes to the next priority.) From there we speak the number, subtract 1 from COUNT, and go back to priority label start. If COUNT is less than or equal to 0, control goes to priority label goodbye, and the call is hung up.

The classic example of conditional branching is affectionately known as the anti-girl-friend logic. If the Caller ID number of the incoming call matches the phone number of the recipient's ex-girlfriend, Asterisk gives a different message than it ordinarily would to any other caller. While somewhat simple and primitive, it's a good example for learning about conditional branching within the Asterisk dialplan.

This example uses the CALLERID function, which allows us to retrieve the Caller ID information on the inbound call. Let's assume for the sake of this example that the victim's phone number is 888-555-1212:

```
exten => 123,1,GotoIf($[${CALLERID(num)} = 8885551212]?reject:allow)
exten => 123,n(allow),Dial(Zap/4)
exten => 123,n,Hangup()
exten => 123,n(reject),Playback(abandon-all-hope)
exten => 123,n,Hangup()
```

In priority 1, we call the GotoIf() application. It tells Asterisk to go to priority label reject if the Caller ID number matches 8885551212, and otherwise to go to priority label allow (we could have simply omitted the label name, causing the GotoIf() to fall through). If the Caller ID number matches, control of the call goes to priority label reject, which plays back an uninspiring message to the undesired caller. Otherwise, the call attempts to dial the recipient on channel Zap/4.

Time-Based Conditional Branching with GotoIfTime()

Another way to use conditional branching in your dialplan is with the GotoIfTime() application. Whereas GotoIf() evaluates an expression to decide what to do, GotoIf

`Time()` looks at the current system time and uses that to decide whether or not to follow a different branch in the dialplan.

The most obvious use of this application is to give your callers a different greeting before and after normal business hours.

The syntax for the `GotoIfTime()` application looks like this:

```
GotoIfTime(times,days_of_week,days_of_month,months?label)
```

In short, `GotoIfTime()` sends the call to the specified *label* if the current date and time match the criteria specified by *times*, *days_of_week*, *days_of_month*, and *months*. Let's look at each argument in more detail:

times

This is a list of one or more time ranges, in a 24-hour format. As an example, 9:00 A.M. through 5:00 P.M. would be specified as `09:00-17:00`. The day starts at 0:00 and ends at 23:59.

It is worth noting that times will properly wrap around. So if you wish to specify the times your office is closed, you might write `18:00-9:00` in the *times* parameter, and it will perform as expected. Note that this technique works as well for the other components of `GotoIfTime`. For example, you can write `sat-sun` to specify the weekend days.

days_of_week

This is a list of one or more days of the week. The days should be specified as `mon`, `tue`, `wed`, `thu`, `fri`, `sat`, and/or `sun`. Monday through Friday would be expressed as `mon-fri`. Tuesday and Thursday would be expressed as `tue&thu`.

Note that you can specify a combination of ranges and single days, as in: `sun-mon&wed&fri-sat`, or, more simply: `wed&fri-mon`.

days_of_month

This is a list of the numerical days of the month. Days are specified by the numbers `1` through `31`. The 7th through the 12th would be expressed as `7-12`, and the 15th and 30th of the month would be written as `15&30`.

months

This is a list of one or more months of the year. The months should be written as `jan-apr` for a range, and separated with ampersands when wanting to include non-sequencial months, such as `jan&mar&jun`. You can also combine them like so: `jan-apr&jun&oct-dec`.

If you wish to match on all possible values for any of these arguments, simply put an * in for that argument.

The *label* argument can be any of the following:

- A priority label within the same extension, such as `time_has_passed`
- An extension and a priority within the same context, such as `123,time_has_passed`
- A context, extension, and priority, such as `incoming,123,time_has_passed`

Now that we've covered the syntax, let's look at a couple of examples. The following example would match from *9:00 A.M. to 5:59 P.M.*, on *Monday through Friday*, on *any day of the month*, in *any month of the year*:

```
exten => s,1,GotoIfTime(09:00-17:59,mon-fri,*,*?open,s,1)
```

If the caller calls during these hours, the call will be sent to the first priority of the s extension in the context named **open**. If the call is made outside of the specified times, it will be sent to the next priority of the current extension. This allows you to easily branch on multiple times, as shown in the next example (note that you should always put your most specific time matches before the least specific ones):

```
; If it's any hour of the day, on any day of the week,
; during the fourth day of the month, in the month of July,
; we're closed
exten => s,1,GotoIfTime(*,*,4,jul?open,s,1)

; During business hours, send calls to the open context
exten => s,n,GotoIfTime(09:00-17:59|mon-fri|*|*?open,s,1)
exten => s,n,GotoIfTime(09:00-11:59|sat|*|*?open,s,1)

; Otherwise, we're closed
exten => s,n,Goto(closed,s,1)
```

If you run into the situation where you ask the question, "But I specified 17:58 and it's now 17:59. Why is it still doing the same thing?" it should be noted that the granularity of the `GotoIfTime()` application is only to a two-minute period. So if you specify 18:00 as the ending time of a period, the system will continue to perform the same way for an additional minute, until 18:01:59.

Voicemail

One of the most popular (or, arguably, unpopular) features of any modern telephone system is voicemail. Naturally, Asterisk has a reasonably flexible voicemail system. Some of the features of Asterisk's voicemail system include:

- Unlimited password-protected voicemail boxes, each containing mailbox folders for organizing voicemail
- Different greetings for busy and unavailable states

- Default and custom greetings
- The ability to associate phones with more than one mailbox and mailboxes with more than one phone
- Email notification of voicemail, with the voicemail optionally attached as a sound file[§]
- Voicemail forwarding and broadcasts
- Message-waiting indicator (flashing light or stuttered dial tone) on many types of phones
- Company directory of employees, based on voicemail boxes

And that's just the tip of the iceberg! In this section, we'll introduce you to the fundamentals of a typical voicemail setup.

The voicemail configuration is defined in the configuration file called *voicemail.conf*. This file contains an assortment of settings that you can use to customize the voicemail system to your needs. Covering all of the available options in *voicemail.conf* would be beyond the scope of this chapter, but the sample configuration file is well documented and quite easy to follow. For now, look near the bottom of the file, where voicemail contexts and voicemail boxes are defined.

Just as dialplan contexts keep different parts of your dialplan separate, voicemail contexts allow you to define different sets of mailboxes that are separate from one another. This allows you to host voicemail for several different companies or offices on the same server. Voicemail contexts are defined in the same way as dialplan contexts, with square brackets surrounding the name of the context. For our examples, we'll be using the [default] voicemail context.

Creating Mailboxes

Inside each voicemail context, we define different mailboxes. The syntax for defining a mailbox is:

```
mailbox => password,name[,email[,pager_email[,options]]]
```

Let's explain what each part of the mailbox definition does:

mailbox
> This is the mailbox number. It usually corresponds with the extension number of the associated set.

password
> This is the numeric password that the mailbox owner will use to access her voicemail. If the user changes her password, the system will update this field in the *voicemail.conf* file.

---

§ No, you really don't have to pay for this—and yes, it really does work.

name

> This is the name of the mailbox owner. The company directory uses the text in this field to allow callers to spell usernames.

email

> This is the email address of the mailbox owner. Asterisk can send voicemail notifications (including the voicemail message itself) to the specified email box.

pager_email

> This is the email address of the mailbox owner's pager or cell phone. Asterisk can send a short voicemail notification message to the specified email address.

options

> This field is a list of options that sets the mailbox owner's time zone and overrides the global voicemail settings. There are nine valid options: attach, serveremail, tz, saycid, review, operator, callback, dialout, and exitcontext. These options should be in *option* = *value* pairs, separated by the pipe character (|). The tz option sets the user's time zone to a time zone previously defined in the [zonemessages] section of *voicemail.conf*, and the other eight options override the global voicemail settings with the same names.

A typical mailbox definition might look something like this:

```
101 => 1234,Joe Public,jpublic@somedomain.com,jpublic@pagergateway.net,
tz=central|attach=yes
```

Continuing with our dialplan from the last chapter, let's set up voicemail boxes for John and Jane. We'll give John a password of **1234** and Jane a password of **4444** (remember, these go in *voicemail.conf*, not in *extensions.conf*):

```
[default]
101 => 1234,John Doe,john@asteriskdocs.org,jdoe@pagergateway.tld
102 => 4444,Jane Doe,jane@asteriskdocs.org,jane@pagergateway.tld
```

Adding Voicemail to the Dialplan

Now that we've created mailboxes for Jane and John, let's allow callers to leave messages for them if they don't answer the phone. To do this, we'll use the VoiceMail() application.

The VoiceMail() application sends the caller to the specified mailbox, so that he can leave a message. The mailbox should be specified as *mailbox* @ *context*, where *context* is the name of the voicemail context. The option letters b or u can be added to request the type of greeting. If the letter b is used, the caller will hear the mailbox owner's *busy* message. If the letter u is used, the caller will hear the mailbox owner's *unavailable* message (if one exists).

Let's use this in our sample dialplan. Previously, we had a line like this in our [internal] context, which allowed us to call John:

```
exten => 101,1,Dial(${JOHN})
```

Next, let's add an unavailable message that the caller will be played if John doesn't answer the phone within 10 seconds. Remember, the second argument to the `Dial()` application is a timeout. If the call is not answered before the timeout expires, the call is sent to the next priority. Let's add a 10-second timeout, and a priority to send the caller to voicemail if John doesn't answer in time:

```
exten => 101,1,Dial(${JOHN},10)
exten => 101,n,VoiceMail(101@default,u)
```

Now, let's change it so that if John is busy (on another call), it'll send us to his voicemail, where we'll hear his busy message. To do this, we will make use of the `${DIALSTATUS}` variable which contains one of several status values (see `core show application dial` at the Asterisk console for a listing of all the possible values):

```
exten => 101,1,Dial(${JOHN},10)
exten => 101,n,GotoIf($["${DIALSTATUS}" = "BUSY"]?busy:unavail)
exten => 101,n(unavail),Voicemail(101@default,u)
exten => 101,n,Hangup()
exten => 101,n(busy),VoiceMail(101@default,b)
exten => 101,n,Hangup()
```

Now callers will get John's voicemail (with the appropriate greeting) if John is either busy or unavailable. A slight problem remains, however, in that John has no way of retrieving his messages. Let's remedy that.

Accessing Voicemail

Users can retrieve their voicemail messages, change their voicemail options, and record their voicemail greetings by using the `VoiceMailMain()` application. In its typical form, `VoiceMailMain()` is called without any arguments. Let's add extension 700 to the [`internal`] context of our dialplan so that internal users can dial it to access their voicemail messages:

```
exten => 700,1,VoiceMailMain()
```

Creating a Dial-by-Name Directory

One last feature of the Asterisk voicemail system we should cover is the dial-by-name directory. This is created with the `Directory()` application. This application uses the names defined in the mailboxes in *voicemail.conf* to present the caller with a dial-by-name directory of the users.

`Directory()` takes up to three arguments: the voicemail context from which to read the names, the optional dialplan context in which to dial the user, and an option string (which is also optional). By default, `Directory()` searches for the user by last name, but passing the `f` option forces it to search by first name instead. Let's add two dial-by-name directories to the [`incoming`] context of our sample dialplan, so that callers can search by either first or last name:

```
exten => 8,1,Directory(default,incoming,f)
exten => 9,1,Directory(default,incoming)
```

If callers press 8, they'll get a directory by first name. If they dial 9, they'll get the directory by last name.

Macros

Macros[||] are a very useful construct designed to avoid repetition in the dialplan. They also help in making changes to the dialplan. To illustrate this point, let's look at our sample dialplan again. If you remember the changes we made for voicemail, we ended up with the following for John's extension:

```
exten => 101,1,Dial(${JOHN},10)
exten => 101,n,GotoIf($["${DIALSTATUS}" = "BUSY"]?busy:unavail)
exten => 101,n(unavail),Voicemail(101@default,u)
exten => 101,n,Hangup()
exten => 101,n(busy),VoiceMail(101@default,b)
exten => 101,n,Hangup()
```

Now imagine you have a hundred users on your Asterisk system—setting up the extensions would involve a lot of copying and pasting. Then imagine that you need to make a change to the way your extensions work. That would involve a lot of editing, and you'd be almost certain to have errors.

Instead, you can define a macro that contains a list of steps to take, and then have all of the phone extensions refer to that macro. All you need to change is the macro, and everything in the dialplan that references that macro will change as well.

 If you're familiar with computer programming, you'll recognize that macros are similar to subroutines in many modern programming languages. If you're not familiar with computer programming, don't worry —we'll walk you through creating a macro.

The best way to appreciate macros is to see one in action, so let's move right along.

[||] Although Macro seems like a general-purpose dialplan subroutine, it has a stack overflow problem that means you should not try to nest Macro calls more than five levels deep. As of this writing, we do not know whether the Macro application will be patched for 1.4, or if it will be rewritten for future versions. If you plan to do a lot of macros within macros (and call complex functions within them), you may run into stability problems. You will know you have a problem with just one test call, so if your dialplan tests out, you're good to go. We also recommend that you take a look at the Gosub and Return applications, as a lot of macro functionality can be implemented without actually using Macro(). Also, please note that we are not suggesting that you don't use Macro(). It is fantastic and works very well; it just doesn't nest efficiently.

Defining Macros

Let's take the dialplan logic we used above to set up voicemail for John and turn it into a macro. Then we'll use the macro to give John and Jane (and the rest of their coworkers) the same functionality.

Macro definitions look a lot like contexts. (In fact, you could argue that they really are small, limited contexts.) You define a macro by placing macro- and the name of your macro in square brackets, like this:

```
[macro-voicemail]
```

Macro names must start with macro-. This distinguishes them from regular contexts. The commands within the macro are built almost identically to anything else in the dialplan; the only limiting factor is that macros use only the s extension. Let's add our voicemail logic to the macro, changing the extension to s as we go:

```
[macro-voicemail]
exten => s,1,Dial(${JOHN},10)
exten => s,n,GotoIf($["${DIALSTATUS}" = "BUSY"]?busy:unavail)
exten => s,n(unavail),Voicemail(101@default,u)
exten => s,n,Hangup()
exten => s,n(busy),VoiceMail(101@default,b)
exten => s,n,Hangup()
```

That's a start, but it's not perfect, as it's still specific to John and his mailbox number. To make the macro generic so that it will work not only for John but also for all of his coworkers, we'll take advantage of another property of macros: arguments. But first, let's see how we call macros in our dialplan.

Calling Macros from the Dialplan

To use a macro in our dialplan, we use the Macro() application. This application calls the specified macro and passes it any arguments. For example, to call our voicemail macro from our dialplan, we can do the following:

```
exten => 101,1,Macro(voicemail)
```

The Macro() application also defines several special variables for our use. They include:

${MACRO_CONTEXT}
> The original context in which the macro was called.

${MACRO_EXTEN}
> The original extension in which the macro was called.

${MACRO_PRIORITY}
> The original priority in which the macro was called.

${ARG n }
> The nth argument passed to the macro. For example, the first argument would be ${ARG1}, the second ${ARG2}, and so on.

As we explained earlier, the way we initially defined our macro was hardcoded for John, instead of being generic. Let's change our macro to use ${MACRO_EXTEN} instead of 101 for the mailbox number. That way, if we call the macro from extension 101 the voice-mail messages will go to mailbox 101, and if we call the macro from extension 102 messages will go to mailbox 102, and so on:

```
[macro-voicemail]
exten => s,1,Dial(${JOHN},10)
exten => s,n,GotoIf($["${DIALSTATUS}" = "BUSY"]?busy:unavail)
exten => s,n(unavail),Voicemail(${MACRO_EXTEN}@default,u)
exten => s,n,Hangup()
exten => s,n(busy),VoiceMail(${MACRO_EXTEN}@default,b)
exten => s,n,Hangup()
```

Using Arguments in Macros

Okay, now we're getting closer to having the macro the way we want it, but we still have one thing left to change; we need to pass in the channel to dial, as it's currently still hardcoded for ${JOHN} (remember that we defined the variable JOHN as the channel to call when we want to reach John). Let's pass in the channel as an argument, and then our first macro will be complete:

```
[macro-voicemail]
exten => s,1,Dial(${ARG1},10)
exten => s,n,GotoIf($["${DIALSTATUS}" = "BUSY"]?busy:unavail)
exten => s,n(unavail),Voicemail(${MCARO_EXTEN}@default,u)
exten => s,n,Hangup()
exten => s,n(busy),VoiceMail(${MCARO_EXTEN}@default,b)
exten => s,n,Hangup()
```

Now that our macro is done, we can use it in our dialplan. Here's how we can call our macro to provide voicemail to John, Jane, and Jack:

```
exten => 101,1,Macro(voicemail,${JOHN})
exten => 102,1,Macro(voicemail,${JANE})
exten => 103,1,Macro(voicemail,${JACK})
```

With 50 or more users, this dialplan will still look neat and organized; we'll simply have one line per user, referencing a macro that can be as complicated as required. We could even have a few different macros for various user types, such as executives, courtesy_phones, call_center_agents, analog_sets, sales_department, and so on.

A more advanced version of the macro might look something like this:

```
[macro-voicemail]
exten => s,1,Dial(${ARG1},20)
exten => s,n,Goto(s-${DIALSTATUS},1)
exten => s-NOANSWER,1,Voicemail(${MACRO_EXTEN},u)
exten => s-NOANSWER,n,Goto(incoming,s,1)
exten => s-BUSY,1,Voicemail(${MACRO_EXTEN},b)
exten => s-BUSY,n,Goto(incoming,s,1)
exten => _s-.,1,Goto(s-NOANSWER,1)
```

This macro depends on a nice side effect of the Dial() application: when you use the Dial() application, it sets the DIALSTATUS variable to indicate whether the call was successful or not. In this case, we're handling the NOANSWER and BUSY cases, and treating all other result codes as a NOANSWER.

Using the Asterisk Database (AstDB)

Having fun yet? It gets even better!

Asterisk provides a powerful mechanism for storing values called the Asterisk database (AstDB). The AstDB provides a simple way to store data for use within your dialplan.

 For those of you with experience using relational databases such as PostgreSQL or MySQL, the Asterisk database is not a traditional relational database. It is a Berkeley DB Version 1 database. There are several ways to store data from Asterisk in a relational database. Check out Chapter 12 for a more about relational databases.

The Asterisk database stores its data in groupings called *families*, with values identified by *keys*. Within a family, a key may be used only once. For example, if we had a family called test, we could store only one value with a key called count. Each stored value must be associated with a family.

Storing Data in the AstDB

To store a new value in the Asterisk database, we use the Set() application,[#] but instead of using it to set a channel variable, we use it to set an AstDB variable. For example, to assign the count key in the test family with the value of 1, write the following:

```
exten => 456,1,Set(DB(test/count)=1)
```

If a key named count already exists in the test family, its value will be overwritten with the new value. You can also store values from the Asterisk command line, by running the command database put *family key value*. For our example, you would type **data base put test count 1**.

Retrieving Data from the AstDB

To retrieve a value from the Asterisk database and assign it to a variable, we use the Set() application again. Let's retrieve the value of count (again, from the test family), assign it to a variable called COUNT, and then speak the value to the caller:

[#] Previous versions of Asterisk had applications called DBput() and DBget() that were used to set values in and retrieve values from the AstDB. If you're using an old version of Asterisk, you'll want to use those applications instead.

```
exten => 456,1,Set(DB(test/count)=1)
exten => 456,n,Set(COUNT=${DB(test/count)})
exten => 456,n,SayNumber(${COUNT})
```

You may also check the value of a given key from the Asterisk command line by running the command `database get family key`. To view the entire contents of the AstDB, use the `database show` command.

Deleting Data from the AstDB

There are two ways to delete data from the Asterisk database. To delete a key, you can use the `DB_DELETE()` application. It takes the path to the key as its arguments, like this:

```
; deletes the key and returns its value in one step
exten => 457,1,Verbose(0, The value was ${DB_DELETE(test/count)})
```

You can also delete an entire key family by using the `DBdeltree()` application. The `DBdeltree()` application takes a single argument—the name of the key family—to delete. To delete the entire test family, do the following:

```
exten => 457,1,DBdeltree(test)
```

To delete keys and key families from the AstDB via the command-line interface, use the `database del key` and `database deltree family` commands, respectively.

Using the AstDB in the Dialplan

There are an infinite number of ways to use the Asterisk database in a dialplan. To introduce the AstDB, we'll show two simple examples. The first is a simple counting example to show that the Asterisk database is persistent (meaning that it survives system reboots). In the second example, we'll use the `BLACKLIST()` function to evaluate whether or not a number is on the blacklist and should be blocked.

To begin the counting example, let's first retrieve a number (the value of the **count** key) from the database and assign it to a variable named COUNT. If the key doesn't exist, DB () will return NULL (no value). In order to verify if a value exists in the database or not, we will introduce the `ISNULL()` function that will verify whether a value was returned, and if not, we will initialize the AstDB with the `Set()` application, where we will set the value in the database to 1. The next priority will send us back to priority 1. This will happen the very first time we dial this extension:

```
exten => 678,1,Set(COUNT=${DB(test/count)})
exten => 678,n,GotoIf($[${ISNULL(${COUNT})}]?:continue)
exten => 678,n,Set(DB(test/count)=1)
exten => 678,n,Goto(1)
exten => 678,n(continue),NoOp()
```

Next, we'll say the current value of COUNT, and then increment COUNT:

```
exten => 678,1,Set(COUNT=${DB(test/count)})
exten => 678,n,GotoIf($[${ISNULL(${COUNT})}]?:continue)
```

```
exten => 678,n,Set(DB(test/count)=1)
exten => 678,n,Goto(1)
exten => 678,n(continue),NoOp()
exten => 678,n,SayNumber(${COUNT})
exten => 678,n,Set(COUNT=$[${COUNT} + 1])
```

Now that we've incremented COUNT, let's put the new value back into the database. Remember that storing a value for an existing key overwrites the previous value:

```
exten => 678,1,Set(COUNT=${DB(test/count)})
exten => 678,n,GotoIf($[${ISNULL(${COUNT})}]?:continue)
exten => 678,n,Set(DB(test/count)=1)
exten => 678,n,Goto(1)
exten => 678,n(continue),NoOp()
exten => 678,n,SayNumber(${COUNT})
exten => 678,n,Set(COUNT=$[${COUNT} + 1])
exten => 678,n,Set(DB(test/count)=${COUNT})
```

Finally, we'll loop back to the first priority. This way, the application will continue counting:

```
exten => 678,1,Set(COUNT=${DB(test/count)})
exten => 678,n,GotoIf($[${ISNULL(${COUNT})}]?:continue)
exten => 678,n,Set(DB(test/count)=1)
exten => 678,n,Goto(1)
exten => 678,n(continue),NoOp()
exten => 678,n,SayNumber(${COUNT})
exten => 678,n,Set(COUNT=$[${COUNT} + 1]
exten => 678,n,Set(DB(test/count)=${COUNT})
exten => 678,n,Goto(1)
```

Go ahead and try this example. Listen to it count for a while, and then hang up. When you dial this extension again, it should continue counting from where it left off. The value stored in the database will be persistent, even across a restart of Asterisk.

In the next example, we'll create dialplan logic around the BLACKLIST() function, which checks to see if the current Caller ID number exists in the blacklist. (The blacklist is simply a family called blacklist in the AstDB.) If BLACKLIST() finds the number in the blacklist, it returns the value 1, otherwise it will return 0. We can use these values in combination with a GotoIf() to control whether the call will execute the Dial() application:

```
exten => 124,1,GotoIf($[${BLACKLIST()}]?blocked,1)
exten => 124,n,Dial(${JOHN})

exten => blocked,1,Playback(privacy-you-are-blacklisted)
exten => blocked,n,Playback(vm-goodbye)
exten => blocked,n,Hangup()
```

To add a number to the blacklist, run the database put blacklist *number* 1 command from the Asterisk command-line interface.

Handy Asterisk Features

Now that we've gone over some more of the basics, let's look at a few popular functions that have been incorporated into Asterisk.

Zapateller()

Zapateller() is a simple Asterisk application that plays a special information tone at the beginning of a call, which causes auto-dialers (usually used by telemarketers) to think that the line has been disconnected. Not only will they hang up, but their systems will flag your number as out of service, which could help you avoid all kinds of telemarketing calls. To use this functionality within your dialplan, simply call the Zapateller() application.

We'll also use the optional nocallerid option so that the tone will be played only when there is no Caller ID information on the incoming call. For example, you might use Zapateller() in the s extension of your [incoming] context, like this:

```
[incomimg]
exten => s,1,Zapateller(nocallerid)
exten => s,n,Playback(enter-ext-of-person)
```

Call Parking

Another handy feature is called *call parking*. Call parking allows you to place a call on hold in a "parking lot," so that it can be taken off hold from another extension. Parameters for call parking (such as the extensions to use, the number of spaces, and so on) are all controlled within the *features.conf* configuration file. The [general] section of the *features.conf* file contains four settings related to call parking:

parkext
: This is the parking lot extension. Transfer a call to this extension, and the system will tell you which parking position the call is in. By default, the parking extension is 700.

parkpos
: This option defines the number of parking slots. For example, setting it to 701-720 creates 20 parking positions, numbered 701 through 720.

context
: This is the name of the parking context. To be able to park calls, you must include this context.

parkingtime
: If set, this option controls how long (in seconds) a call can stay in the parking lot. If the call isn't picked up within the specified time, the extension that parked the call will be called back.

You must restart Asterisk after editing *features.conf*, as the file is read only on startup. Running the `reload` command will not cause the *features.conf* file to be read.

Also note that because the user needs to be able to transfer the calls to the parking lot extension, you should make sure you're using the t and/or T options to the `Dial()` application.

So, let's create a simple dialplan to show off call parking:

```
[incoming]
include => parkedcalls

exten => 103,1,Dial(SIP/Bob,,tT)
exten => 104,1,Dial(SIP/Charlie,,tT)
```

To illustrate how call parking works, say that Alice calls into the system and dials extension 103 to reach Bob. After a while, Bob transfers the call to extension 700, which tells him that the call from Alice has been parked in position 701. Bob then dials Charlie at extension 104, and tells him that Alice is at extension 701. Charlie then dials extension 701 and begins to talk to Alice. This is a simple and effective way of allowing callers to be transferred between users.

The t and T arguments to `Dial()` are not needed on all channel types. For example, many SIP phones implement this via a softkey or hardkey and utilize SIP signaling.

Conferencing with MeetMe()

Last but not least, let's cover setting up an audio conference bridge with the `MeetMe()` application.[*] This application allows multiple callers to converse together, as if they were all in the same physical location. Some of the main features include:

- The ability to create password-protected conferences
- Conference administration (mute conference, lock conference, kick participants)
- The option of muting all but one participant (useful for company announcements, broadcasts, etc.)
- Static or dynamic conference creation

Let's walk through setting up a basic conference room. The configuration options for the MeetMe conferencing system are found in *meetme.conf*. Inside the configuration file, you define conference rooms and optional numeric passwords. (If a password is

---

[*] In the world of legacy PBXes, this type of functionality is very expensive. Either you have to pay big bucks for a dial-in service, or you have to add an expensive conferencing bridge to your proprietary PBX.

defined here, it will be required to enter all conferences using that room.) For our example, let's set up a conference room at extension 600. First, we'll set up the conference room in *meetme.conf*. We'll call it **600**, and we won't assign a password at this time:

```
[rooms]
conf => 600
```

Now that the configuration file is complete, we'll need to restart Asterisk so that it can reread the *meetme.conf* file. Next, we'll add support for the conference room to our dialplan with the `MeetMe()` application. `MeetMe()` takes three arguments: the name of the conference room (as defined in *meetme.conf*), a set of options, and the password the user must enter to join this conference. Let's set up a simple conference using room **600**, the **i** option (which announces when people enter and exit the conference), and a password of **54321**:

```
exten => 600,1,MeetMe(600,i,54321)
```

That's all there is to it! When callers enter extension 600, they will be prompted for the password. If they correctly enter **54321**, they will be added to the conference. See Appendix B for a list of all the options supported by the `MeetMe()` application.

Another useful application is `MeetMeCount()`. As its name suggests, this application counts the number of users in a particular conference room. It takes up to two arguments: the conference room in which to count the number of participants, and optionally a variable name to assign the count to. If the variable name is not passed as the second argument, the count is read to the caller:

```
exten => 601,1,Playback(conf-thereare)
exten => 601,n,MeetMeCount(600)
exten => 601,n,Playback(conf-peopleinconf)
```

If you pass a variable as the second argument to `MeetMeCount()`, the count is assigned to the variable, and playback of the count is skipped. You might use this to limit the number of participants, like this:

```
; limit the conference room to 10 participants
exten => 600,1,MeetMeCount(600,CONFCOUNT)
exten => 600,n,GotoIf($[${CONFCOUNT} <= 10]?meetme:conf_full,1)
exten => 600,n(meetme),MeetMe(600,i,54321)

exten => conf_full,1,Playback(conf-full)
```

Isn't Asterisk fun?

Conclusion

In this chapter, we've covered a few more of the many applications in the Asterisk dialplan, and hopefully we've given you the seeds from which you can explore the creation of your own dialplans. As with the previous chapter, we invite you to go back and reread any sections that require clarification.

The following chapters take us away from Asterisk for a bit, in order to talk about some of the technologies that all telephone systems use. We'll be referring to Asterisk a lot, but much of what we want to discuss are things that are common to many telecom systems.

## CHAPTER 7

Understanding Telephony

*Utility is when you have one telephone, luxury is when
you have two, opulence is when you have three—and
paradise is when you have none.*

—Doug Larson

We're now going to take a break from Asterisk for a chapter or two, because we want
to spend some time discussing the technologies with which your Asterisk system will
need to interface. In this chapter, we are going to talk about some of the technologies
of the traditional telephone network—especially those that people most commonly
want to connect to Asterisk. (We'll discuss Voice over IP in the next chapter.)

While tomes could be written about the technologies in use in telecom networks, the
material in this chapter was chosen based on our experiences in the community, which
helped us to define the specific items that might be most useful. Although this knowledge may not be strictly required in order to configure your Asterisk system, it will be
of great benefit when interconnecting to systems (and talking with people) from the
world of traditional telecommunications.

Analog Telephony

The purpose of the Public Switched Telephone Network (PSTN) is to establish and
maintain audio connections between two endpoints in order to carry speech.

Although humans can perceive sound vibrations in the range of 20–20,000 Hz,[*] most
of the sounds we make when speaking tend to be in the range of 250–3,000 Hz. Since
the purpose of the telephone network is to transmit the sounds of people speaking, it

---

[*] If you want to play around with what different frequencies look like on an oscilloscope, grab a copy of Sound
Frequency Analyzer, from Reliable Software. It's a really simple and fun way to visualize what sounds "look"
like. The spectrograph gives a good picture of the complex harmonics our voices can generate, as well as an
appreciation for the background sounds that always surround us. You should also try the delightfully
annoying NCH Tone Generator, from NCH Swift Sound.

was designed with a bandwidth of somewhere in the range of 300–3,500 Hz. This limited bandwidth means that some sound quality will be lost (as anyone who's had to listen to music on hold can attest to), especially in the higher frequencies.

Parts of an Analog Telephone

An analog phone is composed of five parts: the ringer, the dial pad, the hybrid (or network), and the hook switch and handset (both of which are considered parts of the hybrid). The ringer, the dial pad, and the hybrid can operate completely independently of one another.

Ringer

When the central office (CO) wants to signal an incoming call, it will connect an alternating current (AC) signal of roughly 90 volts to your circuit. This will cause the bell in your telephone to produce a ringing sound. (In electronic telephones, this ringer may be a small electronic warbler rather than a bell. Ultimately, a ringer can be anything that is capable of reacting to the ringing voltage; for example, strobe lights are often employed in noisy environments such as factories.)

 Ringing voltage can be hazardous. Be very careful to take precautions when working with an in-service telephone line.

Many people confuse the AC voltage that triggers the ringer with the direct current (DC) voltage that powers the phone. Remember that a ringer needs an alternating current in order to oscillate (just as a church bell won't ring if you don't supply the movement), and you've got it.

In North America, the number of ringers you can connect to your line is dependent on the Ringer Equivalence Number (REN) of your various devices. (The REN must be listed on each device.) The total REN for all devices connected to your line cannot exceed 5.0. An REN of 1.0 is equivalent to an old-fashioned analog set with an electromechanical ringer. Some electronic phones have an REN of 0.3 or even less. If you connect too many devices that require too much current, you will find that none of them will be able to ring.

Dial pad

When you place a telephone call, you need some way of letting the network know the address of the party you wish to reach. The dial pad is the portion of the phone that provides this functionality. In the early days of the PSTN, dial pads were in fact rotary devices that used pulses to indicate digits. This was a rather slow process, so the telephone companies eventually introduced touch-tone dialing. With touch-tone—also known as Dual-Tone Multi Frequency (DTMF)—dialing, the dial pad consists of 12 buttons. Each button has two frequencies assigned to it (see Table 7-1).

Table 7-1. DTMF digits

	1209 Hz	1336 Hz	1477 Hz	1633 Hz [a]
697 Hz	1	2	3	A
770 Hz	4	5	6	B
852 Hz	7	8	9	C
941 Hz	*	0	#	D

[a] Notice that this column contains letters that are not typically present as keys on a telephone dial pad. They are part of the DTMF standard nonetheless, and any proper telephone contains the electronics required to create them, even if it doesn't contain the buttons themselves. (These buttons actually do exist on some telephones, which are mostly used in military and government applications.)

When you press a button on your dial pad, the two corresponding frequencies are transmitted down the line. The far end can interpret these frequencies and note which digit was pressed.

Hybrid (or network)

The hybrid is a type of transformer that handles the need to combine the signals transmitted and received across a single pair of wires in the PSTN and two pairs of wires in the handset. One of the functions the hybrid performs is regulating *sidetone*, which is the amount of your transmitted signal that is returned to your earpiece; its purpose is to provide a more natural-sounding conversation. Too much sidetone, and your voice will sound too loud; too little, and you'll think the line has gone dead.

Hook switch (or switch hook).   This device signals the state of the telephone circuit to the CO. When you pick up your telephone, the hook switch closes the loop between you and the CO, which is seen as a request for a dial tone. When you hang up, the hook switch opens the circuit, which indicates that the call has ended.[†]

The hook switch can also be used for signaling purposes. Some electronic analog phones have a button labeled Link that causes an event called a *flash*. You can perform a flash manually by depressing the hook switch for a duration of between 200 and 1,200 milliseconds. If you leave it down for longer than that, the carrier may assume you've hung up. The purpose of the Link button is to handle this timing for you. If you've ever used call waiting or three-way calling on an analog line, you have performed a hook switch flash for the purpose of signaling the network.

[†] When referring to the state of an analog circuit, people often speak in terms of "off-hook" and "on-hook." When your line is "off-hook," your telephone is "on" a call. If your phone is "on-hook," the telephone is essentially "off," or idle.

Ring   Tip

Figure 7-1. Tip and Ring

Handset. The handset is composed of the transmitter and receiver. It performs the conversion between the sound energy humans use and the electrical energy the telephone network uses.

Tip and Ring

In an analog telephone circuit, there are two wires. In North America, these wires are referred to as Tip and Ring.[‡] This terminology comes from the days when telephone calls were connected by live operators sitting at cord boards. The plugs that they used had two contacts—one located at the tip of the plug and the other connected to the ring around the middle (Figure 7-1).

The Tip lead is the positive polarity wire. In North America, this wire is typically green and provides the return path. The Ring wire is the negative polarity wire. In North America, this wire is normally red. For modern Cat 5 and 6 cables, the Tip is usually the white wire, and Ring is the coloured wire. When your telephone is on-hook, this wire will have a potential of −48V DC with respect to Tip. Off-hook, this voltage drops to roughly −7V DC.

Digital Telephony

Analog telephony is almost dead.

In the PSTN, the famous Last Mile is the final remaining piece of the telephone network still using technology pioneered well over a hundred years ago.[§]

One of the primary challenges when transmitting analog signals is that all sorts of things can interfere with those signals, causing low volume, static, and all manner of other undesired effects. Instead of trying to preserve an analog waveform over distances that may span thousands of miles, why not simply measure the characteristics of the original

---

[‡] They may have other names elsewhere in the world (such as "A" and "B").

[§] "The Last Mile" is a term that was originally used to describe the only portion of the PSTN that had not been converted to fiber optics: the connection between the central office and the customer. The Last Mile is more than that, however, as it also has significance as a valuable asset of the traditional phone companies; they own a connection into your home. The Last Mile is becoming more and more difficult to describe in technical terms, as there are now so many ways to connect the network to the customer. As a thing of strategic value to telecom, cable, and other utilities, its importance is obvious.

sound and send that information to the far end? The original waveform wouldn't get there, but all the information needed to reconstruct it would.

This is the principle of all digital audio (including telephony): sample the characteristics of the source waveform, store the measured information, and send that data to the far end. Then, at the far end, use the transmitted information to generate a completely new audio signal that has the same characteristics as the original. The reproduction is so good that the human ear can't tell the difference.

The principle advantage of digital audio is that the sampled data can be mathematically checked for errors all along the route to its destination, ensuring that a perfect duplicate of the original arrives at the far end. Distance no longer affects quality, and interference can be detected and eliminated.

Pulse-Code Modulation

There are several ways to digitally encode audio, but the most common method (and the one used in telephony systems) is known as Pulse-Code Modulation (PCM). To illustrate how this works, let's go through a few examples.

Digitally encoding an analog waveform

The principle of PCM is that the amplitude[||]of the analog waveform is sampled at specific intervals so that it can later be re-created. The amount of detail that is captured is dependent both on the bit resolution of each sample and on how frequently the samples are taken. A higher bit resolution and a higher sampling rate will provide greater accuracy, but more bandwidth will be required to transmit this more detailed information.

To get a better idea of how PCM works, consider the waveform displayed in Figure 7-2.

To digitally encode the wave, it must be sampled on a regular basis, and the amplitude of the wave at each moment in time must be measured. The process of slicing up a waveform into moments in time and measuring the energy at each moment is called *quantization*, or *sampling*.

The samples will need to be taken frequently enough and will need to capture enough information to ensure that the far end can re-create a sufficiently similar waveform. To achieve a more accurate sample, more bits will be required. To explain this concept, we will start with a very low resolution, using four bits to represent our amplitude. This will make it easier to visualize both the quantization process itself and the effect that resolution has on quality.

[||] Amplitude is essentially the power or strength of the signal. If you have ever held a skipping rope or garden hose and given it a whip, you have seen the resultant wave. The taller the wave, the greater the amplitude.

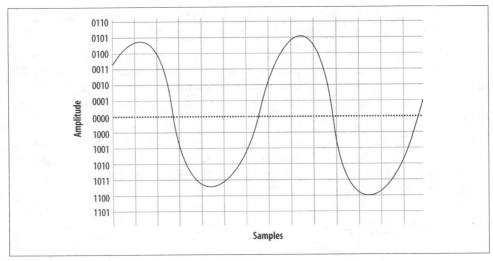

Figure 7-2. A simple sinusoidal (sine) wave

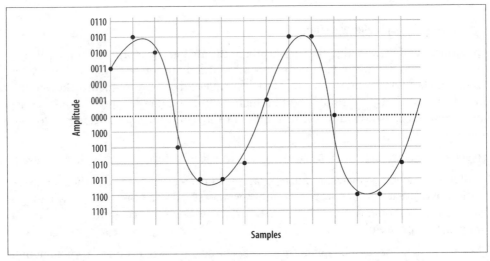

Figure 7-3. Sampling our sine wave using four bits

Figure 7-3 shows the information that will be captured when we sample our sine wave at four-bit resolution.

At each time interval, we measure the amplitude of the wave and record the corresponding intensity—in other words, we sample it. You will notice that the four-bit resolution limits our accuracy. The first sample has to be rounded to **0011**, and the next quantization yields a sample of **0101**. Then comes **0100**, followed by **1001**, **1011**, and so forth. In total, we have 14 samples (in reality, several thousand samples must be taken per second).

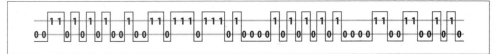

Figure 7-4. PCM encoded waveform

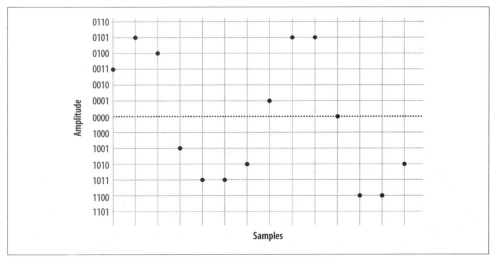

Figure 7-5. Plotted PCM signal

If we string together all the values, we can send them to the other side as:

    0011 0101 0100 1001 1011 1011 1010 0001 0101 0101 0000 1100 1100 1010

On the wire, this code might look something like Figure 7-4.

When the far end's digital-to-analog (D/A) converter receives this signal, it can use the information to plot the samples, as shown in Figure 7-5.

From this information, the waveform can be reconstructed (see Figure 7-6).

As you can see if you compare Figure 7-2 with Figure 7-6, this reconstruction of the waveform is not very accurate. This was done intentionally, to demonstrate an important point: the quality of the digitally encoded waveform is affected by the resolution and rate at which it is sampled. At too low a sampling rate, and with too low a sample resolution, the audio quality will not be acceptable.

Increasing the sampling resolution and rate

Let's take another look at our original waveform, this time using five bits to define our quantization intervals (Figure 7-7).

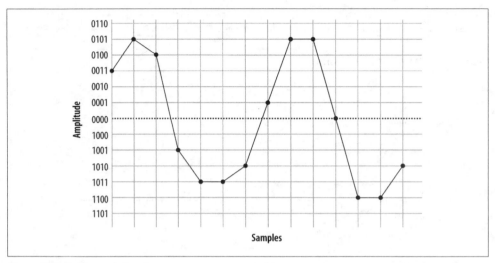

Figure 7-6. Delineated signal

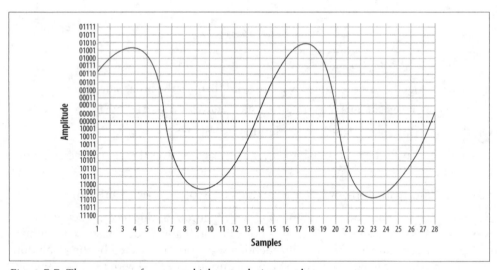

Figure 7-7. The same waveform, on a higher-resolution overlay

 In reality, there is no such thing as five-bit PCM. In the telephone network, PCM samples are encoded using eight bits.[#]

[#] Other digital audio methods may employ 16 bits or more.

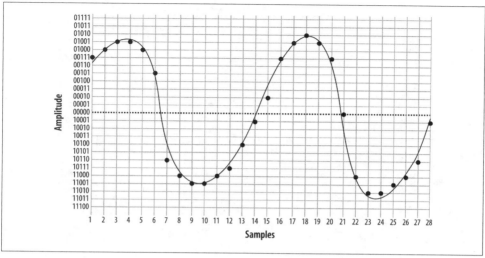

Figure 7-8. The same waveform at double the resolution

We'll also double our sampling frequency. The points plotted this time are shown in Figure 7-8.

We now have twice the number of samples, at twice the resolution. Here they are:

```
00111 01000 01001 01001 01000 00101 10110 11000 11001 11001 11000 10111
10100 10001 00010 00111 01001 01010 01001 00111 00000 11000 11010 11010
11001 11000 10110 10001
```

When received at the other end, that information can now be plotted as shown in Figure 7-9.

From this information, the waveform shown in Figure 7-10 can then be generated.

As you can see, the resultant waveform is a far more accurate representation of the original. However, you can also see that there is still room for improvement.

Note that 40 bits were required to encode the waveform at 4-bit resolution, while 156 bits were needed to send the same waveform using 5-bit resolution (and also doubling the sampling rate). The point is, there is a tradeoff: the higher the quality of audio you wish to encode, the more bits required to do it, and the more bits you wish to send (in real time, naturally), the more bandwidth you will need to consume.

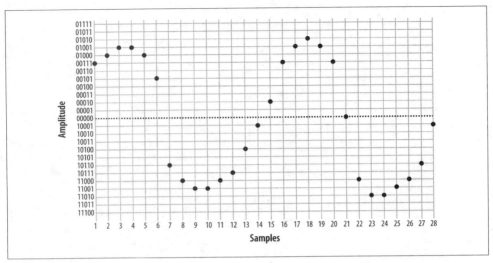

Figure 7-9. Five-bit plotted PCM signal

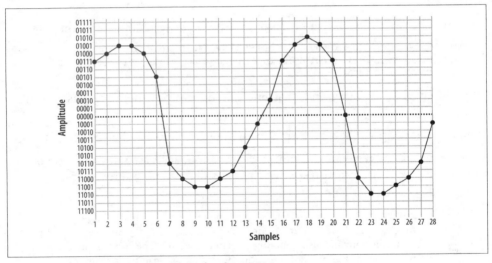

Figure 7-10. Waveform delineated from five-bit PCM

Nyquist's Theorem

So how much sampling is enough? That very same question was considered in the 1920s by an electrical engineer (and AT&T/Bell employee) named Harry Nyquist. Nyquist's Theorem states: "When sampling a signal, the *sampling frequency* must be greater than twice the bandwidth of the input signal in order to be able to reconstruct the original perfectly from the sampled version."*

In essence, what this means is that to accurately encode an analog signal you have to sample it twice as often as the total bandwidth you wish to reproduce. Since the telephone network will not carry frequencies below 300 Hz and above 4,000 Hz, a sampling frequency of 8,000 samples per second will be sufficient to reproduce any frequency within the bandwidth of an analog telephone. Keep that 8,000 samples per second in mind; we're going to talk about it more later.

Logarithmic companding

So, we've gone over the basics of quantization, and we've discussed the fact that more quantization intervals (i.e., a higher sampling rate) give better quality but also require more bandwidth. Lastly, we've discussed the minimum sample rate needed to accurately measure the range of frequencies we wish to be able to transmit (in the case of the telephone, it's 8,000 Hz). This is all starting to add up to a fair bit of data being sent on the wire, so we're going to want to talk about companding.

Companding is a method of improving the dynamic range of a sampling method without losing important accuracy. It works by quantizing higher amplitudes in a much coarser fashion than lower amplitudes. In other words, if you yell into your phone, you will not be sampled as cleanly as you will be when speaking normally. Yelling is also not good for your blood pressure, so it's best to avoid it.

Two companding methods are commonly employed: μlaw[†] in North America, and alaw in the rest of the world. They operate on the same principles but are otherwise not compatible with each other.

Companding divides the waveform into *cords*, each of which has several *steps*. Quantization involves matching the measured amplitude to an appropriate step within a cord. The value of the band and cord numbers (as well as the sign—positive or negative) becomes the signal. The following diagrams will give you a visual idea of what companding does. They are not based on any standard, but rather were made up for the purpose of illustration (again, in the telephone network, companding will be done at an eight-bit, not five-bit, resolution).

Figure 7-11 illustrates five-bit companding. As you can see, amplitudes near the zero-crossing point will be sampled far more accurately than higher amplitudes (either positive or negative). However, since the human ear, the transmitter, and the receiver will also tend to distort loud signals, this isn't really a problem.

---

[*] Nyquist published two papers, "Certain Factors Affecting Telegraph Speed" (1924) and "Certain Topics in Telegraph Transmission Theory" (1928), in which he postulated what became known as Nyquist's Theorem. Proven in 1949 by Claude Shannon ("Communication in the Presence of Noise"), it is also referred to as the Nyquist-Shannon sampling theorem.

[†] μlaw is often referred to as "ulaw" because, let's face it, how many of us have μ keys on our keyboards? μ is in fact the Greek letter Mu; thus, you will also see μlaw written (more correctly) as "Mu-law." When spoken, it is correct to confidently say "Mew-law," but if folks look at you strangely, and you're feeling generous, you can help them out and tell them it's "ulaw." Many people just don't appreciate trivia.

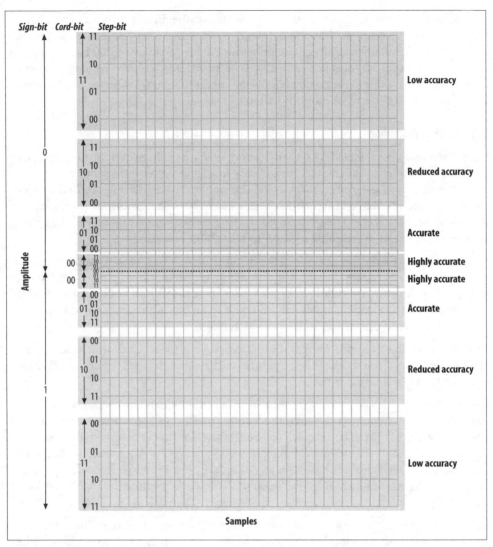

Figure 7-11. Five-bit companding

A quantized sample might look like Figure 7-12. It yields the following bit stream:

```
00000 10011 10100 10101 01101 00001 00011 11010 00010 00001 01000 10011
10100 10100 00101 00100 00101 10101 10011 10001 00011 00001 00000 10100
10010 10101 01101 10100 00101 11010 00100 00000 01000
```

Aliasing

If you've ever watched the wheels on a wagon turn backward in an old Western movie, you've seen the effects of aliasing. The frame rate of the movie cannot keep up with the rotational frequency of the spokes, and a false rotation is perceived.

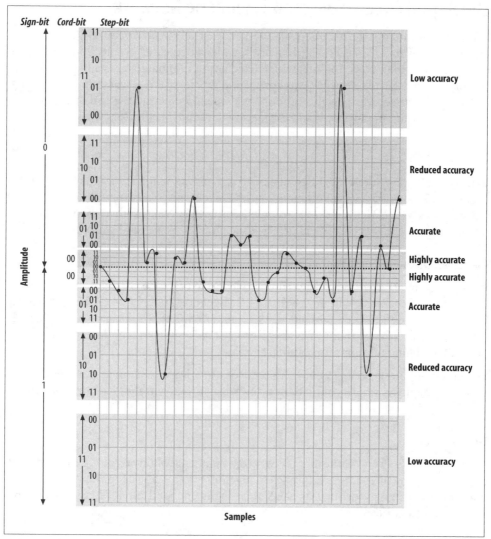

Figure 7-12. Quantized and companded at 5-bit resolution

In a digital audio system (which the modern PSTN arguably is), aliasing always occurs if frequencies that are greater than one-half the sampling rate are presented to the analog-to-digital (A/D) converter. In PSTN, that includes any audio frequencies above 4,000 Hz (half the sampling rate of 8,000 Hz). This problem is easily corrected by passing the audio through a low-pass filter[‡] before presenting it to the A/D converter.[§]

[‡] A low-pass filter, as its name implies, allows through only frequencies that are lower than its cut-off frequency. Other types of filters are high-pass filters (which remove low frequencies) and band-pass filters (which filter out both high and low frequencies).

The Digital Circuit-Switched Telephone Network

For over a hundred years, telephone networks were exclusively circuit-switched. What this meant was that for every telephone call made, a dedicated connection was established between the two endpoints, with a fixed amount of bandwidth allocated to that circuit. Creating such a network was costly, and where distance was concerned, using that network was costly as well. Although we are all predicting the end of the circuit-switched network, many people still use it every day, and it really does work rather well.

Circuit Types

In the PSTN, there are many different sizes of circuits serving the various needs of the network. Between the central office and a subscriber, one or more analog circuits, or a few dozen channels delivered over a digital circuit, generally suffice. Between PSTN offices (and with larger customers), fiber-optic circuits are generally used.

The humble DS-0—the foundation of it all

Since the standard method of digitizing a telephone call is to record an 8-bit sample 8,000 times per second, we can see that a PCM-encoded telephone circuit will need a bandwidth of eight times 8,000 bits per second, or 64,000 bps. This 64 Kbps channel is referred to as a DS-0 (that's "Dee-Ess-Zero"). The DS-0 is the fundamental building block of all digital telecommunications circuits.

Even the ubiquitous analog circuit is sampled into a DS-0 as soon as possible. Sometimes this happens where your circuit terminates at the central office, and sometimes well before.[||]

T-carrier circuits

The venerable T1 is one of the more recognized digital telephony terms. A T1 is a digital circuit consisting of 24 DS-0s multiplexed together into a 1.544 Mbps bitstream.[#] This bit stream is properly defined as a *DS-1*. Voice is encoded on a T1 using the μlaw companding algorithm.

[§] If you ever have to do audio recordings for a system, you might want to take advantage of the band-pass filter that is built into most telephone sets. Doing a recording using even high-end recording equipment can pick up all kinds of background noise that you don't even hear until you downsample, at which point the background noise produces aliasing (which can sound like all kinds of weird things). Conversely, the phone records in the correct format already, so the noise never enters the audio stream. Having said all that, no matter what you use to do recordings, avoid environments that have a lot of background noise. Typical offices can be a lot noisier than you'd think, as HVAC equipment can produce noise that we don't even realize is there.

[||] Digital telephone sets (including IP sets) do the analog-to-digital conversion right at the point where the handset plugs into the phone, so the DS-0 is created right at the phone set.

[#] The 24 DS-0s use 1.536 Mbps, and the remaining .008 Mbps is used by framing bits.

The European version of the T1 was developed by the European Conference of Postal and Telecommunications Administrations[*] (CEPT), and was first referred to as a *CEPT-1*. It is now called an *E1*.

The E1 is comprised of 32 DS-0s, but the method of PCM encoding is different: E1s use alaw companding. This means that connecting between an E1-based network and a T1-based network will always require a transcoding step. Note that an E1, although it has 32 channels, is also considered a *DS-1*. It is likely that E1 is far more widely deployed, as it is used everywhere in the world except North American and Japan.

The various other T-carriers (T2, T3, and T4) are multiples of the T1, each based on the humble DS-0. Table 7-2 illustrates the relationships between the different T-carrier circuits.

Table 7-2. T-carrier circuits

Carrier	Equivalent data bitrate	Number of DS-0s	Data bitrate
T1	24 DS-0s	24	1.544 Mbps
T2	4 T1s	96	6.312 Mbps
T3	7 T2s	672	44.736 Mbps
T4	6 T3s	4,032	274.176 Mbps

At densities above T3, it is very uncommon to see a T-carrier circuit. For these speeds, optical carrier (OC) circuits may be used.

SONET and OC circuits

The Synchronous Optical Network (SONET) was developed out of a desire to take the T-carrier system to the next technological level: fiber optics. SONET is based on the bandwidth of a T3 (44.736 Mbps), with a slight overhead making it 51.84 Mbps. This is referred to as an *OC-1* or *STS-1*. As Table 7-3 shows, all higher-speed OC circuits are multiples of this base rate.

Table 7-3. OC circuits

Carrier	Equivalent data bitrate	Number of DS-0s	Data bitrate
OC-1	1 DS-3 (plus overhead)	672	51.840 Mbps
OC-3	3 DS-3s	2,016	155.520 Mbps
OC-12	12 DS-3s	8,064	622.080 Mbps
OC-48	48 DS-3s	32,256	2488.320 Mbps
OC-192	192 DS-3s	129,024	9953.280 Mbps

[*] Conférence Européenne des Administrations des Postes et des Télécommunications.

SONET was created in an effort to standardize optical circuits, but due to its high cost, coupled with the value offered by many newer schemes, such as Dense Wave Division Multiplexing (DWDM), there is some controversy surrounding its future.

Digital Signaling Protocols

As with any circuit, it is not enough for the circuits used in the PSTN to just carry (voice) data between endpoints. Mechanisms must also be provided to pass information about the state of the channel between each endpoint. (Disconnect and answer supervision are two examples of basic signaling that might need to take place; Caller ID is an example of a more complex form of signaling.)

Channel Associated Signaling (CAS)

Also known as robbed-bit signaling, CAS is what you will use to transmit voice on a T1 when ISDN is not available. Rather than taking advantage of the power of the digital circuit, CAS simulates analog channels. CAS works by stealing bits from the audio stream for signaling purposes. Although the effect on audio quality is not really noticeable, the lack of a powerful signaling channel limits your flexibility.

When configuring a CAS T1, the signaling options at each end must match. E&M (Ear & Mouth or recEive & transMit) signaling is generally preferred, as it offers the best supervision. Having said that, in an Asterisk environment the most likely reason for you to use CAS would be for a channel bank, which means you are most likely going to have to use FXS signaling.

CAS is very rarely used on PSTN circuits anymore, due to the superiority of ISDN-PRI. One of the limitations of CAS is that it does not allow the dynamic assignment of channels to different functions. Also, Caller ID information (which may not even be supported) has to be sent as part of the audio stream. CAS is commonly used on the T1 link in channel banks.

ISDN

The Integrated Services Digital Network (ISDN) has been around for more than 20 years. Because it separates the channels that carry the traffic (the bearer channels, or B-channels) from the channel that carries the signaling information (the D-channel), ISDN allows for the delivery of a much richer set of features than CAS. In the beginning, ISDN promised to deliver much the same sort of functionality that the Internet has given us, including advanced capabilities for voice, video, and data transfer.

Unfortunately, rather than ratifying a standard and sticking to it, the respective telecommunications manufacturers all decided to add their own tweaks to the protocol, in the belief that their versions were superior and would eventually dominate the market. As a result, getting two ISDN-compliant systems to connect to each other was often a painful and expensive task. The carriers who had to implement and support this expensive technology, in turn, priced it so that it was not rapidly adopted. Currently, ISDN is rarely used for much more than basic trunking—in fact, the acronym ISDN has become a joke in the industry: "It Still Does Nothing."

Having said that, ISDN has become quite popular for trunking, and it is now (mostly) standards-compliant. If you have a PBX with more than a dozen lines connected to the PSTN, there's a very good chance that you'll be running an ISDN-*PRI* (Primary Rate Interface) circuit. Also, in places where DSL and cable access to the Internet are not available (or are too expensive), an ISDN-*BRI* (Basic Rate Interface) circuit might provide you with an affordable 128 Kbps connection. In much of North America, the use of BRI for Internet connectivity has been deprecated in favor of DSL and cable modems (and it is never used for voice), but in many European countries it has almost totally replaced analog circuits.

ISDN-BRI/BRA. Basic Rate Interface (or Basic Rate Access) is the flavor of ISDN, and is designed to service small endpoints such as workstations.

The BRI flavor of the ISDN specification is often referred to simply as "ISDN," but this can be a source of confusion, as ISDN is a protocol, not a type of circuit (not to mention that PRI circuits are also correctly referred to as ISDN!).

A Basic Rate ISDN circuit consists of two 64 Kbps B-channels controlled by a 16-Kbps D-channel, for a total of 144 Kbps.

Basic Rate ISDN has been a source of much confusion during its life, due to problems with standards compliance, technical complexity, and poor documentation. Still, many European telecos have widely implemented ISDN-BRI, and thus it is more popular in Europe than in North America.

ISDN-PRI/PRA. The Primary Rate Interface (or Primary Rate Access) flavor of ISDN is used to provide ISDN service over larger network connections. A Primary Rate ISDN circuit uses a single DS-0 channel as a signaling link (the D-channel); the remaining channels serve as B-channels.

In North America, Primary Rate ISDN is commonly carried on one or more T1 circuits. Since a T1 has 24 channels, a North American PRI circuit typically consists of 23 B-channels and 1 D-channel. For this reason, PRI is often referred to as 23B+D.[†]

In Europe, a 32-channel E1 circuit is used, so a Primary Rate ISDN circuit is referred to as 30B+D (the final channel is used for synchronization).

[†] PRI is actually quite a bit more flexible than that, as it is possible to span a single PRI circuit across multiple T1 spans. This can give rise, for example, to a 47B+D circuit (where a single D-channel serves two T1s) or a 46B+2D circuit (where primary and backup D-channels serve a pair of T1s). You will sometimes see PRI described as nB+nD, because the number of B- and D-channels is, in fact, quite variable. For this reason, you should never refer to a T1 carrying PRI as "a PRI." For all you know, the PRI circuit spans multiple T1s, as is common in larger PBX deployments.

Primary Rate ISDN is very popular, due to its technical benefits and generally competitive pricing at higher densities. If you believe you will require more than a dozen or so PSTN lines, you should look into Primary Rate ISDN pricing.

From a technical perspective, ISDN-PRI is always preferable to CAS.

Signaling System 7

Signaling System 7 (SS7) is the signaling system used by carriers. It is conceptually similar to ISDN, and it is instrumental in providing a mechanism for the carriers to transmit the additional information ISDN endpoints typically need to pass. However, the technology of SS7 is different from that of ISDN; one big difference is that SS7 runs on a completely separate network from the actual trunks that carry the calls.

SS7 support in Asterisk is on the horizon, as there is much interest in making Asterisk compatible with the carrier networks. An open source version of SS7 (*http:// www.openss7.org*) exists, but work is still needed for full SS7 compliance, and as of this writing it is not known whether this version will be integrated with Asterisk. Another promising source of SS7 support comes from Sangoma Technologies, which offers SS7 functionality in many of its products.

It should be noted that adding support for SS7 in Asterisk is not going to be as simple as writing a proper driver. Connecting equipment to an SS7 network will not be possible without that equipment having passed extremely rigorous certification processes. Even then, it seems doubtful that any traditional carrier is going to be in a hurry to allow such a thing to happen, mostly for strategic and political reasons.

Packet-Switched Networks

In the mid-1990s, network performance improved to the point where it became possible to send a stream of media information in real time across a network connection. Because the media stream is chopped up into segments, which are then wrapped in an addressing envelope, such connections are referred to as *packet-based*. The challenge, of course, is to send a flood of these packets between two endpoints, ensuring that the packets arrive in the same order in which they were sent, in less than 150 milliseconds, with none lost. This is the essence of Voice over IP.

Conclusion

This chapter has explored the technologies currently in use in the PSTN. In the next chapter, we will discuss protocols for VoIP: the carrying of telephone connections across IP-based networks. These protocols define different mechanisms for carrying telephone conversations, but their significance is far greater than just that. Bringing the telephone network into the data network will finally erase the line between telephones and computers, which holds the promise of a revolutionary evolution in the way we communicate.

## CHAPTER 8

Protocols for VoIP

The Internet is a telephone system that's gotten uppity.

—Clifford Stoll

The telecommunications industry spans over 100 years, and Asterisk integrates most —if not all—of the major technologies that it has made use of over the last century. To make the most out of Asterisk, you need not be a professional in all areas, but understanding the differences between the various codecs and protocols will give you a greater appreciation and understanding of the system as a whole.

This chapter explains Voice over IP and what makes VoIP networks different from the traditional circuit-switched voice networks that were the topic of the last chapter. We will explore the need for VoIP protocols, outlining the history and potential future of each. We'll also look at security considerations and these protocols' abilities to work within topologies such as Network Address Translation (NAT). The following VoIP protocols will be discussed (some more briefly than others):

- IAX
- SIP
- H.323
- MGCP
- Skinny/SCCP
- UNISTIM

Codecs are the means by which analog voice can be converted to a digital signal and carried across the Internet. Bandwidth at any location is finite, and the number of simultaneous conversations any particular connection can carry is directly related to the type of codec implemented. In this chapter, we'll also explore the differences between the following codecs in regards to bandwidth requirements (compression level) and quality:

- G.711
- G.726
- G.729A
- GSM
- iLBC
- Speex
- MP3

We will then conclude the chapter with a discussion of how voice traffic can be routed reliably, what causes echo and how to deal with it, and how Asterisk controls the authentication of inbound and outbound calls.

The Need for VoIP Protocols

The basic premise of VoIP is the packetization[*] of audio streams for transport over Internet Protocol-based networks. The challenges to accomplishing this relate to the manner in which humans communicate. Not only must the signal arrive in essentially the same form that it was transmitted in, but it needs to do so in less than 150 milliseconds. If packets are lost or delayed, there will be degradation to the quality of the communications experience, meaning that two people will have difficulty in carrying on a conversation.

The transport protocols that collectively are called "the Internet" were not originally designed with real-time streaming of media in mind. Endpoints were expected to resolve missing packets by waiting longer for them to arrive, requesting retransmission, or, in some cases, considering the information to be gone for good and simply carrying on without it. In a typical voice conversation, these mechanisms will not serve. Our conversations do not adapt well to the loss of letters or words, nor to any appreciable delay between transmittal and receipt.

The traditional PSTN was designed specifically for the purpose of voice transmission, and it is perfectly suited to the task from a technical standpoint. From a flexibility standpoint, however, its flaws are obvious to even people with a very limited understanding of the technology. VoIP holds the promise of incorporating voice communications into all of the other protocols we carry on our networks, but due to the special demands of a voice conversation, special skills are needed to design, build, and maintain these networks.

The problem with packet-based voice transmission stems from the fact that the way in which we speak is totally incompatible with the way in which IP transports data.

---

[*] This word hasn't quite made it into the dictionary, but it is a term that is becoming more and more common. It refers to the process of chopping a steady stream of information into discrete chunks (or *packets*), suitable for delivery independently of one another.

Speaking and listening consist of the relaying of a stream of audio, whereas the Internet protocols are designed to chop everything up, encapsulate the bits of information into thousands of packages, and then deliver each package in whatever way possible to the far end. Clearly, some way of dealing with this is required.

VoIP Protocols

The mechanism for carrying a VoIP connection generally involves a series of signaling transactions between the endpoints (and gateways in between), culminating into two persistent media streams (one for each direction) that carry the actual conversation. There are several protocols in existence to handle this. In this section, we will discuss some of those that are important to VoIP in general and to Asterisk specifically.

IAX (The "Inter-Asterisk eXchange" Protocol)

If you claim to be one of the folks in the know when it comes to Asterisk, your test will come when you have to pronounce the name of this protocol. It would seem that you should say "eye-ay-ex", but this hardly rolls off the tongue very well.[†] Fortunately, the proper pronunciation is in fact "eeks."[‡] IAX is an open protocol, meaning that anyone can download and develop for it, but it is not yet a standard of any kind.[§]

It is expected that IAX2 will become an IETF protocol soon. IAX2 is currently in draft status with the IETF, and it is widely expected to become an official protocol in a few years' time.

In Asterisk, IAX is supported by the *chan_iax2.so* module.

History

The IAX protocol was developed by Digium for the purpose of communicating with other Asterisk servers (hence the Inter-Asterisk eXchange protocol). It is very important to note that IAX is not at all limited to Asterisk. The standard is open for anyone to use, and it is supported by many other open source telecom projects, as well as by several hardware vendors. IAX is a transport protocol (much like SIP) that uses a single UDP port (4569) for both the channel signaling and media streams. As discussed later in this chapter, this makes it easier to manage when behind NATed firewalls.

IAX also has the unique ability to trunk multiple sessions into one dataflow, which can be a tremendous bandwidth advantage when sending a lot of simultaneous channels to a remote box. Trunking allows multiple media streams to be represented with a

---

[†] It sounds like the name of a Dutch football team.

[‡] Go ahead. Say it. Now that sounds much better, doesn't it?

[§] Officially, the current version is IAX2, but all support for IAX1 has been dropped, so whether you say "IAX" or "IAX2," it is expected that you are talking about the same version.

single datagram header, that will lower the overhead associated with individual channels. This helps to lower latency and reduce the processing power and bandwidth required, allowing the protocol to scale much more easily with a large number of active channels between endpoints. If you have a large quantity of IP calls to pass between two endpoints, you should take a close look at IAX trunking.

Future

Since IAX was optimized for voice, it has received some criticism for not better supporting video—but in fact, IAX holds the potential to carry pretty much any media stream desired. Because it is an open protocol, future media types are certain to be incorporated as the community desires them.

Security considerations

IAX includes the ability to authenticate in three ways: plain text, MD5 hashing, and RSA key exchange. This, of course, does nothing to encrypt the media path or headers between endpoints. Many solutions include using a Virtual Private Network (VPN) appliance or software to encrypt the stream in another layer of technology, which requires the endpoints to pre-establish a method of having these tunnels configured and operational. However, IAX is now also able to encrypt the streams between endpoints with dynamic key exchange at call setup (using the configuration option encryption=aes128), allowing the use of automatic key rollover.

IAX and NAT

The IAX2 protocol was deliberately designed to work from behind devices performing NAT. The use of a single UDP port for both signaling and transmission of media also keeps the number of holes required in your firewall to a minimum. These considerations have helped make IAX one of the easiest protocols (if not the easiest) to implement in secure networks.

SIP

The Session Initiation Protocol (SIP) has taken the telecommunications industry by storm. SIP has pretty much dethroned the once-mighty H.323 as the VoIP protocol of choice—certainly at the endpoints of the network. The premise of SIP is that each end of a connection is a peer; the protocol negotiates capabilities between them. What makes SIP compelling is that it is a relatively simple protocol, with a syntax similar to that of other familiar protocols such as HTTP and SMTP. SIP is supported in Asterisk with the *chan_sip.so* module.[||]

History

SIP was originally submitted to the Internet Engineering Task Force (IETF) in February of 1996 as "draft-ietf-mmusic-sip-00." The initial draft looked nothing like the SIP we

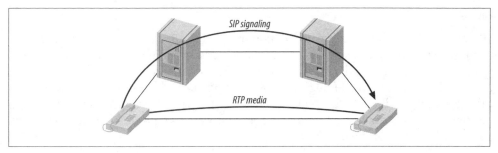

Figure 8-1. The SIP trapezoid

know today and contained only a single request type: a call setup request. In March of 1999, after 11 revisions, SIP RFC 2543 was born.

At first, SIP was all but ignored, as H.323 was considered the protocol of choice for VoIP transport negotiation. However, as the buzz grew, SIP began to gain popularity, and while there may be a lot of different factors that accelerated its growth, we'd like to think that a large part of its success is due to its freely available specification.

SIP is an application-layer signaling protocol that uses the well-known port 5060 for communications. SIP can be transported with either the UDP or TCP transport-layer protocols. Asterisk does not currently have a TCP implementation for transporting SIP messages, but it is possible that future versions may support it (and patches to the code base are gladly accepted). SIP is used to "establish, modify, and terminate multimedia sessions such as Internet telephony calls."[#]

SIP does not transport media between endpoints.

RTP is used to transmit media (i.e., voice) between endpoints. RTP uses high-numbered, unprivileged ports in Asterisk (10,000 through 20,000, by default).

A common topology to illustrate SIP and RTP, commonly referred to as the "SIP trapezoid," is shown in Figure 8-1. When Alice wants to call Bob, Alice's phone contacts her proxy server, and the proxy tries to find Bob (often connecting through his proxy). Once the phones have started the call, they communicate directly with each other (if possible), so that the data doesn't have to tie up the resources of the proxy.

SIP was not the first, and is not the only, VoIP protocol in use today (others include H. 323, MGCP, IAX, and so on), but currently it seems to have the most momentum with hardware vendors. The advantages of the SIP protocol lie in its wide acceptance and architectural flexibility (and, we used to say, simplicity!).

[‖] Having just called SIP simple, it should be noted that it is by no means lightweight. It has been said that if one were to read all of the IETF RFCs that are relevant to SIP, one would have more than 3,000 pages of reading to do. SIP is quickly earning a reputation for being far too bloated, but that does nothing to lessen its popularity.

[#] RFC 3261, SIP: Session Initiation Protocol, p. 9, Section 2.

Future

SIP has earned its place as the protocol that justified VoIP. All new user and enterprise products are expected to support SIP, and any existing products will now be a tough sell unless a migration path to SIP is offered. SIP is widely expected to deliver far more than VoIP capabilities, including the ability to transmit video, music, and any type of real-time multimedia. While its use as a ubiquitous general-purpose media transport mechanism seems doubtful, SIP is unarguably poised to deliver the majority of new voice applications for the next few years.

Security considerations

SIP uses a challenge/response system to authenticate users. An initial `INVITE` is sent to the proxy with which the end device wishes to communicate. The proxy then sends back a 407 Proxy Authorization Request message, which contains a random set of characters referred to as a *nonce*. This nonce is used along with the password to generate an MD5 hash, which is then sent back in the subsequent `INVITE`. Assuming the MD5 hash matches the one that the proxy generated, the client is then authenticated.

Denial of Service (DoS) attacks are probably the most common type of attack on VoIP communications. A DoS attack can occur when a large number of invalid `INVITE` requests are sent to a proxy server in an attempt to overwhelm the system. These attacks are relatively simple to implement, and their effects on the users of the system are immediate. SIP has several methods of minimizing the effects of DoS attacks, but ultimately they are impossible to prevent.

SIP implements a scheme to guarantee that a secure, encrypted transport mechanism (namely Transport Layer Security, or TLS) is used to establish communication between the caller and the domain of the callee. Beyond that, the request is sent securely to the end device, based upon the local security policies of the network. Note that the encryption of the media (that is, the RTP stream) is beyond the scope of SIP itself and must be dealt with separately.

More information regarding SIP security considerations, including registration hijacking, server impersonation, and session teardown, can be found in Section 26 of SIP RFC 3261.

SIP and NAT

Probably the biggest technical hurdle SIP has to conquer is the challenge of carrying out transactions across a NAT layer. Because SIP encapsulates addressing information in its data frames, and NAT happens at a lower network layer, the addressing information is not automatically modified and, thus, the media streams will not have the correct addressing information needed to complete the connection when NAT is in place. In addition to this, the firewalls normally integrated with NAT will not consider the incoming media stream to be part of the SIP transaction, and will block the connection. Newer firewalls and Session Border Controllers are SIP-aware, but this is still considered a shortcoming in this protocol, and it causes no end of trouble to network professionals needing to connect SIP endpoints using existing network infrastructure.

H.323

This International Telecommunication Union (ITU) protocol was originally designed to provide an IP transport mechanism for video conferencing. It has become the standard in IP-based video-conferencing equipment, and it briefly enjoyed fame as a VoIP protocol as well. While there is much heated debate over whether SIP or H.323 (or IAX) will dominate the VoIP protocol world, in Asterisk, H.323 has largely been deprecated in favor of IAX and SIP. H.323 has not enjoyed much success among users and enterprises, although it might still be the most widely used VoIP protocol among carriers.

The three versions of H.323 supported in Asterisk are handled by the modules *chan_h323.so* (supplied with Asterisk), *chan_oh323.so* (available as a free add-on), and *chan_ooh323.so* (supplied in asterisk-addons).

 You have probably used H.323 without even knowing it—Microsoft's NetMeeting client is arguably the most widely deployed H.323 client.

History

H.323 was developed by the ITU in May of 1996 as a means to transmit voice, video, data, and fax communications across an IP-based network while maintaining connectivity with the PSTN. Since that time, H.323 has gone through several versions and annexes (which add functionality to the protocol), allowing it to operate in pure VoIP networks and more widely distributed networks.

Future

The future of H.323 is a subject of debate. If the media is any measure, it doesn't look good for H.323; it hardly ever gets mentioned (certainly not with the regularity of SIP). H.323 is often regarded as technically superior to SIP, but, as with so many other technologies, that sort of thing is seldom the deciding factor in whether technology enjoys success. One of the factors that makes H.323 unpopular is its complexity—although many argue that the once-simple SIP is starting to suffer from the same problem.

H.323 still carries by far the majority of worldwide carrier VoIP traffic, but as people become less and less dependent on traditional carriers for their telecom needs, the future of H.323 becomes more difficult to predict with any certainty. While H.323 may not be the protocol of choice for new implementations, we can certainly expect to have to deal with H.323 interoperability issues for some time to come.

Security considerations

H.323 is a relatively secure protocol and does not require many security considerations beyond those that are common to any network communicating with the Internet. Since H.323 uses the RTP protocol for media communications, it does not natively support encrypted media paths. The use of a VPN or other encrypted tunnel between endpoints is the most common way of securely encapsulating communications. Of course, this has the disadvantage of requiring the establishment of these secure tunnels between endpoints, which may not always be convenient (or even possible). As VoIP becomes used more often to communicate with financial institutions such as banks, we're likely to require extensions to the most commonly used VoIP protocols to natively support strong encryption methods.

H.323 and NAT

The H.323 standard uses the Internet Engineering Task Force (IETF) RTP protocol to transport media between endpoints. Because of this, H.323 has the same issues as SIP when dealing with network topologies involving NAT. The easiest method is to simply forward the appropriate ports through your NAT device to the internal client.

To receive calls, you will always need to forward TCP port 1720 to the client. In addition, you will need to forward the UDP ports for the RTP media and RTCP control streams (see the manual for your device for the port range it requires). Older clients, such as Microsoft NetMeeting, will also require TCP ports forwarded for H.245 tunneling (again, see your client's manual for the port number range).

If you have a number of clients behind the NAT device, you will need to use a *gatekeeper* running in proxy mode. The gatekeeper will require an interface attached to the private IP subnet and the public Internet. Your H.323 client on the private IP subnet will then register to the gatekeeper, which will proxy calls on the clients' behalf. Note that any external clients that wish to call you will also be required to register with the proxy server.

At this time, Asterisk can't act as an H.323 gatekeeper. You'll have to use a separate application, such as the open source OpenH323 Gatekeeper (*http://www.gnugk.org*).

MGCP

The Media Gateway Control Protocol (MGCP) also comes to us from the IETF. While MGCP deployment is more widespread than one might think, it is quickly losing ground to protocols such as SIP and IAX. Still, Asterisk loves protocols, so naturally it has rudimentary support for it.

MGCP is defined in RFC 3435.[*] It was designed to make the end devices (such as phones) as simple as possible, and have all the call logic and processing handled by

---

[*] RFC 3435 obsoletes RFC 2705.

media gateways and call agents. Unlike SIP, MGCP uses a centralized model. MGCP phones cannot directly call other MGCP phones; they must always go through some type of controller.

Asterisk supports MGCP through the *chan_mgcp.so* module, and the endpoints are defined in the configuration file *mgcp.conf*. Since Asterisk provides only basic call agent services, it cannot emulate an MGCP phone (to register to another MGCP controller as a user agent, for example).

If you have some MGCP phones lying around, you will be able to use them with Asterisk. If you are planning to put MGCP phones into production on an Asterisk system, keep in mind that the community has moved on to more popular protocols, and you will therefore need to budget your software support needs accordingly. If possible (for example, with Cisco phones), you should upgrade MGCP phones to SIP.

Proprietary Protocols

Finally, let's take a look at two proprietary protocols that are supported in Asterisk.

Skinny/SCCP

The Skinny Client Control Protocol (SCCP) is proprietary to Cisco VoIP equipment. It is the default protocol for endpoints on a Cisco Call Manager PBX.[†] Skinny is supported in Asterisk, but if you are connecting Cisco phones to Asterisk, it is generally recommended that you obtain SIP images for any phones that support it and connect via SIP instead.

UNISTIM

Support for Nortel's proprietary VoIP protocol, UNISTIM, means that Asterisk is the first PBX in history to natively support proprietary IP terminals from the two biggest players in VoIP—Nortel and Cisco. UNISTIM support is totally experimental, and does not work well enough to put into production, but the fact that somebody took the trouble to do this demonstrates the power of the Asterisk platform.

Codecs

Codecs are generally understood to be various mathematical models used to digitally encode (and compress) analog audio information. Many of these models take into account the human brain's ability to form an impression from incomplete information. We've all seen optical illusions; likewise, voice-compression algorithms take advantage of our tendency to interpret what we *believe* we should hear, rather than what we

---

[†] Cisco has recently announced that it will be migrating toward SIP in its future products.

actually hear.[‡] The purpose of the various encoding algorithms is to strike a balance between efficiency and quality.[§]

Originally, the term *codec* referred to a COder/DECoder: a device that converts between analog and digital. Now, the term seems to relate more to COmpression/ DECompression.

Before we dig into the individual codecs, take a look at Table 8-1—it's a quick reference that you may want to refer back to.

Table 8-1. Codec quick reference

Codec	Data bitrate (Kbps)	License required?
G.711	64 Kbps	No
G.726	16, 24, 32, or 40 Kbps	No
G.729A	8 Kbps	Yes (no for passthrough)
GSM	13 Kbps	No
iLBC	13.3 Kbps (30-ms frames) or 15.2 Kbps (20-ms frames)	No
Speex	Variable (between 2.15 and 22.4 Kbps)	No

G.711

G.711 is the fundamental codec of the PSTN. In fact, if someone refers to PCM (discussed in the previous chapter) with respect to a telephone network, you are allowed to think of G.711. Two companding methods are used: μlaw in North America and alaw in the rest of the world. Either one delivers an 8-bit word transmitted 8,000 times per second. If you do the math, you will see that this requires 64,000 bits to be transmitted per second.

Many people will tell you that G.711 is an uncompressed codec. This is not exactly true, as companding is considered a form of compression. What is true is that G.711 is the base codec from which all of the others are derived.

G.711 imposes minimal (almost zero) load on the CPU.

[‡] "Aoccdrnig to rsereach at an Elingsh uinervtisy, it deosn't mttaer in waht oredr the ltteers in a wrod are, the olny iprmoetnt tihng is taht frist and lsat ltteres are in the rghit pclae. The rset can be a toatl mses and you can sitll raed it wouthit a porbelm. Tihs is bcuseae we do not raed ervey lteter by istlef, but the wrod as a wlohe." (The source of this quote is unknown–see *http://www.bisso.com/ujg_archives/000228.html*.) We do the same thing with sound–if there is enough information, our brain can fill in the gaps.

[§] On an audio CD, quality is far more important than saving bandwidth, so the audio is quantized at 16 bits (times 2, as it's stereo), with a sampling rate of 44,100 Hz. Considering that the CD was invented in the late 1970s, this was quite impressive stuff back then. The telephone network does not require this level of quality (and needs to optimize bandwidth), so telephone signals are encoded using 8 bits, at a sampling frequency of 8,000 Hz.

G.726

This codec has been around for some time (it used to be G.721, which is now obsolete), and it is one of the original compressed codecs. It is also known as Adaptive Differential Pulse-Code Modulation (ADPCM), and it can run at several bitrates. The most common rates are 16 Kbps, 24 Kbps, and 32 Kbps. As of this writing, Asterisk currently supports only the ADPCM-32 rate, which is far and away the most popular rate for this codec.

G.726 offers quality nearly identical to G.711, but it uses only half the bandwidth. This is possible because rather than sending the result of the quantization measurement, it sends only enough information to describe the difference between the current sample and the previous one. G.726 fell from favor in the 1990s due to its inability to carry modem and fax signals, but because of its bandwidth/CPU performance ratio it is now making a comeback. G.726 is especially attractive because it does not require a lot of computational work from the system.

G.729A

Considering how little bandwidth it uses, G.729A delivers impressive sound quality. It does this through the use of Conjugate-Structure Algebraic-Code-Excited Linear Prediction (CS-ACELP).[||] Because of patents, you can't use G729A without paying a licensing fee; however, it is extremely popular and is, thus, well supported on many different phones and systems.

To achieve its impressive compression ratio, this codec requires an equally impressive amount of effort from the CPU. In an Asterisk system, the use of heavily compressed codecs will quickly bog down the CPU.

G.729A uses 8 Kbps of bandwidth.

GSM

GSM is the darling codec of Asterisk. This codec does not come encumbered with a licensing requirement the way that G.729A does, and it offers outstanding performance with respect to the demand it places on the CPU. The sound quality is generally considered to be of a lesser grade than that produced by G.729A, but much of this comes down to personal opinion; be sure to try it out. GSM operates at 13 Kbps.

[||] CELP is a popular method of compressing speech. By mathematically modeling the various ways humans make sounds, a codebook of sounds can be built. Rather than sending an actual sampled sound, a code corresponding to the sound is determined. CELP codecs take this information (which by itself would produce a very robot-like sound) and attempt to add the personality back in. (Of course, there is much more to it than that.) Jason Woodward's Speech Coding page (*http://www-mobile.ecs.soton.ac.uk/speech_codecs/*) is a source of helpful information for the non-mathematically inclined. This is fairly heavy stuff, though, so wear your thinking cap.

iLBC

The Internet Low Bitrate Codec (iLBC) provides an attractive mix of low bandwidth usage and quality, and it is especially well suited to sustaining reasonable quality on lossy network links.

Naturally, Asterisk supports it (and support elsewhere is growing), but it is not as popular as the ITU codecs and, thus, may not be compatible with common IP telephones and commercial VoIP systems. IETF RFCs 3951 and 3952 have been published in support of iLBC, and iLBC is on the IETF standards track.

Because iLBC uses complex algorithms to achieve its high levels of compression, it has a fairly high CPU cost in Asterisk.

While you are allowed to use iLBC without paying royalty fees, the holder of the iLBC patent, Global IP Sound (GIPS), wants to know whenever you use it in a commercial application. The way you do that is by downloading and printing a copy of the iLBC license, signing it, and returning it to GIPS. If you want to read about iLBC and its license, you can do so at *http://www.ilbcfreeware.org*.

iLBC operates at 13.3 Kbps (30 ms frames) and 15.2 Kbps (20 ms frames).

Speex

Speex is a variable bitrate (VBR) codec, which means that it is able to dynamically modify its bitrate to respond to changing network conditions. It is offered in both narrowband and wideband versions, depending on whether you want telephone quality or better.

Speex is a totally free codec, licensed under the *Xiph.org* variant of the BSD license.

An Internet draft for Speex is available, and more information about Speex can be found at its home page (*http://www.speex.org*).

Speex can operate at anywhere from 2.15 to 22.4 Kbps, due to its variable bitrate.

MP3

Sure thing, MP3 is a codec. Specifically, it's the Moving Picture Experts Group Audio Layer 3 Encoding Standard.[#] With a name like that, it's no wonder we call it MP3! In Asterisk, the MP3 codec is typically used for Music on Hold (MoH). MP3 is not a telephony codec, as it is optimized for music, not voice; nevertheless, it's very popular with VoIP telephony systems as a method of delivering Music on Hold.

[#] If you want to learn all about MPEG audio, do a web search for Davis Pan's paper titled "A Tutorial on MPEG/Audio Compression."

Be aware that music cannot usually be broadcast without a license. Many people assume that there is no legal problem with connecting a radio station or CD as a Music on Hold source, but this is very rarely true.

Quality of Service

Quality of Service, or *QoS* as it's more popularly termed, refers to the challenge of delivering a time-sensitive stream of data across a network that was designed to deliver data in an ad hoc, best-effort sort of way. Although there is no hard rule, it is generally accepted that if you can deliver the sound produced by the speaker to the listener's ear within 150 milliseconds, a normal flow of conversation is possible. When delay exceeds 300 milliseconds, it becomes difficult to avoid interrupting each other. Beyond 500 milliseconds, normal conversation becomes increasingly awkward and frustrating.

In addition to getting it there on time, it is also essential to ensure that the transmitted information arrives intact. Too many lost packets will prevent the far end from completely reproducing the sampled audio, and gaps in the data will be heard as static or, in severe cases, entire missed words or sentences. Even packet loss of 5 percent can severely impede a VoIP network.

TCP, UDP, and SCTP

If you're going to send data on an IP-based network, it will be transported using one of the three transport protocols discussed here.

Transmission Control Protocol

The Transmission Control Protocol (TCP) is almost never used for VoIP, for while it does have mechanisms in place to ensure delivery, it is not inherently in any hurry to do so. Unless you have an extremely low-latency interconnection between the two endpoints, TCP will tend to cause more problems than it solves.

The purpose of TCP is to guarantee the delivery of packets. In order to do this, several mechanisms are implemented, such as packet numbering (for reconstructing blocks of data), delivery acknowledgment, and re-requesting lost packets. In the world of VoIP, getting the packets to the endpoint quickly is paramount—but 20 years of cellular telephony has trained us to tolerate a few lost packets.[*]

---

[*] The order of arrival is important in voice communication, because the audio will be processed and sent to the caller ASAP. However, with a jitter buffer the order of arrival isn't as important, as it provides a small window of time in which the packets can be reordered before being passed on to the caller.

TCP's high processing overhead, state management, and acknowledgment of arrival work well for transmitting large amounts of data, but they simply aren't efficient enough for real-time media communications.

User Datagram Protocol

Unlike TCP, the User Datagram Protocol (UDP) does not offer any sort of delivery guarantee. Packets are placed on the wire as quickly as possible and released into the world to find their way to their final destinations, with no word back as to whether they got there or not. Since UDP itself does not offer any kind of guarantee that the data will arrive,[†] it achieves its efficiency by spending very little effort on what it is transporting.

> TCP is a more "socially responsible" protocol because the bandwidth is more evenly distributed to clients connecting to a server. As the percentage of UDP traffic increases, it is possible that a network could become overwhelmed.

Stream Control Transmission Protocol

Approved by the IETF as a proposed standard in RFC 2960, SCTP is a relatively new transport protocol. From the ground up, it was designed to address the shortcomings of both TCP and UDP, especially as related to the types of services that used to be delivered over circuit-switched telephony networks.

Some of the goals of SCTP were:

- Better congestion-avoidance techniques (specifically, avoiding Denial of Service attacks)
- Strict sequencing of data delivery
- Lower latency for improved real-time transmissions

By overcoming the major shortcomings of TCP and UDP, the SCTP developers hoped to create a robust protocol for the transmission of SS7 and other types of PSTN signaling over an IP-based network.

Differentiated Service

Differentiated service, or DiffServ, is not so much a QoS mechanism as a method by which traffic can be flagged and given specific treatment. Obviously, DiffServ can help to provide QoS by allowing certain types of packets to take precedence over others.

[†] Keep in mind that the upper-layer protocols or applications can implement their own packet-acknowledgment systems.

While this will certainly increase the chance of a VoIP packet passing quickly through each link, it does not guarantee anything.

Guaranteed Service

The ultimate guarantee of QoS is provided by the PSTN. For each conversation, a 64 Kbps channel is completely dedicated to the call; the bandwidth is guaranteed. Similarly, protocols that offer guaranteed service can ensure that a required amount of bandwidth is dedicated to the connection being served. As with any packetized networking technology, these mechanisms generally operate best when traffic is below maximum levels. When a connection approaches its limits, it is next to impossible to eliminate degradation.

MPLS

Multiprotocol Label Switching (MPLS) is a method for engineering network traffic patterns independent of layer-3 routing tables. The protocol works by assigning short labels (MPLS frames) to network packets, which routers then use to forward the packets to the MPLS egress router, and ultimately to their final destinations. Traditionally, routers make an independent forwarding decision based on an IP table lookup at each hop in the network. In an MPLS network, this lookup is performed only once, when the packet enters the MPLS cloud at the ingress router. The packet is then assigned to a stream, referred to as a Label Switched Path (LSP), and identified by a label. The label is used as a lookup index in the MPLS forwarding table, and the packet traverses the LSP independent of layer-3 routing decisions. This allows the administrators of large networks to fine-tune routing decisions and make the best use of network resources. Additionally, information can be associated with a label to prioritize packet forwarding.

RSVP

MPLS contains no method to dynamically establish LSPs, but you can use the Reservation Protocol (RSVP) with MPLS. RSVP is a signaling protocol used to simplify the establishment of LSPs and to report problems to the MPLS ingress router. The advantage of using RSVP in conjunction with MPLS is the reduction in administrative overhead. If you don't use RSVP with MPLS, you'll have to go to every single router and configure the labels and each path manually. Using RSVP makes the network more dynamic by distributing control of labels to the routers. This enables the network to become more responsive to changing conditions, because it can be set up to change the paths based on certain conditions, such as a certain path going down (perhaps due to a faulty router). The configuration within the router will then be able to use RSVP to distribute new labels to the routers in the MPLS network, with no (or minimal) human intervention.

Best Effort

The simplest, least expensive approach to QoS is not to provide it at all—the "best effort" method. While this might sound like a bad idea, it can in fact work very well. Any VoIP call that traverses the public Internet is almost certain to be best-effort, as QoS mechanisms are not yet common in this environment.

Echo

You may not realize it, but echo has been a problem in the PSTN for as long as there have been telephones. You probably haven't often experienced it, because the telecom industry has spent large sums of money designing expensive echo cancellation devices. Also, when the endpoints are physically close—e.g., when you phone your neighbor down the street—the delay is so minimal that anything you transmit will be returned back so quickly that it will be indistinguishable from the sidetone[‡] normally occurring in your telephone. So the fact of the matter is that there is echo on your local calls much of the time, but you cannot perceive it with a regular telephone because it happens almost instantaneously. It may be helpful to understand this if you consider that when you stand in a room and speak, everything you say echos back to you off of the walls and ceiling (and possibly floor if it's not carpeted), but does not cause any problems because it happens so fast you do not perceive a delay.

The reason that VoIP telephone systems such as Asterisk can experience echo is that the addition of a VoIP telephone introduces a slight delay. It takes a few milliseconds for the packets to travel from your phone and the server (and vice versa). Suddenly there is an appreciable delay, which allows you to perceive the echo that was always there, but never had a delay before.

Why Echo Occurs

Before we discuss measures to deal with echo, let's first take a look at why echo occurs in the analog world.

If you hear echo, it's not your phone that's causing the problem; it's the far end of the circuit. Conversely, echo heard on the far end is being generated at your end. Echo can be caused by the fact that an analog local loop circuit has to transmit and receive on the same pair of wires. If this circuit is not electrically balanced, or if a low-quality telephone is connected to the end of the circuit, signals it receives can be reflected back, becoming part of the return transmission. When this reflected circuit gets back to you, you will hear the words you spoke just moments before. Humans will perceive an echo

---

[‡] As discussed in Chapter 7, sidetone is a function in your telephone that returns part of what you say back to your own ear, to provide a more natural-sounding conversation.

beyond a certain amount of delay (possibly as low as 20 milliseconds for some people). This echo will become annoying as the delay increases.

In a cheap telephone, it is possible for echo to be generated in the body of the handset. This is why some cheap IP phones can cause echo even when the entire end-to-end connection does not contain an analog circuit.[§] In the VoIP world, echo is usually introduced either by an analog circuit somewhere in the connection, or by a cheap endpoint reflecting back some of the signal (e.g., feedback through a hands-free or poorly designed handset or headset). The greater the latency on the network, the more annoying this echo can be.

Managing Echo on Zaptel Channels

In the *zconfig.h* configuration file, you can choose from one of several echo-canceller algorithms, with the default being MARK2. Experiment with the various echo cancellers on your network to determine the best one for your environment. Asterisk also has an option in the *zconfig.h* file to make the echo cancellation more aggressive. You can enable it by uncommenting the following line:

```
#define AGGRESSIVE_SUPPRESSOR
```

Note that aggressive echo cancellation can create a walkie-talkie, half-duplex effect. It should be enabled only if all other methods of reducing echo have failed.

Enable echo cancellation for Zaptel interfaces in the *zapata.conf* file. The default configuration enables echo cancellation with echocancel=yes. echocancelwhenbridged=yes will enable echo cancellation for TDM bridged calls. While bridged calls should not require echo cancellation, this may improve call quality.

When echo cancellation is enabled, the echo canceller learns of echo on the line by listening for it for the duration of the call. Consequently, echo may be heard at the beginning of a call and eventually lessen after a period of time. To avoid this situation, you can employ a method called *echo training*, which will mute the line briefly at the beginning of a call, and then send a tone from which the amount of echo on the line can be determined. This allows Asterisk to deal with the echo more quickly. Echo training can be enabled with echotraining=yes.

Hardware Echo Cancellation

The most effective way to handle echo cancellation is not in software. If you are planning on deploying a good quality system, spend the extra money and purchase cards for the system that have onboard hardware echo cancellation. These cards are somewhat more expensive, but they quickly pay for themselves in terms of reduced load on the CPU, as well as reduced load on you due to less user complaints.

[§] Actually, the handset in any phone, be it traditional or VoIP, is an analog connection.

Asterisk and VoIP

It should come as no surprise that Asterisk loves to talk VoIP. But in order to do so, Asterisk needs to know which function it is to perform: that of client, server, or both. One of the most complex and often confusing concepts in Asterisk is the naming scheme of inbound and outbound authentication.

Users and Peers and Friends—Oh My!

Connections that authenticate to us, or that we authenticate, are defined in the *iax.conf* and *sip.conf* files as *users* and *peers*. Connections that do both may be defined as *friends*. When determining which way the authentication is occurring, it is always important to view the direction of the channels from Asterisk's viewpoint, as connections are being accepted and created by the Asterisk server.

Users

A connection defined as a `user` is any system/user/endpoint that we allow to connect to us. Keep in mind that a `user` definition does not provide a method with which to call that user; the `user` type is used simply to create a channel for incoming calls.[||] A `user` definition will require a context name to be defined to indicate where the incoming authenticated call will enter the dialplan (in *extensions.conf*).

Peers

A connection defined as a `peer` type is an outgoing connection. Think of it this way: *users* place calls to us, while we place calls to our *peers*. Since peers do not place calls to us, a `peer` definition does not typically require the configuration of a context name. However, there is one exception: if calls that originate from your system are returned to your system in a loopback, the incoming calls (which originate from a SIP proxy, not a user agent) will be matched on the `peer` definition. The `default` context should handle these incoming calls appropriately, although it's preferable for contexts to be defined for them on a per-*peer* basis.[#]

In order to know where to send a call to a host, we must know its location in relation to the Internet (that is, its IP address). The location of a *peer* may be defined either statically or dynamically. A dynamic *peer* is configured with `host=dynamic` under the peer definition heading. Because the IP address of a dynamic *peer* may change

---

[||] In SIP, this is not *always* the case. If the endpoint is a SIP proxy service (as opposed to a user agent), Asterisk will authenticate based on the `peer` definition, matching the IP address and port in the `Contact` field of the SIP header against the hostname (and port, if specified) defined for the peer (if the port is not specified, the one defined in the `[general]` section will be used). See the discussion of the SIP `insecure` option in Appendix A for more on this subject.

[#] For more information on this topic, see the discussion of the SIP `context` option in Appendix A.

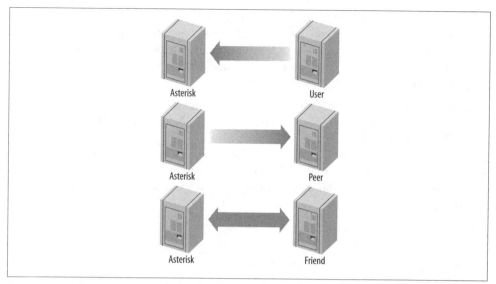

Figure 8-2. Call origination relationships of users, peers, and friends to Asterisk

constantly, it must register with the Asterisk box to let it know what its IP address is, so calls can successfully be routed to it. If the remote end is another Asterisk box, the use of a `register` statement is required, as discussed below.

Friends

Defining a type as a `friend` is a shortcut for defining it as both a `user` and a `peer`. However, connections that are both a `user` and a `peer` aren't always defined this way, because defining each direction of call creation individually (using both a `user` and a `peer` definition) allows more granularity and control over the individual connections.

Figure 8-2 shows the flow of authentication control in relation to Asterisk.

register Statements

A `register` statement is a way of telling a remote peer where your Asterisk box is in relation to the Internet. Asterisk uses `register` statements to authenticate to remote providers when you are employing a dynamic IP address, or when the provider does not have your IP address on record. There are situations when a `register` statement is not required, but to demonstrate when a `register` statement *is* required, let's look at an example.

Say you have a remote peer that is providing DID services to you. When someone calls the number +1-800-555-1212, the call goes over the physical PSTN network to your service provider and into its Asterisk server, possibly over its T1 connection. This call is then routed to your Asterisk server via the Internet.

Your service provider will have a definition in either its *sip.conf* or *iax.conf* configuration file (depending on whether you are connecting with the SIP or IAX protocol, respectively) for your Asterisk server. If you receive calls only from this provider, you would define them as a user (if it was another Asterisk system, you might be defined in its system as a `peer`).

Now let's say that your box is on your home Internet connection, with a dynamic IP address. Your service provider has a static IP address (or perhaps a fully qualified domain name), which you place in your configuration file. Since you have a dynamic address, your service provider specifies `host=dynamic` in its configuration file. In order to know where to route your +1-800-555-1212 call, your service provider needs to know where you are located in relation to the Internet. This is where the `register` statement comes into use.

The `register` statement is a way of authenticating and telling your `peer` where you are. In the `[general]` section of your configuration file, you place a statement similar to this:

```
register => username:secret@my_remote_peer
```

You can verify a successful register with the use of the `iax2 show registry` and `sip show registry` commands at the Asterisk console.

VoIP Security

In this book we can barely scratch the surface of the complex matter of VoIP security; therefore before we dig in, we want to steer you in the direction of the VoIP Security Alliance (*http://www.voipsa.org*). This fantastic resource contains an excellent mailing list, white papers, howtos, and a general compendium of all matters relating to VoIP security. Just as email has been abused by the selfish and criminal, so too will voice. The fine folks at VoIPSA are doing what they can to ensure that we address these challenges now, before they become an epidemic. In the realm of books on the subject, we recommend the most excellent *Hacking Exposed VoIP* by David Endler and Mark Collier (McGraw-Hill Osborne Media). If you are responsible for deploying any VoIP system, you need to be aware of this stuff.

Spam over Internet Telephony (SPIT)

We don't want to think about this, but we know it's coming. The simple fact is that there are people in this world who lack certain social skills, and, coupled with a kind of mindless greed, these folks think nothing of flooding the Internet with massive volumes of email. These same types of characters will similarly think little of doing the same with voice. We already know what it's like to get flooded with telemarketing calls; try to imagine what happens when it costs the telemarketer almost nothing to send voice spam. Regulation has not stopped email spam, and it will probably not stop voice spam, so it will be up to us to prevent it.

Encrypting Audio with Secure RTP

If you can sniff the packets coming out of an Asterisk system, you can extract the audio from the RTP streams. This data can be fed offline to a speech processing system, which can listen for keywords such as "credit card number" or "PIN", and present that data to someone who has an interest in it. The stream can also be evaluated to see if there are DTMF tones embedded in it, which is dangerous because many services ask for password and credit card information input via the dialpad. In business, strategic information could also be gleaned from being able to capture and evaluate audio.

Using Secure RTP can combat this problem by encrypting the RTP streams; however, Asterisk does not support SRTP as of this writing. Work is under way to provide SRTP support (a patch exists in the trunk release, but it is not known as of this writing whether this will be back-ported to 1.4).

Spoofing

In the traditional telephone network, it is very difficult to successfully adopt someone else's identity. Your activities can (and will) be traced back to you, and the authorities will quickly put an end to the fun. In the world of IP, it is much easier to remain anonymous. As such, it is no stretch to imagine that hordes of enterprising criminals will only be too happy to make calls to your credit card company or bank, pretending to be you. If a trusted mechanism is not discovered to combat spoofing, we will quickly learn that we cannot trust VoIP calls.

What Can Be Done?

The first thing to keep in mind when considering security on a VoIP system is that VoIP is based on network protocols, and needs be evaluated from that perspective. This is not to say that traditional telecom security should be ignored, but we need to pay attention to the underlying network.

Basic network security

One of the most effective things that can be done is to secure access to the voice network. The use of firewalls and VLANs are examples of how this can be achieved. By default, the voice network should be accessible only to those things that have a need. For example, if you do not have any softphones in use, do not allow client PCs access to the voice network.

Segregating voice and data traffic. Unless there is a need to have voice and data on the same network, there may be some value in keeping them separate (this can have other benefits as well, such as simplifying QoS configurations). It is not unheard of to build the internal voice network on a totally separate LAN, using existing CAT3 cabling and terminating on inexpensive network switches. It can be less expensive too.

DMZ.   Placing your VoIP system in a DMZ can provide an additional layer of protection for your LAN, while still allowing connectivity for relevant applications. Should your VoIP system be compromised, it will be much more difficult to use it to launch an attack on the rest of your network, since it is not trusted. Regardless of whether you deploy within a DMZ, any abnormal traffic coming out of the system should be suspect.

Server hardening.   Hardening your Asterisk server is critical. Not only are there performance benefits to doing this (running nonessential processes can eat up valuable CPU and RAM), the elimination of anything not required will reduce the chance that an exploited vulnerability in the operating system can be used to gain access and launch an attack on other parts of your network.

Running Asterisk as non-*root* is an essential part of system hardening. See Chapter 11 for more information.

Encryption

Even though Asterisk does not yet fully support SRTP, it is still possible to encrypt VoIP traffic. For example, between sites a VPN could be employed. Consideration should be given to the performance cost of this, but in general this can be a very effective way to secure VoIP traffic and it is relatively simple to implement.

Physical security

Physical security should not be ignored. All terminating equipment (such as switches, routers, and the PBX itself) should be secured in an environment that can only be accessed by authorized persons. At the user end (such as under desks), it can be more difficult to deliver physical security, but if the network responds only to devices that it is familiar with (such as restricting DHCP to devices whose MAC is known), unauthorized intrusion can be mitigated somewhat.

Conclusion

If you listen to the buzz in the telecom industry, you might think that VoIP is the future of telephony. But to Asterisk, VoIP is more a case of "been there, done that." For Asterisk, the future of telephony is much more exciting. We'll take a look at that vision a bit later, in Chapter 15. In the next chapter, we are going to delve into one of the more revolutionary and powerful concepts of Asterisk: AGI, the Asterisk Gateway Interface.

## CHAPTER 9

The Asterisk Gateway Interface (AGI)

*Even he, to whom most things that most people would
think were pretty smart were pretty dumb, thought it
was pretty smart.*

—Douglas Adams, *The Salmon of Doubt*

The Asterisk Gateway Interface, or AGI, provides a standard interface by which external programs may control the Asterisk dialplan. Usually, AGI scripts are used to do advanced logic, communicate with relational databases (such as PostgreSQL or MySQL), and access other external resources. Turning over control of the dialplan to an external AGI script enables Asterisk to easily perform tasks that would otherwise be difficult or impossible.

This chapter covers the fundamentals of AGI communication. It will not teach you how to be a programmer—rather, we'll assume that you're already a competent programmer, so that we can show you how to write AGI programs. If you don't know how to do computer programming, this chapter probably isn't for you, and you should skip ahead to the next chapter.

Over the course of this chapter, we'll write a sample AGI program in each of the Perl, PHP, and Python programming languages. Note, however, that because Asterisk provides a standard interface for AGI scripts, these scripts can be written in almost any modern programming language. We've chosen to highlight Perl, PHP, and Python because they're the languages most commonly used for AGI programming.

Fundamentals of AGI Communication

Instead of releasing an API for programming, AGI scripts communicate with Asterisk over communications channels (*file handles*, in programming parlance) known as STDIN, STDOUT, and STDERR. Most computer programmers will recognize these channels, but just in case you're not familiar with them, we'll cover them here.

What Are STDIN, STDOUT, and STDERR?

STDIN , STDOUT , and STDERR are channels by which programs in Unix-like environments receive information from and send information to external programs. STDIN, or *standard input*, is the information that is sent to the program, either from the keyboard or from another program. In our case, information coming from Asterisk itself comes in on the program's STDIN file handle. STDOUT, or *standard output*, is the file handle that the AGI script uses to pass information back to Asterisk. Finally, the AGI script can use the STDERR (*standard error*) file handle to write error messages to the Asterisk console.

Let's sum up these three communications concepts:

* An AGI script reads from STDIN to get information from Asterisk.
* An AGI script writes data to STDOUT to send information to Asterisk.
* An AGI script may write data to STDERR to send debug information to the Asterisk console.

> At this time, writing to STDERR from within your AGI script writes the information only to the *first* Asterisk console—that is, the first Asterisk console started with the -c parameters.
>
> This is rather unfortunate, and will hopefully be remedied soon by the Asterisk developers.
>
> If you're using the *safe_asterisk* program to start Asterisk (which you probably are), it starts a remote console on TTY9. (Try pressing Ctrl-Alt-F9, and see if you get an Asterisk command-line interface.) This means that all of the AGI debug information will print on only that remote console. You may want to disable this console in *safe_asterisk* to allow you to see the debug information in another console. (You may also want to disable that console for security reasons, as you might not want just anyone to be able to walk up to your Asterisk server and have access to a console without any kind of authentication.)

The Standard Pattern of AGI Communication

The communication between Asterisk and an AGI script follows a predefined pattern. Let's enumerate the steps, and then we'll walk through one of the sample AGI scripts that come with Asterisk.

When an AGI script starts, Asterisk sends a list of variables and their values to the AGI script. The variables might look something like this:

```
agi_request: test.py
agi_channel: Zap/1-1
agi_language: en
agi_callerid:
agi_context: default
```

```
agi_extension: 123
agi_priority: 2
```

After sending these variables, Asterisk sends a blank line. This is the signal that Asterisk is done sending the variables, and it is time for the AGI script to control the dialplan.

At this point, the AGI script sends commands to Asterisk by writing to STDOUT. After the script sends each command, Asterisk sends a response that the AGI script should read. These actions (sending commands to Asterisk and reading the responses) can continue for the duration of the AGI script.

You may be asking yourself what commands you can use from within your AGI script. Good question—we'll cover the basic commands shortly.*

Calling an AGI Script from the Dialplan

In order to work properly, your AGI script must be executable. To use an AGI script inside your dialplan, simply call the AGI() application, with the name of the AGI script as the argument, like this:

```
exten => 123,1,Answer()
exten => 123,2,AGI(agi-test.agi)
```

AGI scripts often reside in the AGI directory (usually located in */var/lib/asterisk/agi-bin*), but you can specify the complete path to the AGI script.

AGI(), EAGI(), DeadAGI(), and FastAGI()

In addition to the AGI() application, there are several other AGI applications suited to different circumstances. While they won't be covered in this chapter, they should be quite simple to figure out once you understand the basics of AGI scripting.

The EAGI() (enhanced AGI) application acts just like AGI() but allows your AGI script to read the inbound audio stream on file descriptor number three.

The DeadAGI() application is also just like AGI(), but it works correctly on a channel that is dead (i.e., a channel that has been hung up). As this implies, the regular AGI() application doesn't work on dead channels.

The FastAGI() application allows the AGI script to be called across the network, so that multiple Asterisk servers can call AGI scripts from a central location.

In this chapter, we'll first cover the sample *agi-test.agi* script that comes with Asterisk (which was written in Perl), then write a weather report AGI program in PHP, and finish up by writing an AGI program in Python to play a math game.

---

* To get a list of available AGI commands, type show agi at the Asterisk command-line interface. You can also refer to Appendix C for an AGI command reference.

Writing AGI Scripts in Perl

Asterisk comes with a sample AGI script called *agi-test.agi*. Let's step through the file while we cover the core concepts of AGI programming. While this particular script is written in Perl, please remember that your own AGI programs may be written in almost any programming language. Just to prove it, we're going to cover AGI programming in a couple of other languages later in the chapter.

Let's get started! We'll look at each section of the code in turn, and describe what it does:

```
#!/usr/bin/perl
```

This line tells the system that this particular script is written in Perl, so it should use the Perl interpreter to execute the script. If you've done much Linux or Unix scripting, this line should be familiar to you. This line assumes, of course, that your Perl binary is located in the */usr/bin/* directory. Change this to match the location of your Perl interpreter.

```
use strict;
```

`use strict` tells Perl to act, well, strict about possible programming errors, such as undeclared variables. While not absolutely necessary, enabling this will help you avoid common programming pitfalls.

```
$|=1;
```

This line tells Perl not to buffer its output—in other words, that it should write any data immediately, instead of waiting for a block of data before outputting it. You'll see this as a recurring theme throughout the chapter.

```
# Set up some variables
my %AGI; my $tests = 0; my $fail = 0; my $pass = 0;
```

You should *always* use unbuffered output when writing AGI scripts. Otherwise, your AGI may not work as expected, because Asterisk may be waiting for the output of your program, while your program thinks it has sent the output to Asterisk and is waiting for a response.

Here, we set up four variables. The first is a hash called `AGI`, which is used to store the variables that Asterisk passes to our script at the beginning of the AGI session. The next three are scalar values, used to count the total number of tests, the number of failed tests, and the number of passed tests, respectively.

```
while(<STDIN>) {
        chomp;
        last unless length($_);
        if (/^agi_(\w+)\:\s+(.*)$/) {
                $AGI{$1} = $2;
        }
}
```

As we explained earlier, Asterisk sends a group of variables to the AGI program at startup. This loop simply takes all of these variables and stores them in the hash named AGI. They can be used later in the program or simply ignored, but they should always be read from STDIN before continuing on with the logic of the program.

```
print STDERR "AGI Environment Dump:\n";
foreach my $i (sort keys %AGI) {
        print STDERR " -- $i = $AGI{$i}\n";
}
```

This loop simply writes each of the values that we stored in the AGI hash to STDERR. This is useful for debugging the AGI script, as STDERR is printed to the Asterisk console.[†]

```
sub checkresult {
        my ($res) = @_;
        my $retval;
        $tests++;
        chomp $res;
        if ($res =~ /^200/) {
                $res =~ /result=(-?\d+)/;
                if (!length($1)) {
                        print STDERR "FAIL ($res)\n";
                        $fail++;
                } else {
                        print STDERR "PASS ($1)\n";
                        $pass++;
                }
        } else {
                print STDERR "FAIL (unexpected result '$res')\n";
                $fail++;
        }
}
```

This subroutine reads in the result of an AGI command from Asterisk and decodes the result to determine whether the command passes or fails.

Now that the preliminaries are out of the way, we can get to the core logic of the AGI script:

```
print STDERR "1.  Testing 'sendfile'...";
print "STREAM FILE beep \"\"\n";
my $result = <STDIN>;
&checkresult($result);
```

This first test shows how to use the STREAM FILE command. The STREAM FILE command tells Asterisk to play a sound file to the caller, just as the Background() application does. In this case, we're telling Asterisk to play a file called *beep.gsm*.[‡]

[†] Actually, to the first spawned Asterisk console (i.e., the first instance of Asterisk called with the -c option). If *safe_asterisk* was used to start Asterisk, the first Asterisk console will be on TTY9, which means that you will not be able to view AGI errors remotely.

[‡] Asterisk automatically selects the best format, based on translation cost and availability, so the file extension is never used in the function.

You will notice that the second argument is passed by putting in a set of double quotes, escaped by backslashes. Without the double quotes to indicate the second argument, this command does not work correctly.

 You must pass *all required arguments* to the AGI commands. If you want to skip a required argument, you must send empty quotes (properly escaped in your particular programming language), as shown above. If you don't pass the required number of arguments, your AGI script will not work.

You should also make sure you pass a line feed (the \n on the end of the print statement) at the end of the command.

After sending the STREAM FILE command, this test reads the result from STDIN and calls the checkresult subroutine to determine if Asterisk was able to play the file. The STREAM FILE command takes three arguments, two of which are required:

- The name of the sound file to play back
- The digits that may interrupt the playback
- The position at which to start playing the sound, specified in number of samples (optional)

In short, this test told Asterisk to play back the file named *beep.gsm*, and then it checked the result to make sure the command was successfully executed by Asterisk.

```
print STDERR "2.  Testing 'sendtext'...";
print "SEND TEXT \"hello world\"\n";
my $result = <STDIN>;
&checkresult($result);
```

This test shows us how to call the SEND TEXT command, which is similar to the SendText() application. This command will send the specified text to the caller, if the caller's channel type supports the sending of text.

The SEND TEXT command takes one argument: the text to send to the channel. If the text contains spaces (as in the previous code block), the argument should be encapsulated with quotes, so that Asterisk will know that the entire text string is a single argument to the command. Again, notice that the quotation marks are escaped, as they must be sent to Asterisk, not used to terminate the string in Perl.

```
print STDERR "3.  Testing 'sendimage'...";
print "SEND IMAGE asterisk-image\n";
my $result = <STDIN>;
&checkresult($result);
```

This test calls the SEND IMAGE command, which is similar to the SendImage() application. Its single argument is the name of an image file to send to the caller. As with the SEND TEXT command, this command works only if the calling channel supports the receiving images.

```
print STDERR "4.  Testing 'saynumber'...";
print "SAY NUMBER 192837465 \"\"\n";
my $result = <STDIN>;
&checkresult($result);
```

This test sends Asterisk the SAY NUMBER command. This command behaves identically to the SayNumber() dialplan application. It takes two arguments:

- The number to say
- The digits that may interrupt the command

Again, since we're not passing in any digits as the second argument, we need to pass in an empty set of quotes.

```
print STDERR "5.  Testing 'waitdtmf'...";
print "WAIT FOR DIGIT 1000\n";
my $result = <STDIN>;
&checkresult($result);
```

This test shows the WAIT FOR DIGIT command. This command waits the specified number of milliseconds for the caller to enter a DTMF digit. If you want the command to wait indefinitely for a digit, use -1 as the timeout. This application returns the decimal ASCII value of the digit that was pressed.

```
print STDERR "6.  Testing 'record'...";
print "RECORD FILE testagi gsm 1234 3000\n";
my $result = <STDIN>;
&checkresult($result);
```

This section of code shows us the RECORD FILE command. This command is used to record the call audio, similar to the Record() dialplan application. RECORD FILE takes seven arguments, the last three of which are optional:

- The filename of the recorded file.
- The format in which to record the audio.
- The digits that may interrupt the recording.
- The timeout (maximum recording time) in milliseconds, or -1 for no timeout.
- The number of samples to skip before starting the recording (optional).
- The word BEEP, if you'd like Asterisk to beep before the recording starts (optional).
- The number of seconds before Asterisk decides that the user is done with the recording and returns, even though the timeout hasn't been reached and no DTMF digits have been entered (optional). This argument must be preceded by s=.

In this particular case, we're recording a file called *testagi* (in the GSM format), with any of the DTMF digits 1 through 4 terminating the recording, and a maximum recording time of 3,000 milliseconds.

```
print STDERR "6a.  Testing 'record' playback...";
print "STREAM FILE testagi \"\"\n";
```

```
my $result = <STDIN>;
&checkresult($result);
```

The second part of this test plays back the audio that was recorded earlier, using the STREAM FILE command. We've already covered STREAM FILE, so this section of code needs no further explanation.

```
print STDERR "================== Complete =====================\n";
print STDERR "$tests tests completed, $pass passed, $fail failed\n";
print STDERR "================================================\n";
```

At the end of the AGI script, a summary of the tests is printed to STDERR, which should end up on the Asterisk console.

In summary, you should remember the following when writing AGI programs in Perl:

- Turn on strict language checking with the use strict command.[§]
- Turn off output buffering by setting $|=1.
- Data from Asterisk is received using a while(<STDIN>) loop.
- Write values with the print command.
- Use the print STDERR command to write debug information to the Asterisk console.

The Perl AGI Library

If you are interested in building your own AGI scripts in Perl, you may want to check out the Asterisk::AGI Perl module written by James Golovich, which is located at *http://asterisk.gnuinter.net*. The Asterisk::AGI module makes it even easier to write AGI scripts in Perl.

Creating AGI Scripts in PHP

We promised we'd cover several languages, so let's go ahead and see what an AGI script in PHP looks like. The fundamentals of AGI programming still apply; only the programming language has changed. In this example, we'll write an AGI script to download a weather report from the Internet and deliver the temperature, wind direction, and wind speed back to the caller:

```
#!/usr/bin/php -q
<?php
```

The first line tells the system to use the PHP interpreter to run this script. The -q option turns off HTML error messages. You should ensure that there aren't any extra lines between the first line and the opening PHP tag, as they'll confuse Asterisk.

```
# change this to match the code of your particular city
# for a complete list of U.S. cities, go to
```

[§] This advice probably applies to any Perl program you might write, especially if you're new to Perl.

```
# http://www.nws.noaa.gov/data/current_obs/
$weatherURL="http://www.nws.noaa.gov/data/current_obs/KMDQ.xml";
```

This tells our AGI script where to go to get the current weather conditions. In this example, we're getting the weather for Huntsville, Alabama. Feel free to visit the web site listed above for a complete list of stations throughout the United States of America.[||]

```
# don't let this script run for more than 60 seconds
set_time_limit(60);
```

Here, we tell PHP not to let this program run for more than 60 seconds. This is a safety net, which will end the script if for some reason it takes more than 60 seconds to run.

```
# turn off output buffering
ob_implicit_flush(false);
```

This command turns off output buffering, meaning that all data will be sent immediately to the AGI interface and will not be buffered.

```
# turn off error reporting, as it will most likely interfere with
# the AGI interface
error_reporting(0);
```

This command turns off all error reporting, as it can interfere with the AGI interface. (You might find it helpful to comment out this line during testing.)

```
# create file handles if needed
if (!defined('STDIN'))
{
    define('STDIN', fopen('php://stdin', 'r'));
}
if (!defined('STDOUT'))
{
    define('STDOUT', fopen('php://stdout', 'w'));
}
if (!defined('STDERR'))
{
    define('STDERR', fopen('php://stderr', 'w'));
}
```

This section of code ensures that we have open file handles for STDIN, STDOUT, and STDERR, which will handle all communication between Asterisk and our script.

```
# retrieve all AGI variables from Asterisk
while (!feof(STDIN))
{
    $temp = trim(fgets(STDIN,4096));
    if (($temp == "") || ($temp == "\n"))
    {
        break;
    }
}
```

[||] We apologize to our readers outside of the United States for using a weather service that only works for U.S. cities. If you can find a good international weather service that provides its data in XML, it shouldn't be too hard to change this AGI script to work with that particular service. Once we find one, we'll update this script for future editions.

```
    $s = split(":",$temp);
    $name = str_replace("agi_","",$s[0]);
    $agi[$name] = trim($s[1]);
}
```

Next, we'll read in all of the AGI variables passed to us by Asterisk. Using the `fgets` command in PHP to read the data from `STDIN`, we'll save each variable in the hash called `$agi`. Remember that we could use these variables in the logic of our AGI script, although we won't in this example.

```
# print all AGI variables for debugging purposes
foreach($agi as $key=>$value)
{
    fwrite(STDERR,"-- $key = $value\n");
    fflush(STDERR);
}
```

Here, we print the variables back out to `STDERR` for debugging purposes.

```
#retrieve this web page
$weatherPage=file_get_contents($weatherURL);
```

This line of code retrieves the XML file from the National Weather Service and puts the contents into the variable called `$weatherPage`. This variable will be used later on to extract the pieces of the weather report that we want.

```
#grab temperature in Fahrenheit
if (preg_match("/<temp_f>([0-9]+)<\/temp_f>/i",$weatherPage,$matches))
{
    $currentTemp=$matches[1];
}
```

This section of code extracts the temperature (in Fahrenheit) from the weather report, using the `preg_match` command. This command uses Perl-compatible regular expressions[#] to extract the needed data.

```
#grab wind direction
if (preg_match("/<wind_dir>North<\/wind_dir>/i",$weatherPage))
{
    $currentWindDirection='northerly';
}
elseif (preg_match("/<wind_dir>South<\/wind_dir>/i",$weatherPage))
{
    $currentWindDirection='southerly';
}
elseif (preg_match("/<wind_dir>East<\/wind_dir>/i",$weatherPage))
{
    $currentWindDirection='easterly';
}
elseif (preg_match("/<wind_dir>West<\/wind_dir>/i",$weatherPage))
{
    $currentWindDirection='westerly';
}
```

[#] The ultimate guide to regular expressions is O'Reilly's *Mastering Regular Expressions*, by Jeffrey E.F. Friedl.

```
elseif (preg_match("/<wind_dir>Northwest<\/wind_dir>/i",$weatherPage))
{
    $currentWindDirection='northwesterly';
}
elseif (preg_match("/<wind_dir>Northeast<\/wind_dir>/i",$weatherPage))
{
    $currentWindDirection='northeasterly';
}
elseif (preg_match("/<wind_dir>Southwest<\/wind_dir>/i",$weatherPage))
{
    $currentWindDirection='southwesterly';
}
elseif (preg_match("/<wind_dir>Southeast<\/wind_dir>/i",$weatherPage))
{
    $currentWindDirection='southeasterly';
}
```

The wind direction is found through the use of preg_match (located in the wind_dir tags) and is assigned to the variable $currentWindDirection.

```
#grab wind speed
if (preg_match("/<wind_mph>([0-9.]+)<\/wind_mph>/i",$weatherPage,$matches))
{
    $currentWindSpeed = $matches[1];
}
```

Finally, we'll grab the current wind speed and assign it to the $currentWindSpeed variable.

```
# tell the caller the current conditions
if ($currentTemp)
{
    fwrite(STDOUT,"STREAM FILE temperature \"\"\n");
    fflush(STDOUT);
    $result = trim(fgets(STDIN,4096));
    checkresult($result);
    fwrite(STDOUT,"STREAM FILE is \"\"\n");
    fflush(STDOUT);
    $result = trim(fgets(STDIN,4096));
    checkresult($result);
    fwrite(STDOUT,"SAY NUMBER $currentTemp \"\"\n");
    fflush(STDOUT);
    $result = trim(fgets(STDIN,4096));
    checkresult($result);
    fwrite(STDOUT,"STREAM FILE degrees \"\"\n");
    fflush(STDOUT);
    $result = trim(fgets(STDIN,4096));
    checkresult($result);
    fwrite(STDOUT,"STREAM FILE fahrenheit \"\"\n");
    fflush(STDOUT);
    $result = trim(fgets(STDIN,4096));
    checkresult($result);
}

if ($currentWindDirection && $currentWindSpeed)
{
```

```
        fwrite(STDOUT,"STREAM FILE with \"\"\n");
        fflush(STDOUT);
        $result = trim(fgets(STDIN,4096));
        checkresult($result);
        fwrite(STDOUT,"STREAM FILE $currentWindDirection \"\"\n");
        fflush(STDOUT);
        $result = trim(fgets(STDIN,4096));
        checkresult($result);
        fwrite(STDOUT,"STREAM FILE wx/winds \"\"\n");
        fflush(STDOUT);
        $result = trim(fgets(STDIN,4096));
        checkresult($result);
        fwrite(STDOUT,"STREAM FILE at \"\"\n";)
        fflush(STDOUT);
        $result = trim(fgets(STDIN,4096));
        checkresult($result);
        fwrite(STDOUT,"SAY NUMBER $currentWindSpeed \"\"\n");
        fflush(STDOUT);
        $result = trim(fgets(STDIN,4096));
        checkresult($result);
        fwrite($STDOUT,"STREAM FILE miles-per-hour \"\"\n");
        fflush(STDOUT);
        $result = trim(fgets(STDIN,4096));
        checkresult($result);
    }
```

Now that we've collected our data, we can send AGI commands to Asterisk (checking the results as we go) that will deliver the current weather conditions to the caller. This will be achieved through the use of the STREAM FILE and SAY NUMBER AGI commands.

We've said it before, and we'll say it again: when calling AGI commands, you must pass in all of the required arguments. In this case, both STREAM FILE and SAY NUMBER commands require a second argument; we'll pass empty quotes escaped by the backslash character.

You should also notice that we call the fflush command each time we write to STDOUT. While this is arguably redundant, there's no harm in ensuring that the AGI command is not buffered and is sent immediately to Asterisk.

```
    function checkresult($res)
    {
        trim($res);
        if (preg_match('/^200/',$res))
        {
            if (! preg_match('/result=(-?\d+)/',$res,$matches))
            {
                fwrite(STDERR,"FAIL ($res)\n");
                fflush(STDERR);
                return 0;
            }
            else
            {
                fwrite(STDERR,"PASS (".$matches[1].")\n");
                fflush(STDERR);
```

```
            return $matches[1];
        }
    }
    else
    {
        fwrite(STDERR,"FAIL (unexpected result '$res')\n");
        fflush(STDERR);
        return -1;
    }
}
```

The checkresult function is identical in purpose to the checkresult subroutine we saw in our Perl example. As its name suggests, it checks the result that Asterisk returns whenever we call an AGI command.

```
?>
```

At the end of the file, we have our closing PHP tag. Don't place any whitespace after the closing PHP tag, as it can confuse the AGI interface.

We've now covered two different languages in order to demonstrate the similarities and differences of programming an AGI script in PHP as opposed to Perl. The following things should be remembered when writing an AGI script in PHP:

- Invoke PHP with the -q switch; it turns off HTML in error messages.
- Turn off the time limit, or set it to a reasonable value (newer versions of PHP automatically disable the time limit when PHP is invoked from the command line).
- Turn off output buffering with the ob_implicit_flush(false) command.
- Open file handles to STDIN, STDOUT, and STDERR (newer versions of PHP may have one or more of these file handles already opened; see the previous code for a slick way of making this work across most versions of PHP).
- Read variables from STDIN using the fgets function.
- Use the fwrite function to write to STDOUT and STDERR.
- Always call the fflush function after writing to either STDOUT or STDERR.

The PHP AGI Library

For advanced AGI programming in PHP, you may want to check out the PHPAGI project at *http://phpagi.sourceforge.net*. It was originally written by Matthew Asham and is being developed by several other members of the Asterisk community.

Writing AGI Scripts in Python

The AGI script we'll be writing in Python, called "The Subtraction Game," was inspired by a Perl program written by Ed Guy and discussed by him at the 2004 AstriCon conference. Ed described his enthusiasm for the power and simplicity of Asterisk when he found he could write a quick Perl script to help his young daughter improve her math skills.

Since we've already written a Perl program using AGI, and Ed has already written the math program in Perl, we figured we'd take a stab at it in Python!

Let's go through our Python script:

```
#!/usr/bin/python
```

This line tells the system to run this script in the Python interpreter. For small scripts, you may consider adding the -u option to this line, which will run Python in unbuffered mode. This is not recommended, however, for larger or frequently used AGI scripts, as it can affect system performance.

```
import sys
import re
import time
import random
```

Here, we import several libraries that we'll be using in our AGI script.

```
# Read and ignore AGI environment (read until blank line)

env = {}
tests = 0;

while 1:
    line = sys.stdin.readline().strip()

    if line == '':
        break
    key,data = line.split(':')
    if key[:4] <> 'agi_':
        #skip input that doesn't begin with agi_
        sys.stderr.write("Did not work!\n");
        sys.stderr.flush()
        continue
    key = key.strip()
    data = data.strip()
    if key <> '':
        env[key] = data

sys.stderr.write("AGI Environment Dump:\n");
sys.stderr.flush()
for key in env.keys():
    sys.stderr.write(" -- %s = %s\n" % (key, env[key]))
    sys.stderr.flush()
```

This section of code reads in the variables that are passed to our script from Asterisk, and saves them into a dictionary named env. These values are then written to STDERR for debugging purposes.

```
def checkresult (params):
    params = params.rstrip()
    if re.search('^200',params):
```

```
        result = re.search('result=(\d+)',params)
        if (not result):
            sys.stderr.write("FAIL ('%s')\n" % params)
            sys.stderr.flush()
            return -1
        else:
            result = result.group(1)
            #debug("Result:%s Params:%s" % (result, params))
            sys.stderr.write("PASS (%s)\n" % result)
            sys.stderr.flush()
            return result
    else:
        sys.stderr.write("FAIL (unexpected result '%s')\n" % params)
        sys.stderr.flush()
        return -2
```

The checkresult function is almost identical in purpose to the checkresult subroutine in the sample Perl AGI script we covered earlier in the chapter. It reads in the result of an Asterisk command, parses the answer, and reports whether or not the command was successful.

```
def sayit (params):
    sys.stderr.write("STREAM FILE %s \"\"\n" % str(params))
    sys.stderr.flush()
    sys.stdout.write("STREAM FILE %s \"\"\n" % str(params))
    sys.stdout.flush()
    result = sys.stdin.readline().strip()
    checkresult(result)
```

The sayit function is a simple wrapper around the STREAM FILE command.

```
def saynumber (params):
    sys.stderr.write("SAY NUMBER %s \"\"\n" % params)
    sys.stderr.flush()
    sys.stdout.write("SAY NUMBER %s \"\"\n" % params)
    sys.stdout.flush()
    result = sys.stdin.readline().strip()
    checkresult(result)
```

The saynumber function is a simple wrapper around the SAY NUMBER command.

```
def getnumber (prompt, timelimit, digcount):
    sys.stderr.write("GET DATA %s %d %d\n" % (prompt, timelimit, digcount))
    sys.stderr.flush()
    sys.stdout.write("GET DATA %s %d %d\n" % (prompt, timelimit, digcount))
    sys.stdout.flush()
    result = sys.stdin.readline().strip()
    result = checkresult(result)
    sys.stderr.write("digits are %s\n" % result)
    sys.stderr.flush()
    if result:
        return result
    else:
        result = -1
```

The getnumber function calls the GET DATA command to get DTMF input from the caller. It is used in our program to get the caller's answers to the subtraction problems.

```
limit=20
digitcount=2
score=0
count=0
ttanswer=5000
```

Here, we initialize a few variables that we'll be using in our program.

```
starttime = time.time()
t = time.time() - starttime
```

In these lines we set the starttime variable to the current time and initialize t to 0. We'll use the t variable to keep track of the number of seconds that have elapsed since the AGI script was started.

```
sayit("subtraction-game-welcome")
```

Next, we welcome the caller to the subtraction game.

```
while ( t < 180 ):

    big = random.randint(0,limit+1)
    big += 10
    subt= random.randint(0,big)
    ans = big - subt
    count += 1

    #give problem:
    sayit("subtraction-game-next");
    saynumber(big);
    sayit("minus");
    saynumber(subt);
    res = getnumber("equals",ttanswer,digitcount);

    if (int(res) == ans) :
        score+=1
        sayit("subtraction-game-good");
    else :
        sayit("subtraction-game-wrong");
        saynumber(ans);

    t = time.time() - starttime
```

This is the heart of the AGI script. We loop through this section of code and give subtraction problems to the caller until 180 seconds have elapsed. Near the top of the loop, we calculate two random numbers and their difference. We then present the problem to the caller, and read in the caller's response. If the caller answers incorrectly, we give the correct answer.

```
pct = float(score)/float(count)*100;
sys.stderr.write("Percentage correct is %d\n" % pct)
sys.stderr.flush()
```

```
sayit("subtraction-game-timesup")
saynumber(score)
sayit("subtraction-game-right")
saynumber(count)
sayit("subtraction-game-pct")
saynumber(pct)
```

After the user is done answering the subtraction problems, she is given her score.

As you have seen, the basics you should remember when writing AGI scripts in Python are:

- Flush the output buffer after every write. This will ensure that your AGI program won't hang while Asterisk is waiting for the buffer to fill and Python is waiting for the response from Asterisk.
- Read data from Asterisk with the `sys.stdin.readline` command.
- Write commands to Asterisk with the `sys.stdout.write` command. Don't forget to call `sys.stdout.flush` after writing.

The Python AGI Library

If you are planning on writing lot of Python AGI code, you may want to check out Karl Putland's Python module, *Pyst*. You can find it at *http://www.sourceforge.net/projects/pyst/*.

Debugging in AGI

Debugging AGI programs, as with any other type of program, can be frustrating. Luckily, there are two advantages to debugging AGI scripts. First, since all of the communications between Asterisk and the AGI program happen over STDIN and STDOUT (and, of course, STDERR), you should be able to run your AGI script directly from the operating system. Second, Asterisk has a handy command for showing all of the communications between itself and the AGI script: `agi debug`.

Debugging from the Operating System

As mentioned above, you should be able to run your program directly from the operating system to see how it behaves. The secret here is to act just like Asterisk does, providing your script with the following:

- A list of variables and their values, such as `agi_test:1`.
- A blank line feed (\n) to indicate that you're done passing variables.
- Responses to each of the AGI commands from your AGI script. Usually, typing **200 response=1** is sufficient.

Trying your program directly from the operating system may help you to more easily spot bugs in your program.

Using Asterisk's agi debug Command

The Asterisk command-line interface has a very useful command for debugging AGI scripts, which is called (appropriately enough) agi debug. If you type **agi debug** at an Asterisk console and then run an AGI, you'll see something like the following:

```
    -- Executing AGI("Zap/1-1", "temperature.php") in new stack
        -- Launched AGI Script /var/lib/asterisk/agi-bin/temperature.php
AGI Tx >> agi_request: temperature.php
AGI Tx >> agi_channel: Zap/1-1
AGI Tx >> agi_language: en
AGI Tx >> agi_type: Zap
AGI Tx >> agi_uniqueid: 1116732890.8
AGI Tx >> agi_callerid: 101
AGI Tx >> agi_calleridname: Tom Jones
AGI Tx >> agi_callingpres: 0
AGI Tx >> agi_callingani2: 0
AGI Tx >> agi_callington: 0
AGI Tx >> agi_callingtns: 0
AGI Tx >> agi_dnid: unknown
AGI Tx >> agi_rdnis: unknown
AGI Tx >> agi_context: incoming
AGI Tx >> agi_extension: 141
AGI Tx >> agi_priority: 2
AGI Tx >> agi_enhanced: 0.0
AGI Tx >> agi_accountcode:
AGI Tx >>
AGI Rx << STREAM FILE temperature ""
AGI Tx >> 200 result=0 endpos=6400
AGI Rx << STREAM FILE is ""
AGI Tx >> 200 result=0 endpos=5440
AGI Rx << SAY NUMBER 67 ""
        -- Playing 'digits/60' (language 'en')
        -- Playing 'digits/7' (language 'en')
AGI Tx >> 200 result=0
AGI Rx << STREAM FILE degrees ""
AGI Tx >> 200 result=0 endpos=6720
AGI Rx << STREAM FILE fahrenheit ""
AGI Tx >> 200 result=0 endpos=8000
        -- AGI Script temperature.php completed, returning 0
```

You'll see three types of lines while your AGI script is running. The first type, prefaced with AGI TX >>, are the lines that Asterisk transmits to your program's STDIN. The second type, prefaced with AGI RX <<, are the commands your AGI program writes back to Asterisk over STDOUT. The third type, prefaced by --, are the standard Asterisk messages presented as it executes certain commands.

To disable AGI debugging after it has been started, simply type **agi no debug** at an Asterisk console.

Using the `agi debug` command will enable you to see the communication between Asterisk and your program, which can be very useful when debugging. Hopefully, these two tips will greatly improve your ability to write and debug powerful AGI programs.

Conclusion

For a developer, AGI is one of the more revolutionary and compelling reasons to choose Asterisk over a closed, proprietary PBX. Still, AGI is only part of the picture. In Chapter 10, we'll explore another powerful programming interface known as the Asterisk Manager Interface.

## CHAPTER 10

Asterisk Manager Interface (AMI) and Adhearsion

> *Do but take care to express yourself in a plain, easy*
> *Manner, in well-chosen, significant and decent Terms,*
> *and to give a harmonious and pleasing Turn to your*
> *Periods: study to explain your Thoughts, and set them*
> *in the truest Light, labouring as much as possible, not*
> *to leave them dark nor intricate, but clear and*
> *intelligible.*
>
> —Miguel de Cervantes, Preface to *Don Quixote*

The Manager Interface

The Asterisk Manager Interface (AMI) is a powerful programmatic interface. It allows external programs to both control and monitor an Asterisk system.[*] This interface is often used to integrate Asterisk with existing business processes and systems, CRM (Customer Relationship Management) software. It can also be used for a wide variety of applications, such as automated dialers and click-to-call systems.

The Asterisk Manager Interface listens for connections on a network port. A client program can then connect to the Asterisk Manager Interface on that port, authenticate itself, and send commands to Asterisk. Asterisk will then respond to the request, as well as update the client program with the status of the system.

To use the Manager, you must define an account in the file */etc/asterisk/manager.conf*. This file will look something like this:

```
[general]
enabled = yes
port = 5038
```

---

[*] Contrast this with the Asterisk Gateway Interface (AGI), which allows Asterisk to launch an external program from the dialplan. The AGI and AMI interfaces are very much complimentary to each other.

```
bindaddr = 0.0.0.0

[oreilly]
secret = notvery
;deny=0.0.0.0/0.0.0.0
;permit=209.16.236.73/255.255.255.0
read = system,call,log,verbose,command,agent,user
write = system,call,log,verbose,command,agent,user
```

In the [general] section, you have to enable the service by setting the parameter enabled = yes. You will need to reload the Manager in order to have this change take effect (module reload manager from the Asterisk console). The TCP port defaults to 5038.

Next, you create a section for each user that will authenticate to the system. For each user, you will specify the username in square brackets ([ ]), followed by the password for that user (secret), any IP addresses you wish to deny access to, any IP addresses you wish to permit access to, and the read and write permissions for that user.

> It is very important to understand that other than a clear text password and the ability to restrict IP addresses, there is no security of any kind on the Manager interface. If you are running Manager on an untrusted network (or have any other complex needs), you should consider using David Troy's excellent AstManProxy to handle all of your connections to the manager API.

Connecting to the Manager Interface

It is important to keep in mind that the Manager interface is designed to be used by programs, not fingers. That's not to say that you can't issue commands to it directly—just don't expect a typical console interface, because that's not what Manager is for.

Commands to Manager are delivered in packages with the following syntax (lines are terminated with CR+LF):[†]

```
Action: <action type>
Key 1: Value 1
Key 2: Value 2
etc ...
Variable: Value
Variable: Value
etc...
```

For example, to authenticate with Manager (which is required if you expect to have any interaction whatsoever), you would send the following:

[†] Carriage Return followed by Line Feed. This is popularly achieved by pressing the Enter key on the keyboard, but there are differences across various OS platforms and programming languages, so if you are having trouble with commands to the interface, it may be worth noting the exact character combination that is required. At the time of writing, Wikipedia had a respectable writeup on this concept (*http://en.wikipedia.org/wiki/Newline*).

```
Action: login
Username: oreilly
Secret: notvery
<CR+LF>
```

An extra CR+LF on a blank line will submit the entire package to Manager.

Once authenticated, you will be able to initiate actions, as well as see events generated by Asterisk. On a busy system, this can get quite complicated and become totally impossible to keep track of with the unaided eye. To turn keep Asterisk from sending events, you can add the Events parameter to your login command, like this:

```
Action: login
Username: oreilly
Secret: notvery
Events: off
<CR+LF>
```

If you're worried about sending your secret across the wire in plain text (which you *should* be), you can also authenticate using an MD5 challenge-response system, which works very similar to HTTP digest authentication. To do this, you first call the Challenge action, which will present you with a challenge token:

```
Action: Challenge
AuthType: MD5

Response: Success
Challenge: 840415273
```

You can then take that challenge token, concatenate the plaintext secret onto the end of it, and calculate the MD5 checksum of the resulting string. The result can then be used to login without passing your secret in plain text.

```
Action: Login
AuthType: MD5
Username: Admin
Key: e7a056e1488882c6c509bbe71a049978

Response: Success
Message: Authentication accepted
```

Sending Commands

Once you've successfully logged into the AMI system, you can send commands to Asterisk by using the other actions. We'll show a few commands here so that you can get a feel for how they work.

Transferring a call

The Redirect action can be used to transfer a call. After logging in, you should send an action like the one below:

```
Action: Redirect
Channel: SIP/John-ae201e78
Context: Lab
Exten: 6001
Priority: 1
ActionID: 2340981650981
```

 For each action you sent over the manager interface, you can send along an arbitrary ActionID value. This will allow you to recognize the responses from Asterisk as being related to your actions. It is strongly recommended that you send a unique ActionID with each of your AMI commands.

This URL transfers the specified channel to another extension and priority in the dialplan. The response to this action is:

```
Response: Success
ActionID: 2340981650981
Message: Redirect Successful
```

Reading a configuration file

To read an Asterisk configuration file via the Manger interface, we can use the the GetConfig action. The GetConfig action returns the contents of a configuration file, or portion thereof. The following command retrieves the contents of the file *users.conf*.

```
Action: GetConfig
Filename: users.conf
ActionID: 9873497149817
```

Asterisk then returns the contents of the *users.conf* file. The response looks like:

```
Response: Success
ActionID: 987397149817
Category-000000: general
Line-000000-000000: fullname=New User
Line-000000-000001: userbase=6000
Line-000000-000002: hasvoicemail=yes
Line-000000-000003: hassip=yes
Line-000000-000004: hasiax=yes
Line-000000-000005: hasmanager=no
Line-000000-000006: callwaiting=yes
Line-000000-000007: threewaycalling=yes
Line-000000-000008: callwaitingcallerid=yes
Line-000000-000009: transfer=yes
Line-000000-000010: canpark=yes
Line-000000-000011: cancallforward=yes
Line-000000-000012: callreturn=yes
Line-000000-000013: callgroup=1
Line-000000-000014: pickupgroup=1
Line-000000-000015: host=dynamic
```

Updating configuration files

It is often useful to be able to update an Asterisk configuration file via the Asterisk Manager interface. The `UpdateConfig` action is used to update one or more settings in a configuration file. For example, to delete a user named 6003 from *users.conf* you could issue the following command:

```
Action: UpdateConfig
Filename: users.conf
Reload: yes
SrcFilename: users.conf
DstFilename: users.conf
Action-00000: delcat
Cat-00000: 6003
ActionID: 5298795987243
```

Obviously, we've only scratched the surface of the many different actions available as part of the Asterisk Manager Interface. For a more detailed listing of the available commands, see Appendix F.

The Flash Operator Panel

The Flash Operator Panel (FOP) is one of the most popular examples of the power of the Manager interface. FOP presents a web-based visual view of your system and allows you control of calls.

FOP is most commonly used to enable a live attendant to view the users in the system and connect calls between them. It can also be used in a call-center environment to provide CRM-triggered screen pops.[‡]

The FOP management interface is shown in Figure 10-1. To grab a copy of FOP, head to *http://www.asternic.org*.

FOP isn't all that difficult to set up, but it does require several steps to configure. The configuration of FOP is beyond the scope of this book, but if you head on over to the FOP web site you will find the latest documentation detailing the installation and configuration process.

FOP has a fantastic community and a popular mailing list. The success of FOP has also been aided by its inclusion in Trixbox.

Asterisk Development with Adhearsion

Recently, a new bit of technology has come along which has the potential to change the way we create dialplans.[§]

[‡] Customer Relationship Management (CRM) is an interface that companies use to help manage customer information and interaction.

Figure 10-1. The Flash Operator Panel management interface

A New Approach to Dialplans

Asterisk has matured both in technological innovation and in popularity, but as one becomes more and more immersed in this wonderful world, one cannot help but bump into limitations. When you handle a lot of complex, enterprise-grade scripting with Asterisk alone, you will face many obstacles using dialplan logic. As flexible and powerful as the dialplan is, as a programming language it is quite odd, and much less flexible than most modern scripting languages. When one needs to provide advanced logic, the dialplan, the GUI, and even the more advanced AEL can become very frustrating.

As you build more and more complexity into your dialplans, some of the following may cause you some head-scratching:

- Conditional looping and branching
- Variables
- Complex data structures
- Database/LDAP integration
- Use of third-party libraries
- Exchanging and distributing VoIP functionality
- Extending the configuration languages
- Poor error handling
- Poor date and time handling

§ We would like to thank Jay Phillips for contributing the ideas and code for this section of the book.

- Pattern matching
- Usage consistency
- Source code organization

Many people addressed these matters by writing the advanced logic in external programs such as Perl and PHP, and connecting to Asterisk via AMI and AGI. Unfortunately, while the desired power was now available, these interactions did not always simplify things for the developer. To the contrary, they often made development more complex. Using existing technologies in Asterisk, but aiming to deliver power and simplicity, Adhearsion has a new approach.

Asterisk Development with Adhearsion

Adhearsion is an open source (LGPL) framework that is designed to improve Asterisk solution development. It rests above an Asterisk system, handling parts or all of the dialplan and, in a few unique ways, manages access to Asterisk with several improved interfaces. Because it runs as a separate daemon process and integrates through the already-present Gateway (AGI) and Manager (AMI) interfaces, configuring a context to use Adhearsion is as simple as adding a few lines to your dialplan or adding a user to *manager.conf*.

Adhearsion primarily uses the highly dynamic, object-oriented Ruby programming language, but has optional support for other languages such as C or Java. In the VoIP world, many things exist as conceptual objects, which means that object-oriented programming can make a lot of sense. Those familiar with Python, Perl, or other scripting languages should have no trouble picking up Ruby, and for those who don't, Ruby is an excellent choice for your first scripting language.

Installing Adhearsion

Ruby software is generally installed through Ruby's package manager (similar to Linux package managers, but for the Ruby platform exclusively). Adhearsion exists as a gem in the standard RubyGems trove so, with Ruby and RubyGems installed, Adhearsion is only one install command away.

Installing Ruby/RubyGems on AsteriskNOW

AsteriskNOW comes standard with Ruby but not RubyGems (for support reasons). Thankfully, RubyGems can be easily installed from the Ruby rPath trove with the following command:

```
conary update rubygems=ruby.rpath.org@rpl:devel
source /etc/profile
```

Installing Ruby/RubyGems on Linux

Most Linux distributions' package managers host a Ruby package, though some do not yet have RubyGems. With your respective distro's preferred software management application, install Ruby 1.8.5 or later and RubyGems if available. If RubyGems is not available in CentOS, you can install Ruby by typing:

```
yum install ruby
```

Next, we need RubyGems. You can get that by navigating to */usr/src/*, and entering:

```
wget http://rubyforge.org/frs/download.php/20585/rubygems-0.9.3.tgz
tar zxvf rubygems-0.9.3.tgz
cd rubygems-0.9.3
ruby setup.rb
```

Installing Ruby/RubyGems on Mac OS X

Ruby actually ships standard with OS X, but you will need to upgrade it and install RubyGems from MacPorts, an OS X package manager. With MacPorts installed, (available from *http://www.macports.org* if you do not already have it) you can install Ruby and RubyGems with the following command:

```
sudo port install ruby rb-rubygems
```

You may also need to add `/opt/local/bin` to your PATH variable in `/etc/profile`.

Ruby/RubyGems on Windows

A fantastic "one-click installer" exists for Windows. This installer will automatically install Ruby, RubyGems, and a few commonly used gems all in a matter of minutes. You can download the installer from *http://rubyforge.org/projects/rubyinstaller*.

Installing Adhearsion from RubyGems

Once you've followed one of the previous set of instructions for your system to fetch Ruby and RubyGems, install Adhearsion with the following command:

```
gem install adhearsion
```

If any dependencies are found, you will probably need to allow them to install for Adhearsion to work correctly.

Exploring a New Adhearsion Project

With Adhearsion installed, you can begin creating and initializing a new Adhearsion project with the newly created `ahn` command, a command-line utility that manages nearly everything in Adhearsion.

A sample command for creating a new Adhearsion project is as follows:

```
ahn create ~/newproject
```

This creates a folder at the destination specified containing the directory and file hierarchy Adhearsion needs to operate. Right away, you should be able to execute the newly created application by running:

```
ahn start ~/newproject
```

To familiarize yourself with the Adhearsion system, take a look through the application's folders and read the accompanying documentation.

Adhearsion dialplan writing

The ability to write dialplans in Adhearsion is typically the first feature newcomers use. Since Ruby permits such fine-grained modification of the language itself at runtime, one of the things Adhearsion does is make aesthetic changes, which are intended to streamline the process of developing dialplans.

Below is an Adhearsion Hello World application:

```
my_first_context {
  play "hello-world"
}
```

Though this is completely valid Ruby syntax, not all Ruby applications look like this. Adhearsion makes the declaration of context names comfortable by interpreting the dialplan script specially. Your scripts will be located in the root folder of your newly created Adhearsion application.

As calls come into Asterisk and subsequently Adhearsion, Adhearsion invokes its own version of the context name from which the AGI request originated. Given this, we should ensure that a context in *extensions.conf* has this same name and forwards calls properly to Adhearsion.

The syntax for directing calls to Adhearsion is as follows:

```
[my_first_context]
exten => _.,1,AGI(agi://127.0.0.1)
```

This catches any pattern dialed and goes off to Adhearsion via AGI to handle the call-processing instructions for us. The IP provided here should of course be replaced with the necessary IP to reach your Adhearsion machine.

Now that you have a basic understanding of how Adhearsion and Asterisk interact, here is a more real-world dialplan example in Adhearsion:

```
internal {
  case extension
  when 10..99
    dial SIP/extension
  when 6000..6020, 7000..7030
    # Join a MeetMe conference with "join"
    join extension
  when _'21XX'
    if Time.now.hour.between? 2, 10
      dial SIP/"berlin-office"/extension[2..4]
```

```
      else speak "The German office is closed"
      end
   when US_NUMBER
      dial SIP/'us-trunk-out'/extension
   when /^\d{11,}$/ # Perl-like regular expression
      # Pass any other long numbers straight to our trunk.
      dial IAX/'intl-trunk-out'/extension
   else
      play %w'sorry invalid extension please-try-again'
      end
}
```

With just this small amount of code we accomplish quite a lot. Even with limited or no knowledge of Ruby, you can probably infer the following things:

- We use a switch-like statement on the "extension" variable (which Adhearsion creates for us) and branch depending on that.

- Dialing a number between 10 and 99 routes us to the SIP peer with the dialed numerical username.

- Any number dialed between 6000 and 6200 or between 7000 and 7030 goes to a MeetMe conference of that same number. This of course requires *meetme.conf* to have these conference numbers configured.

- The _'21XX' option comes straight from Asterisk's pattern style. Prepending a String with an underscore in Adhearsion secretly invokes a method that returns a Ruby regular expression. In a Ruby "case" statement, regular expressions can be used in a "when" statement to check against a pattern. The end effect should be very familiar to those with *extensions.conf* writing experience.

- Adhearsion's syntax for representing channels also comes straight from Asterisk's traditional format. SIP/123 can be used verbatim to represent the SIP peer 123. If a trunk were involved, SIP/trunkname/username would act as you would expect.

- The speak() method abstracts an underlying text-to-speech engine. This can be configured to use most of the popular engines.

- A full-blown Perl-like regular can be used in a when statement to perform more sophisticated pattern matching if Asterisk's patterns do not suffice.

- Adhearsion defines a few constants that may be useful to someone writing dialplans. The US_NUMBER constant here is a regular expression for matching an American number.

- If you find the need to play several files in sequence, play() accepts an Array of filenames. By luck, Ruby has a convenient way of creating an Array of Strings.

This is of course just a simple example and covers only the absolute basics of Adhearsion's dialplan authoring capabilities.

Database integration

Though immensely successful in the web development space for serving dynamic content, database integration has always been an underutilized possibility for driving dynamic voice applications with Asterisk. Most Asterisk applications that do accomplish database integration outsource the complexity to a PHP or Perl AGI script because the *extensions.conf* or AEL grammars are simply impractical for the level of sophistication required.

Adhearsion uses a database integration library, called ActiveRecord, developed by the makers of the Ruby on Rails framework. With ActiveRecord, the end user seldom, if ever, writes SQL statements. Instead, the developer accesses the database just like any Ruby object. Because Ruby allows such flexible dynamism, access to the database looks and feels quite natural. Additionally, ActiveRecord abstracts the differences between database management systems, making your database access implementation agnostic.

Without going too much into the internals of ActiveRecord and more sophisticated uses of it, let us consider the following simple MySQL schema:

```
CREATE TABLE groups (
  `id` int(11) DEFAULT NULL auto_increment PRIMARY KEY,
  `description` varchar(255) DEFAULT NULL,
  `hourly_rate` decimal DEFAULT NULL
);

CREATE TABLE customers (
  `id` int(11) DEFAULT NULL auto_increment PRIMARY KEY,
  `name` varchar(255) DEFAULT NULL,
  `phone_number` varchar(10) DEFAULT NULL,
  `usage_this_month` int(11) DEFAULT 0,
  `group_id` int(11) DEFAULT NULL
);
```

In practice we would obviously store much more information about the customer and keep the service usage information in a database-driven call detail record, but this degree of simplicity helps demonstrate ActiveRecord fundamentals more effectively.

To connect Adhearsion to this database, one simply specifies the database access information in a YAML configuration file like so:

```
adapter: mysql
host: localhost
database: adhearsion
username: root
password: pass
```

This tells Adhearsion how to connect to the database, but how we access information in the tables depends on how we model our ActiveRecord objects. Since an object is an instance of a class, we write a class definition to wrap around each table. We define simple properties and relationships in the class with the superclass's methods.

Here are two classes we may use with the aforementioned tables:

```ruby
class Customer < ActiveRecord::Base
  belongs_to :group

  validates_presence_of   :name, :phone_number
  validates_uniqueness_of :phone_number
  validates_associated    :group

  def total_bill
self.group.hourly_rate * self.usage_this_month / 1.hour
  end
end

class Group < ActiveRecord::Base
  has_many :customers
  validates_presence_of :description, :hourly_rate
end
```

From just this small amount of information, ActiveRecord can make a lot of logical inferences. When these classes interpret, ActiveRecord assumes the table names to be customers and groups respectively by lowercasing the classes' names and making them plural. If this convention is not desired, the author can easily override it. Additionally, at interpretation time, ActiveRecord actually peeks into the database's columns and makes available many new dynamically created methods.

The belongs_to and has_many methods in this example define relationships between Customers and Groups. Notice again how ActiveRecord uses pluralization to make the code more expressive in the has_many :customers line. From this example, we also see several validations—policies that ActiveRecord will enforce. When creating a new Customer we must provide a name and phone_number at the bare minimum. No two phone numbers can conflict. Every Customer must have a Group. Every Group must have a description and hourly_rate. These help both the developer and the database stay on track.

Also, notice the total_bill method in the Customer class. On any Customer object we extract from the database, we can call this method, which multiplies the hourly_rate value of the group to which the Customer belongs by the Customer's own phone usage this month (in seconds).

Here are a few examples that may clarify the usefulness of having Ruby objects abstract database logic:

```ruby
everyone = Customer.find :all

jay = Customer.find_by_name "Jay Phillips"
jay.phone_number # Performs a SELECT statement
jay.total_bill   # Performs arithmetic on several SELECT statements
jay.group.customers.average :usage_this_month

jay.group.destroy
jay.group = Group.create :description => "New cool group!",
                         :hourly_rate => 1.23
jay.save
```

Because the database integration here becomes much more natural, Asterisk dialplans becomes much more expressive as well. Below is a sample dialplan of a service provider that imposes a time limit on outgoing calls using information from the database. To remain simple:

```
# Let's assume we're offering VoIP service to customers
# whom we can identify with their callerid.

service {
  # The line of code below performs an SQL SELECT
  # statement on our database. The find_by_phone_number()
  # method was created automatically because ActiveRecord
  # found a phone_number column in the database. Adhearsion
  # creates the "callerid" variable for us.
  caller = Customer.find_by_phone_number callerid

  usage = caller.usage_this_month
  if usage >= 100.hours
    play "sorry-cant-let-you-do-that"
  else
    play %w'to-hear-your-account-balance press-1
            otherwise wait-moment'
    choice = wait_for_digit 3.seconds

    p choice
    if choice == 1
      charge = usage / 60.0 * caller.group.hourly_rate
      play %W"your-account will-reflect-charge-of $#{charge}
              this month for #{usage / 60} minutes and
              #{usage % 60} seconds"
    end

    # We can also write back to the "usage_this_month"
    # property of "caller". When the time method finishes,
    # the database will be updated for this caller.
    caller.usage_this_month += time do
      # Code in this block is timed.
      dial IAX/'main-trunk'/extension
    end
    caller.save
  end
}
```

Robust database integration like this through Adhearsion brings new ease to developing for and managing a PBX. Centrally persistent information allows Asterisk to integrate with other services cleanly while empowering more valuable services whose needs are beyond that of traditional Asterisk development technologies.

Distributing and reusing code

Because an Adhearsion application resides within a single folder, completely copying the VoIP application is as simple as zipping the files. For one of the first times in the Asterisk community, users can easily exchange and build upon one another's successful application. In fact, open sourcing individual Adhearsion applications is greatly encouraged.

Additionally, on a more localized level, users can reuse Adhearsion framework extensions, called helpers, or roll their own. Helpers range from entire sub-frameworks like the Micromenus framework for integrating with on-phone micro-browsers to adding a trivial new dialplan method that returns a random quote by Oscar Wilde.

Below is a simple Adhearsion helper written in Ruby. It creates a new method that will exist across the entire framework, including the dialplan. For simplicity's sake, the method downloads an XML document at a specified HTTP URL and converts it to a Ruby Hash object (Ruby's associative array type):

```
def remote_parse url
  Hash.from_xml open(url).read
end
```

Note that these three lines can work as the entire contents of a helper file. When Adhearsion boots, it executes the script in a way that makes any methods or classes defined available anywhere in the system.

For some issues, particularly ones of scaling Adhearsion, it may be necessary to profile out bottlenecks to the king of efficiency: C. Below is a sample Adhearsion helper that returns the factorial of a number given:

```
int fast_factorial(int input) {
  int fact = 1, count = 1;
  while(count <= input) {
    fact *= count++;
  }
  return fact;
}
```

Again, the code here can exist as the entire contents of a helper file. In this case, because it is written in C, it should have the name *factorial.alien.c*. This tells Adhearsion to invoke its algorithm to read the file, add in the standard C and Ruby language development headers, compile it, cache the shared object, load it into the interpreter, and then wrap the C method in a Ruby equivalent. Below is a sample dialplan that simply speaks back the factorial of six using this C helper:

```
fast_test {
  num = fast_factorial 6
  play num
}
```

Note that the C method becomes a first-class Ruby method. Ruby number objects passed to the method are converted to C's "int" primitive, and the return value is converted back to a Ruby number object.

Helpers promise robust albeit simple extensibility to a VoIP engineer's toolbox, but, best of all, useful helpers can be traded and benefit the entire community.

Integrate with Your Desk Phone Using Micromenus

With increasing competition between modern IP-enabled desk phone manufacturers, the microbrowser feature has snuck in relatively unnoticed and underutilized. The principle is simple: physical desk phones construct interactive menus on a phone by pulling XML over HTTP or the like. Contending interests, however, lead this technology amiss: every vendor's XML differs, microbrowsers are often quirky, and available features vary vastly.

The Micromenus framework exists as an Adhearsion helper and aims to abstract the differences between vendors' phones. For this very particular development domain (i.e., creating menus), Micromenus use a very simple Ruby-based "Domain Specific Language" to program logic cleanly and independent of any phone brands.

Below is a simple example Micromenu:

```
image 'company-logo'
item "Call an Employee" do
  # Creates a list of employees as callable links from a database.
  Employee.find(:all).each do |someone|
    # Simply select someone to call that person on your phone.
    call someone.extension, someone.full_name
  end
end
item "Weather Information" do
  call "Hear the weather report" do
    play weather_report("Portland, OR")
  end
  item "Current: " + weather("Portland, OR")[:current][:temp]
end
item "System Uptime: " + `uptime`
```

A list item displays in two ways. If given only a text String, Micromenus renders only a text element. If the arguments contain a do/end block of nested information, that text becomes a link to a sub-page rendering that nested content.

A call item also has two uses, each producing a link which, when selected, initiates a call. When call receives no do/end block, it simulates physically dialing the number given as the first argument. When a do/end block exists and all calls route through Adhearsion, selecting that item executes the dialplan functionality within the block. This is a great example of how handling the dialplans and on-screen microbrowsers within the same framework pays off well.

From this example we can see a few other neat things about Micromenus:

- Micromenus support sending images. If the requesting phone does not support images, the response will have no mention of them.
- All Adhearsion helpers work here, too. We use the weather helper in this example.
- The Micromenus sub-framework can use Adhearsion's database integration.

- Ruby can execute a command inside backticks and return the result as a String. We report the uptime here with it.

This example of course assumes that you have configured your application's database integration properly and have an Employee class mapping to a table with an exten sion and full_name column.

Because Micromenus simply render varying responses over HTTP, a normal web browser can make a request to the Micromenus' server, too. For these more high-tech endpoints, Micromenus render an attractive interface with Ajax loading, DHTML effects, and draggable windows.

In the interest of "adhering" people together, Micromenus exist as another option to make your Adhearsion VoIP applications more robust.

Integrating with a Web Application

Though Adhearsion by design can integrate with virtually any application, including PHP or Java Servlets, Ruby on Rails makes for a particularly powerful partner. Rails is a web development framework getting a lot of press lately for all the right reasons. Its developers use Ruby to its full potential, showing how meta-programming really does eliminate unnecessary developer work. Rails' remarkable code clarity and application of the Don't Repeat Yourself (DRY) principle served as a strong inspiration to the inception of Adhearsion as it exists today.

Starting with Adhearsion version 0.8.0, Adhearsion's application directory drops in place atop an existing Rails application, sharing data automatically. If you have needs to develop a complex web interface to VoIP functionality, consider this deadly duo.

Using Java

Eyebrows around the world raised when Sun announced their hiring of the two core developers of the JRuby interpreter project, Charles Nutter and Thomas Enebo, in September 2006. JRuby is a Ruby interpreter written in Java instead of C. Because JRuby can compile parts of a Ruby application to Java bytecode, JRuby is actually outperforming the C implementation of Ruby 1.8 in many different benchmarks and promises to outperform Ruby 1.8 in all cases in the near future.

A Ruby application running in JRuby has the full benefit of not only Ruby libraries but any Java library as well. Running Adhearsion in JRuby brings the dumbfounding assortment of third-party Java libraries to PBX dialplan writing. If your corporate environment requires tight integration with other Java technologies, embedding Adhearsion in a J2EE stack may offer the flexibility needed.

More Information

For more information about the fast-moving Adhearsion community, including complete walkthroughs, see Adhearsion's official web site at *http://adhearsion.com*, Adhearsion's official blog at *http://blog.adhearsion.com*, the web site of Adhearsion's parent consulting company *http://codemecca.com*, or for help learning Ruby, see *http://jicksta.com*.

## CHAPTER 11

The Asterisk GUI Framework

...I was constructing a lighthouse while all the others were making ships.

—Charles Simic

This chapter introduces the components that comprise the GUI and help it work with Asterisk. It describes the installation of the web server and the GUI components for those who are not using the AsteriskNOW distribution. It also shows you how to modify the GUI to suit your purposes. Technical information is also provided so that developers wishing to create their own GUI or application can utilize the web server and GUI components. We'd like to thank the folks at Digium for writing this chapter, especially the code examples, which they developed and tested.

Why a GUI for Asterisk?

Since the beginning, Asterisk has been a phone system for the brave. In the early days it took guts and more than a bit of tenacity to make Asterisk do your bidding. Those willing to accept the learning curve, wade into the config files, and fight for their calls were rewarded with a powerful, flexible phone system (as well as a very marketable skill set). However, the mass market was not, and is not, ready to script extensions, manage peers, and handle the other tasks that are the crux of Asterisk administration.

Since the early pre-1.0 days, people have tried to tame the mighty Asterisk with config file generators tied to databases and managed via a range of graphical user interfaces (GUIs). The most successful of these did a fine job of creating an Asterisk-based application, but none of them provided the full flexibility that the raw scripting environment offers. By replacing the digital haiku of the dialplan with a limited list of options, the resulting system is reduced from Asterisk to an Asterisk-based system. Not a bad thing, but not the whole enchilada.[*]

---

[*] In fact, two of the authors of this book once attempted to write the ultimate Asterisk GUI. Lucky for you, they abandoned the project and began writing Asterisk documentation instead!

In order for a GUI to be *the* Asterisk GUI, it would have to leave intact the manually scripted configuration files that have been the *lingua franca* of Asterisk since the dawn of time. It would have to provide a simple, graphical means of configuration without compromising the underlying Asterisk software or irrevocably fixing decisions that should be left open to the end user. It would also have to provide advanced functionality without taxing the computer or stealing valuable resources from the core goal of processing calls.

Coinciding with the release of Asterisk 1.4, Digium launched the Asterisk GUI project. The GUI was originally conceived as a component of Digium's Asterisk embedded appliance. The appliance, sold as the Asterisk Appliance Developers Kit (AADK) as well as in a standalone configuration, is a small, solid-state Linux computer with optional analog (and potentially digital) interfaces. The GUI was built using a flexible and expandable framework that placed as much of the display and validation logic as possible on the client computer. It also took into consideration the need to preserve handwritten config files while providing an automated means of editing them. The resulting framework is known as AJAM (a play on the popular "Web 2.0" technology known as Ajax), which means Asynchronous JavaScript and Asterisk Manager. The core AJAM code, a series of AJAM-enabled web pages, and an extension to the Asterisk manager work together to form the Asterisk GUI framework.

What Is the GUI?

The Asterisk GUI is the interface that comes with the AsteriskNOW distribution or can be added to an existing Asterisk installation. The default interface is geared toward the user who wants to use Asterisk as a PBX for a small business with fairly typical telecom needs. It can best be thought of as a sample of what can be done using AJAM; think of it as a beta interface that can be expected to evolve according to the desires of the community. This has caused a lot of excitement in the Asterisk community, because the underlying technologies behind the GUI raise the bar on what a PBX interface can become. It also enables you to build your own interfaces that are tuned to your unique requirements.

Mark Spencer Talks About the GUI

Asterisk is a powerful telephony platform. However, that power is only as valuable as its ability to be used by a particular target user. There is a lot of value to having graphical interfaces (GUIs) for Asterisk. Most GUIs are specifically designed to support a particular task. For example, some GUIs are designed specifically for voicemail systems. Others are specifically targeted to the hospitality industry. There is some demand to have a GUI that targets Asterisk generally, but there is a natural trade-off between the ease of use and simplicity of a GUI, versus the number of available features. For example, the GUI that a seasoned systems administrator might require would likely be different than that of an office administrator who is only responsible for simple moves, adds, and changes to the system. Given this wide ranging demand, Digium developed a GUI framework called (uncreatively) the Asterisk GUI. Rather than developing a single GUI, Digium developed different GUIs and a framework to trivialize the creation and modification of GUIs for different segments.

A second goal was to make sure that the GUI interacted with Asterisk's traditional configuration methods in a way that did not preclude someone from using them. Most GUIs for Asterisk use an intermediate configuration format or database, then spit out configs for Asterisk to use. Unfortunately that means that any option that is not presented within the GUI cannot be "manually" set in the configuration files. By contrast, the Asterisk GUI actually modifies your traditional Asterisk configuration files, meaning that your changes in the GUI and your changes to the files themselves can co-exist and even flow back and forth. As an example, if you change the caller ID for a user in *users.conf* then refresh the GUI, you'll see the change in the GUI as well. Likewise if you change it in the GUI and reload the file, you'll see the change in the file. If you add new settings that are not presented in the GUI (for example if you add nat=yes to a particular entry in *users.conf*, then change the caller ID in the GUI, you'll see that the nat=yes line will remain in the file even though the caller ID change goes through. Comments are also generally preserved across GUI edits. This means that not only is the GUI no longer required to display all possible configurations, since esoteric ones can be set manually. This also means that when someone starts by using the Asterisk GUI and then outgrows it, there is a natural path for them to be able to start creating more sophisticated functions without abandoning the GUI with which they're familiar.

Using the GUI

When you first log in to a newly created GUI, the system walks you through a wizard that lets you set up the basic elements of your phone system.

 The GUI may not be able to detect all types of TDM interfaces, and thus may report that it cannot find any cards even though you have some installed. It is expected that the GUI will eventually be able to detect and manage any cards that use the Zaptel interface, but this functionality is going to be complex, and is still in development at this time.

The wizard walks you through some basic settings such as extension length and dialing rules. We are not going to get into detail on how the default GUI works. It is in constant development, and what we write about here is not likely to be what you will experience when you read this.

GUI elements

The standard GUI that comes with AsteriskNOW (or can be downloaded via SVN) has a standard set of elements that represent the sorts of things a typical small office PBX might want. The menu items are currently:

- Users
- Conferencing
- Voicemail

- Call Queues
- Service Providers
- Calling Rules
- Incoming Calls
- Voice Menus
- Record a Menu
- Active Channels
- Graphs
- System Info
- Backup
- Options

Architecture of the Asterisk GUI

Before we get too far into exploring the Asterisk GUI (or developing your own), it's important to understand the flow of information between the client (the web browser) and Asterisk. Since these interfaces are Ajax applications, there are a lot of pieces that aren't immediately obvious. The flow of control goes something like this:

- The browser "surfs to" the URL for your management application.
- Asterisk's web server sends the browser an HTML page, libraries, and the application itself (which is written in JavaScript and makes heavy use of Ajax).
- The user interacts with the browser; as needed, the JavaScript application sends commands back to the web server. These commands are in the form of URLs that request some action from the Asterisk server itself.
- The web server interprets the URLs. If the user has logged in successfully, it sends a command (an action) to Asterisk via the Asterisk Manager Interface (AMI), described in Chapter 10.
- Asterisk executes the action and the results (a status code and possibly data) to the web server.
- The web server sends Asterisk's response back to the JavaScript application on the browser.
- The JavaScript application updates the browser's display.

While it may sound a little complicated at first glance, don't be intimidated. It's a very flexible and powerful architecture that can be used for a myriad of applications, not just an Asterisk GUI. For now, however, we'll concentrate on enhancing the Asterisk GUI. Let's begin by configuring the underlying pieces, and then move on to installing and modifying the Asterisk GUI.

Components of the Asterisk GUI

Let's take a closer look at some of the key components of the Asterisk GUI. We'll use these components later in the chapter to modify the Asterisk GUI.

Asterisk Manager Interface

As explained in Chapter 10, the Asterisk Manager Interface allows external programs to control Asterisk. The Manager interface is the heart of the Asterisk GUI, as it does all of the heavy lifting.

Manager over HTTP and the Asterisk web server

The web server built into Asterisk allows manager commands to be sent to Asterisk via HTTP, instead of creating a socket connection directly to the Manager interface. This makes it much simpler for a web application to issue AMI commands to Asterisk using the Asynchronous JavaScript Asterisk Manager (AJAM), which we will cover shortly. The web server can also be configured to serve static content, such as HTML files and images.[†]

AJAM and JavaScript

The AJAM framework uses JavaScript and XML to asynchronously send commands to Asterisk, and to update the information displayed in the web browser.

Installing the Asterisk GUI

If you didn't install AsteriskNOW, you need to download and install the Asterisk GUI files. Once the files are downloaded, you simply compile and install them as part of your Asterisk installation.

  You need Asterisk 1.4 or later in order use the Asterisk GUI.

[†] You may be asking yourself, "Why embed a web server inside of Asterisk? Why not just use an external web server?" While you *can* use an external web server to serve up the Asterisk GUI, it's beyond the scope of this chapter, as the security model behind Ajax permits Ajax requests only to the same domain, port, and protocol that sent the HTML page. This is often referred to as the same-origin policy.

You can get the latest version of the GUI files by checking them out of Digium's Subversion repository.[‡] If you have Subversion installed on your computer, you can downloaded the GUI code by using the following command:

```
# cd /usr/src   # or wherever you prefer to download source code to
# svn co http://svn.digium.com/svn/asterisk-gui/trunk asterisk-gui
```

Installing the GUI is simple as this:

```
# cd asterisk-gui
# ./configure
# make
# make install
# make samples
```

After running the previous commands, the GUI files are installed and part of your Asterisk distribution.

Setting up httpd.conf and manager.conf

Configuring the Asterisk web server to process AJAM requests involves several simple steps. In the /etc/asterisk/http.conf file, add (or un-comment) the following:

```
[general]
enabled=yes
enablestatic=yes       ; without this, you can only send AMI commands, not display
                       ; html content

bindaddr=0.0.0.0       ; address you want the Asterisk HTTP server to respond on
bindport=8088          ; port you want the Asterisk HTTP server to respond on
prefix=asterisk        ; will form part of the URI, similar to a directory name
```

Now that we've got *httpd.conf* set up, we can serve up content to a browser. To allow the web client to send commands to Asterisk, we have to make some changes to the Asterisk Manger Interface (AMI). We do this by adding a few lines to the [general] section of *manager.conf*, and by adding a user account with the config permission set. Open up *manager.conf* and edit it to match the following:

```
[general]
enabled=yes      ; you may already have AMI enabled if you are using it for other things
webenabled=yes   ; this enables the interaction between the Asterisk web server and AMI

[asterisk_http] ; you can name the user whatever you want
secret = gooey
read = system,call,log,verbose,command,agent,user,config
write = system,call,log,verbose,command,agent,user,config
```

Save the changes and restart Asterisk. You should be able to connect to Asterisk's web server through the following URI:

[‡] There is currently no way to download the GUI via FTP. That situation may change at any time, so feel free to check the Asterisk web site for updated information.

```
http://localhost:8088/asterisk/static/ajamdemo.html
```

If for some reason you're having problems getting to that demo page, go back to the *asterisk-gui* source code directory and run:

```
# make checkconfig
```

And that's it! Asterisk is now web-enabled. Now, let's move on to the actual work of developing with the Asterisk GUI.

Developing for the Asterisk GUI

Once you've installed the files for the Asterisk GUI, you can begin to play with developing for the GUI. Over the next few sections, we'll walk through setting up the various components and putting them together to enhance and expand the capabilities of the GUI.

Issuing Manager Commands over HTTP

The Asterisk GUI issues commands to Asterisk by calling specially crafted URLs to the Asterisk web server. This section provides examples of some commonly used commands (actions) and the corresponding web server responses. These AMI URLs have the following general structure:

```
http://hostname:8088/asterisk/rawman?action=command&....parameter=value pairs...
http://hostname:8088/asterisk/manager?action=command&....parameter=value pairs...
http://hostname:8088/asterisk/mxml?action=command&....parameter=value pairs...
```

The difference between the `rawman`, `manager` and `mxml` URLs is important. The web server exports three different views of the AMI interface. If you use a `rawman` URL, the server returns a series of keyword/value pairs in the HTTP response. If you use a `manager` URL, the server returns the result formatted as HTML. In a similar style, if you use a `mxml` URL, the server returns the results formatted in XML. For modern Ajax-style applications, the `rawman` and `mxml` forms are probably more useful.[§]

The actions that can be sent to the server, along with their parameters, are the ordinary manager commands described in Appendix F. Note that the LOGIN and CHALLENGE actions are unique in that they aren't sent to Asterisk itself, but are processed by the Manager interface to authenticate the user. If the user hasn't authenticated correctly, the server returns an error response rather than sending the action to Asterisk for processing.

Let's look at some commonly used actions, and see how we can use them to control the server.

[§] By similar reasoning, the `manager` form is much easier for humans to use for debugging purposes.

LOGIN

The LOGIN command authenticates credentials for the Manager interface's HTML view. Once you are logged in, Asterisk stores a cookie on your browser (valid for the length of the httptimeout setting). This cookie is used to connect to the same session. The URL:

http://localhost:8088/asterisk/rawman?action=login&username=asterisk_http &secret=gooey

sends a login command to the web server that includes the credentials. If successful, the web server responds with:

```
Response: Success
Message:  Authentication accepted
```

This, of course, is a very simplistic way for a login to work. Sending the username and secret password in a URL is bad practice, though it's very useful during development. A more appropriate way to handle the login, and an example of more complex command processing, is to use a challenge/response sequence. Issue a request like this:

http://localhost:8088/asterisk/rawman?action=challenge&AuthType=md5

The CHALLENGE command initiates a challenge/response sequence that can be used to log in a user. The server responds by sending a challenge (an arbitrary string) in the response:

```
Response: Success
Challenge: 113543555
```

Your application answers the challenge by computing the MD5 hash of the challenge concatenated with the user's password. Here's how a user might manually calculate the MD5 hash:

```
# echo -n 113543555gooey | md5sum
       50a0f43ad4c9d99a39f1061cf7301d9a   -
```

You can then use the calculated hash as the login key in a URL like this:

http://localhost:8088/asterisk/rawman?action=login&username=asterisk_http &authtype=md5&key=50a0f43ad4c9d99a39f1061cf7301d9a

 For security reasons, the login action must take place within five seconds of the challenge action. Note also that cookies must be enabled for the challenge/response to work, as the cookie ensures that the login action uses the same manager session ID as the challenge action.

If you use a manager URL to request the challenge (instead of using rawman), the response will be formatted as HTML:

```
<title>Asterisk&trade; Manager Interface</title>
<body bgcolor="#ffffff">
<table align=center bgcolor="#f1f1f1" width="500">
<tr><td colspan="2" bgcolor="#f1f1ff"><h1>  Manager Tester</h1></td></tr>
```

```
<tr><td>Response</td><td>Success</td></tr>
<tr><td>Challenge</td><td>113543555</td></tr>
</table>
</body>
```

Similarly, if you use the mxml view instead, you will receive a response formatted as XML:

```
<Ajax-response>
  <response type='object' id='unknown'>
    <generic response='Success' challenge='113543555' />
  </response>
</Ajax-response>
```

Other than the formatting, there are no other differences between the three types of responses. For most applications, digging the challenge out of the keyword/value pairs will be much simpler than using rawman or mxml in a situation like this, where you don't need to display the HTML to the user.

Transferring a call

The REDIRECT action can be used to transfer a call. Simply generate a URL such as:

http://localhost:8088/asterisk/rawman?action=redirect&channel=SIP/John-ae201e78 &priority=1&exten=6001

This URL transfers the specified channel to another extension and priority in the dialplan. The response to this action is:

```
Response: Success
Message: Redirect Successful
```

Reading a configuration file

The GETCONFIG command returns the contents of a configuration file, or portion thereof. The URL:

http://localhost:8088/asterisk/rawman?action=getconfig&filename=users.conf

returns the contents of the *users.conf* file. The Asterisk GUI uses this functionality to present the current Asterisk configuration to the end user. The response looks like this:

```
Response: Success
  Category-000000: general
  Line-000000-000000: fullname=New User
  Line-000000-000001: userbase=6000
  Line-000000-000002: hasvoicemail=yes
  Line-000000-000003: hassip=yes
  Line-000000-000004: hasiax=yes
  Line-000000-000005: hasmanager=no
  Line-000000-000006: callwaiting=yes
  Line-000000-000007: threewaycalling=yes
  Line-000000-000008: callwaitingcallerid=yes
  Line-000000-000009: transfer=yes
  Line-000000-000010: canpark=yes
  Line-000000-000011: cancallforward=yes
```

```
Line-000000-000012: callreturn=yes
Line-000000-000013: callgroup=1
Line-000000-000014: pickupgroup=1
Line-000000-000015: host=dynamic
Category-000001: 6007
Line-000001-000000: fullname=Bill Savage
Line-000001-000001: secret=1234
Line-000001-000002: email=bsavage@digium.com
Line-000001-000003: cid_number=6001
Line-000001-000004: zapchan=
Line-000001-000005: context=numberplan-custom-1
Line-000001-000006: hasvoicemail=yes
Line-000001-000007: hasdirectory=no
Line-000001-000008: hassip=yes
Line-000001-000009: hasiax=yes
Line-000001-000010: hasmanager=no
Line-000001-000011: callwaiting=yes
Line-000001-000012: threewaycalling=yes
Line-000001-000013: mailbox=6007
Line-000001-000014: hasagent=yes
Line-000001-000015: group=
```

Updating configuration files using UPDATECONFIG

The UPDATECONFIG action is used to update one or more settings in a configuration file. For example, to delete a user you should use a URL like this:

*http://localhost:8088/asterisk/rawman?action=updateconfig&reload=yes&srcfilename
=users.conf&dstfilename=users.conf&Action-000000=delcat&Cat-000000=6003&
Var-000000=&Value-000000=*

Error response

A user must be logged in to the web server before any other commands can be issued. Any of the commands we've discussed will return an error response if the user is not authenticated. If it's not given by an authenticated user, this URI *http://localhost:8088/ asterisk/rawman?action=ping* returns this response to indicate an error:

```
Response: Error
Message:  Authentication Required
```

Ajax, AJAM, and Asterisk

As an acronym, Ajax stands for *Asynchronous JavaScript and XML*. While the term includes the words asynchronous and XML, this does not mean that you can make only asynchronous requests, nor are you required to use XML. Some authors describe Ajax as simply a combination of HTML, JavaScript, DHTML, and DOM. The next generation browsers, such as Mozilla/Firefox, use an XMLHttpRequest (a JavaScript object) to send an asynchronous request to the server. The request is made in the background and processed by the server.

Back on the browser, the result is handled by a callback: whatever the server returns can be stored and used to update the page being displayed. For Internet Explorer 5 or later, the XMLHttp ActiveX object serves the same purpose.

Form processing in a traditional web application

HTML forms are usually submitted by using a submit button (`type=submit`). When the user clicks the submit button, processing stops, and doesn't continue until the server returns a new page:

```
<FORM action="login.php" method="POST">
  <input type=text name="username">
  <input type=password name="password">
  <input type=submit>
</FORM>
```

Before going any further with Ajax or JavaScript, let's take a look at how a traditional web application works. Traditional web applications use HTML's `<FORM>` element to define a form in which all of the parameters a user needs to send to the server are defined. In addition the `action="login.php"` informs the browser where to send all of these variables. The `method="POST"` tells the browser how to send these variables to the server.

Form processing in an Ajax application

An Ajax application uses JavaScript to send the contents of a form to the server. If you have made the request asynchronously, your JavaScript code doesn't wait for the server to respond. This also means that you can let the users continue to interact with the page, even though a request may be taking place in the background. This can be dangerous and, thus, you may want to restrict certain actions until a request has completed. The browser, by default, gives no visual indication that a request is being made in the background. It is your responsibility to inform the user of the progress of a request. Here's the code for submitting the contents of the username and password fields via Ajax:

```
<script language="javascript" type="text/javascript">
function submitform(){
  var uname = document.getElementById("username").value;
  var pwd = document.getElementById("password").value;
  // xmlHttp = new ActiveXObject("Msxml2.XMLHTTP"); // IE 7
  // xmlHttp = new ActiveXObject("Microsoft.XMLHTTP"); // IE 5
  xmlHttp = new XMLHttpRequest(); // Mozilla or Firefox

  var url = "/rawman?action=login&username=" + escape(uname) + "&secret=" + escape(pwd);

  xmlHttp.open("GET", url, true);
  xmlHttp.onreadystatechange = dosomething;
  // dosomething() would be another JavaScript function
  xmlHttp.send(null);
}
</script>
```

The getElementById() method reads the value of the username and the password fields. This code then gets an XMLHttpRequest object, which it uses to send these values back to the server. Note that the kind of object you need depends on whether your users are using Internet Explorer 7, 5, or Mozilla/Firefox. It's fairly easy to write code to handle all of these situations, or to use a library like Prototype to handle platform independence for you. The username and password are encoded in a URL and sent to the server. The call to xmlHttp.onreadystatechange registers a handler to process the result that the server returns to us.

This code only deals with making the XMLHttp request, and it tells the browser to call the dosomething() function when there is a response from the server. Here's a dosomething() function that handles this response:

```
<script language="javascript" type="text/javascript">
  function dosomething() {
    if (xmlHttp.readyState == 4) {
      var login_response = xmlHttp.responseText;
    }
  }
</script>
```

 Make sure that each XMLHttp step has completed (with either success or failure) before performing the next one.

This function is called whenever there's a change in the state of the HTTP request. The if statement saves the response only if the request's readyState is 4, which means that the request has completed. The JavaScript variable login_response now contains the response of the login page.

Note that this is very far from being production-ready code. In particular, the simplistic username and password handling is appropriate for testing, but would be a serious security problem in a production system—even if the application is used only on a private network. To build more robust and secure password handling, use the challenge/response system presented earlier. If you want to learn more about writing Ajax web applications, we recommend *Head Rush Ajax* by Brett McLaughlin (O'Reilly).

The Prototype framework

Prototype (*http://prototypejs.org*) is a JavaScript framework released under an MIT-style license. Prototype can make your job extremely easy while developing an Ajax application. It provides many ways to make your code shorter and clearer. For example, in the submitform function, the call to document.getElementById() can be replaced by the $() function. Likewise, the call to value to get the DOM element's content can be replaced with a call to $F(). Thus, document.getElementById("username").value becomes simply $F('username'); the result is code that's much simpler and more readable.

Prototype also makes it easy to make XMLHttp requests in an elegant manner. Using Prototype's `Ajax` object, the `submitform()` function can be rewritten as:

```
<script language="javascript" type="text/javascript">
function submitform(){
    var url = '/rawman';
    var pars = 'username=' + escape($F('username')) + '&secret=' + escape($F('password'));

var myAjax = new Ajax.Request( url,
  { method: 'get',
    parameters: pars,
    onComplete: dosomething
  });

}
</script>
```

Not only is this code much shorter, you don't have to write browser-specific code in your web pages; Prototype takes care of the differences between Mozilla/Firefox and the various versions of Internet Explorer. Furthermore, it takes care of testing the request's `readyState`, so you leave that annoying `if` statement out of your handler. Prototype has lots of built-in functions, some of which have been extensively used in the Asterisk framework. There's no room to discuss them here, but for more information, see the Short Cuts *Prototype Quick Reference* by Scott Raymond and *Prototype and Scriptaculous: Taking the Pain Out of JavaScript* by Chris Angus, both from O'Reilly.

Customization of the GUI

Now that we've explored the different pieces that form the foundation of the Asterisk GUI, we have what we need to be able to explore the GUI itself and modify it to fit our needs. To get to the Asterisk GUI, go to the following address in your browser: *http://localhost:8088/asterisk/static/config/cfgbasic.html*.

Looking at Figure 11-1 might lead you to conclude that the Asterisk GUI is simply one more Asterisk GUI in an already crowded space. Nothing could be further from the truth. This GUI doesn't just allow you to tweak it, it practically begs you to. In this section we are going to discuss how you can modify the GUI and use AJAM to build your own extensions to the GUI. In order to benefit the most from this information, you need some HTML and JavaScript knowledge.

The GUI home page is named *cfgbasic.html*. All other pages are loaded into the iframe contained within the *cfgbasic.html* page. By default, *cfgbasic.html* loads *home.html* into the main frame.

For most changes to the GUI, you'll eventually need to modify *cfgbasic.html*, which is the login screen.

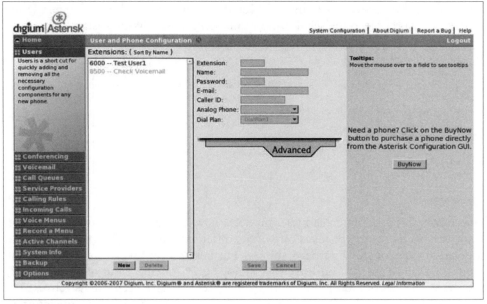

Figure 11-1. A screenshot of the Asterisk GUI

Adding a new tab to the GUI

As an example of customizing the Asterisk GUI, let's create a new tab that displays the contents of *extensions.conf*. First, we need to create a file and put it in the */var/lib/asterisk/static-html/config* directory. In this example, we'll name the file *test.html*:

```
<script src="scripts/prototype.js"></script>
<script src="scripts/astman.js"></script>
<script>
function localAjaxinit() {
        parent.loadscreen(this);
        makerequest('g','extensions.conf', '' , function(t){
                $('ExtensionsDotConf').innerHTML = "<PRE>" + t + "</PRE>";
        });
}
</script>
<body onload="localAjaxinit()" bgcolor="EFEFEF">
        <div id="ExtensionsDotConf"></div>
</body>
```

This code simply displays the configuration of the *extensions.conf* file. Obviously it's a very simple example, but it shows the fundamentals of creating a new page for the Asterisk GUI. Let's walk through the example step by step.

The first line tells the browser to load the Prototype library. The second line tells the browser to load the `astman.js` file, which contains much of the code designed to interact with the Manager interface.

Next, we define a function called `localAjaxinit`. The `localAjaxinit` function first tells this page's parent (*cfgbasic.html* in this case) to run the `loadscreen` function, passing in this page as the parameter. This causes the main GUI screen to load our new *test.html* inside the iframe. The second thing we do inside the `localAjaxinit` function is to use the `makerequest` function. The `makerequest` function is defined in *astman.js* and makes it very convenient to make requests to the web server.[‖]

The first parameter of the `makerequest` function specifies what type of request is being made. It can be set to any of the following:

`'g'`

Use the `GetConfig` action to retrieve the configuration from the configuration file specified in the second parameter.

`'u'`

Use the `UpdateConfig` action to update the configuration in the configuration file specified in the second parameter. The third parameter to the function specifies the configuration data that should be updated.

`''`

If the first parameter to the `makerequest` function is a set of single quotes, then the custom action specified in the third parameter will be sent.

The fourth parameter is the callback function that will be called with the response to the Ajax request.

Examples of Using makerequest

As an example, the following code snippet shows three different ways to use the `makerequest` function. In the first, we'll get the configuration data from *users.conf*. In the second, we'll update *musiconhold.conf* and change the value of the `random` setting in the `default` class. Last but not least, we'll call the `Ping` action. Each of them sets a callback function named t that simply replaces the contents of the `div` with the response of the Ajax call.

```
makerequest( 'g', 'users.conf', '' ,
    function(t) { $('ExtensionsDotConf').innerHTML = "<PRE>" + t +
    "</PRE>"; } );
makerequest( 'u', 'musiconhold.conf',

'&Action-000000=update&Cat-000000=default&Var-000000=random&Value-
    000000=yes' ,
    function(t) { $('ExtensionsDotConf').innerHTML = "<PRE>" + t +
    "</PRE>"; } );

makerequest( '', '', 'action=Ping' ,
    function(t) { $('ExtensionsDotConf').innerHTML = "<PRE>" + t +
    "</PRE>"; } );
```

[‖] In reality, `makerequest` is a simple wrapper around a call to Prototype's `Ajax.Request` method.

The rest of our *test.html* simply contains an HTML body with a `div` element, which is where we'll place the configuration data when we get it. Note that the HTML body tag has an `onload` attribute, which causes the browser to execute the `localAjaxinit` function once the page has finished loading.

Now that we've created a new page, we need to edit *cfgbasic.html* to add this page as a panel in the GUI. Open *cfgbasic.html* and search for a JavaScript function named `returnpanels` and insert this code in the list of panels, where you would like your panel to appear:

```
newpanel( ["Test", "test.html", "Test"]);
```

Now reload the GUI in your browser. You should see a new tab on the lefthand side named `Test` that displays the configuration values for *extensions.conf* when clicked.

While there's a lot more to learn about the AJAM interface and the Asterisk GUI, this example should show just how easy it is to add new functionality to the GUI. In this next example, we'll show how simple it is to expose a setting from the configuration files in the GUI.

Exposing configuration settings in the GUI

As explained earlier, one of the unique benefits of the Asterisk GUI over the other graphical frontends to Asterisk is that it updates the configuration files in place, taking special care not to overwrite or erase any extra settings you might have in your configuration files. To show just how easy it is to expose new settings in the GUI, we'll add a simple checkbox to the GUI to make it possible to set the `nat` setting in *users.conf*.

If you open the GUI and click on the tab labeled `Users`, the GUI loads the file named *users.html* in the iframe. Let's open up *users.html* (usually located in */var/lib/asterisk/static-http/config*) and begin modifying it to add our checkbox.

First, search near the top of the file where a variable named `fieldnames` is defined. This variable contains a list of all of the field names that will be set by this page of the GUI. Simply add `nat` to the end of the list, or add the following line directly below the current definition for `fieldnames`.

```
fieldnames.push('nat');
```

This tells the Asterisk GUI that we want to be able to see the value of `nat` and to be able to set it as well. In order to see or set the value, however, we need to add an element to the HTML form. To do that, search *users.html* for the IAX checkbox, and add the following lines between it and the CTI checkbox.

```
<tr>
  <td align=right><input type='checkbox' id='nat'></td>
  <td>NAT</td>
</tr>
```

Simply reload the page, and that's all there is to it. With just a few lines of additional code, we're able to expose the `nat` setting to the GUI. It couldn't be much simpler!

As you're developing for the Asterisk GUI, you'll probably find that debugging Ajax and JavaScript code can be somewhat difficult at times. We strongly suggest you make use of an extension to Mozilla Firefox named Firebug that greatly simplifies the task of debugging Ajax, Java-Script, and HTML. Check it out at *http://www.getfirebug.com*. There is also a scaled-down version for Internet Explorer known as Firebug Lite, which is available for download at the same web site.

For More Information

Over the course of this chapter, we've introduced you to the Asterisk GUI and the AJAM framework. We've covered the architecture of how the GUI works, and how to modify the GUI. If you would like more information on developing a graphical interface for Asterisk, please refer to the GUI Developers Guide located at *http://asterisknow.org/developers/gui-guide*.

## CHAPTER 12

Relational Database Integration

*Few things are harder to put up with than the annoyance
of a good example.*

—Mark Twain

Introduction

In this chapter we are going to explore integrating some Asterisk features and functions into a database. There are several databases available for Linux, but we have chosen to limit our discussion to PostgreSQL. While we acknowledge MySQL is also an extremely popular database, we had to choose one, and our experience with PostgreSQL tipped the scale in its favor. All that having been said, what we are actually going to be doing is discussing the ODBC connector, so as long has you have familiarity with getting your favorite database ODBC-ready, the substance of this chapter will apply to you.

Integrating Asterisk with databases is one of the fundamental elements allowing clustering of Asterisk into a large, distributed system. By utilizing the power of the database, dynamically changing data can convey information across an array of Asterisk systems. Our newest favorite Asterisk function is func_odbc, which we will cover later in this chapter.

While not all Asterisk deployments will require a relational database, understanding how to harness them opens the lid of a treasure chest full of new ways to design your telecom solution.

Installing the Database

The first thing to do is to install the PostgreSQL database server:[*]

```
# yum install -y postgresql-server
```

---

[*] On a large, busy system you will want to install this on a completely separate box from your Asterisk system.

Then start the database, which will take a few seconds to initialize for the first time:

```
# service postgresql start
```

Next, create a user called *asterisk*, which we will use to connect to and manage the database. Run the following commands:

```
# su - postgres
$ createuser -P
Enter name of user to add: asterisk
Enter password for new user:
Enter it again:
Shall the new role be a superuser? (y/n) n
Shall the new user be allowed to create databases? (y/n) y
Shall the new user be allowed to create more new users? (y/n) n
CREATE USER
```

By default, PostgreSQL does not listen on the TCP/IP connection, which Asterisk will be using. We need to modify the */var/lib/pgsql/data/postgresql.conf* file in order to allow Asterisk to make IP connections to the database. To do this, simply remove the comment from the beginning of the **tcpip_socket** and **port** parameters. Be sure to change the **tcpip_socket** option from **false** to **true**.

```
tcpip_socket = true max_connections = 100
        # note: increasing max_connections costs about 500 bytes of shared
        # memory per connection slot, in addition to costs from shared_buffers
        # and max_locks_per_transaction.
#superuser_reserved_connections = 2
port = 5432
```

Now, edit the */var/lib/pgsql/data/pg_hba.conf* file in order to allow the *asterisk* user we just created to connect to the PostgreSQL server over the TCP/IP socket. At the end of the file, replace everything below # Put your actual configuration here with the following:

```
host    all    asterisk    127.0.0.1    255.255.255.255    md5
local   all    asterisk                                    trust
```

Now we can create the database that we will use throughout this chapter. We're going to create a database called **asterisk** and set the owner to our *asterisk* user.

```
$ createdb --owner=asterisk asterisk
CREATE DATABASE
```

Restart the PostgreSQL server after exiting from the *postgres* user back to root:

```
$ exit
# service postgresql restart
```

We can verify our connection to the PostgreSQL server via TCP/IP like so:

```
# psql -h 127.0.0.1 -U asterisk Password:
Welcome to psql 7.4.16, the PostgreSQL interactive terminal.

Type:  \copyright for distribution terms
       \h for help with SQL commands
```

```
\? for help on internal slash commands
\g or terminate with semicolon to execute query
\q to quit
```

```
asterisk=>
```

Double-check your configuration as discussed earlier if you get the following error, which means connections via the TCP/IP socket are not allowed:

```
psql: could not connect to server: Connection refused
Is the server running on host "127.0.0.1" and accepting
TCP/IP connections on port 5432?
```

Installing and Configuring ODBC

The ODBC connector is a database abstraction layer that makes it possible for Asterisk to communicate with a wide range of databases without requiring the developers to create a separate database connector for every database Asterisk wants to support. This saves a lot of development effort and code maintenance. There is a slight performance cost to this because we are adding another application layer between Asterisk and the database. However, this can be mitigated with proper design and is well worth it when you need powerful, flexible database capabilities in your Asterisk system.

Before we install the connector in Asterisk, we have to install ODBC into Linux itself. To install the ODBC drivers, simply run the command:

```
# yum install -y unixODBC unixODBC-devel libtool-ltdl libtool-ltdl-devel
```

 See Chapter 3 for the matrix of packages you should have installed.

We need to install the *unixODBC-devel* package because it is used by Asterisk to create the ODBC modules we will be using throughout this chapter.

Verify that you have the PostgreSQL ODBC driver configured in the */etc/odbcinst.ini* file. It should look something like this:

```
[PostgreSQL]
Description     = ODBC for PostgreSQL
Driver          = /usr/lib/libodbcpsql.so
Setup           = /usr/lib/libodbcpsqlS.so
FileUsage       = 1
```

Verify the system is able to see the driver by running the following command. It should return the label name PostgreSQL if all is well.

```
# odbcinst -q -d
[PostgreSQL]
```

Next, configure the */etc/odbc.ini* file, which is used to create an identifier that Asterisk will use to reference this configuration. If at any point in the future you need to change the database to something else, you simply need to reconfigure this file, allowing Asterisk to continue to point to the same place.†

```
[asterisk-connector]
Description       = PostgreSQL connection to 'asterisk' database
Driver            = PostgreSQL
Database          = asterisk
Servername        = localhost
UserName          = asterisk
Password          = welcome
Port              = 5432
Protocol          = 7.4
ReadOnly          = No
RowVersioning     = No
ShowSystemTables  = No
ShowOidColumn     = No
FakeOidIndex      = No
ConnSettings      =
```

Let's verify that we can connect to our database using the `isql` application. The `isql` application will not perform the connect as the *root* user, and must be run as the same owner as the database. Since the owner of the **asterisk** database under PostgreSQL is the *asterisk* user, we must create a Linux account with the same name. In Chapter 14, we will take advantage of this user to run Asterisk as non-*root*.

```
# su - asterisk
$ echo "select 1" | isql -v asterisk-connector
+---------------------------------------+
| Connected!                            |
|                                       |
| sql-statement                         |
| help [tablename]                      |
| quit                                  |
|                                       |
+---------------------------------------+
SQL> +------------+
| ?column?   |
+------------+
| 1          |
+------------+
SQLRowCount returns 1
1 rows fetched
$ exit
```

With *unixODBC* installed, configured, and verified to work, we need to recompile Asterisk so that the ODBC modules are created and installed. Change back to your

---

† Yes, this is excessively verbose. The only entries you really need are `Driver`, `Database`, and `Servername`. Even the `Username` and `Password` are specified elsewhere, as seen later.

Asterisk sources directory and run the ./*configure* script so it knows you have installed *unixODBC*.

```
# cd /usr/src/asterisk-1.4
# make distclean
# ./configure
# make menuselect
# make install
```

 Almost everything in this chapter is turned on by default. You will want to run make menuselect to verify that the ODBC related modules are enabled. These include *cdr_odbc*, *func_odbc*, *func_realtime*, *pbx_realtime*, *res_config_odbc*, *res_odbc*. For voicemail stored in an ODBC database, be sure to select ODBC_STORAGE from the Voicemail Build Options menu. You can verify the modules exist in the */usr/lib/asterisk/modules/* directory.

Configuring res_odbc for Access to Our Database

ODBC connections are configured in the *res_odbc.conf* file located in */etc/asterisk*. The *res_odbc.conf* file sets the parameters that the various Asterisk modules will use to connect to the database.[‡]

Modify the *res_odbc.conf* file:

```
[asterisk]
enabled => yes
dsn => asterisk-connector
username => asterisk
password => welcome
pooling => no
limit => 0
pre-connect => yes
```

The dsn option points at the database connection we configured in */etc/odbc.ini*, and the pre-connect option tells Asterisk to open up and maintain a connection to the database when loading the *res_odbc.so* module. This lowers some of the overhead that would come from repeatedly setting up and tearing down the connection to the database.

Once you've configured *res_odbc.conf*, start Asterisk and verify the database connection with the odbc show CLI command:

```
*CLI> odbc show
Name: asterisk
DSN: asterisk-connector
```

[‡] The pooling and limit options are quite useful for MS SQL Server and Sybase databases. These permit you to establish multiple connections (up to limit connections) to a database while ensuring that each connection has only one statement executing at once (this is due to a limitation in the protocol used by these database servers).

```
Pooled: no
Connected: yes
```

Using Realtime

The Asterisk Realtime Architecture (ARA) is a method of storing the configuration files (that would normally be found in */etc/asterisk*) and their configuration options in a database table. There are two types of realtime; *static* and *dynamic*. The static version is similar to the traditional method of reading a configuration file, except that the data is read from the database instead. The dynamic realtime method is used for things such as user and peer objects (SIP, IAX2), and voicemail which loads and updates the information as it is required. Changes to static information requires a reload just as if you had changed the text file on the system, but dynamic information is polled by Asterisk as needed and requires no reload. Realtime is configured in the *extconfig.conf* file located in the */etc/asterisk* directory. This file tells Asterisk what to load from the database and where to load it from, allowing certain files to be loaded from the database and other files to be loaded from the standard configuration files.

Static Realtime

Static realtime is used when you want to store the configuration that you would normally place in the configuration files in */etc/asterisk* but want to load from a database. The same rules that apply to flat files on your system still apply when using static realtime, such as requiring you to either run the reload command from the Asterisk CLI, or to reload the module associated with the configuration file (i.e., `module reload chan_sip.so`).

Using the preload Directive

Most files are able to be loaded via static realtime, but a few files are not able to be loaded using this method. These include *asterisk.conf*, *extconfig.conf*, and *logger.conf*. Additionally, the files *manager.conf*, *cdr.conf*, and *rtp.conf* cannot be loaded from static realtime unless the storage drivers are loaded before the main Asterisk core initializes (this is because the configuration files need to be loaded by realtime before the module goes to read its configuration). Since we are using ODBC in this chapter, we would need to add the following lines to *modules.conf*:

```
; /etc/asterisk/modules.conf
preload => res_odbc.so
preload => res_config_odbc.so
```

When using static realtime, we tell Asterisk which files we want to load from the database using the following syntax in the *extconfig.conf* file:

```
; /etc/asterisk/extconfig.conf
filename.conf => driver,database[,table]
```

 If the table name is not specified, then Asterisk will use the name of the file instead.

The static realtime module uses a specifically formatted table to read the configuration of static files in from the database. You can define the table for static realtime in PostgreSQL as follows:

```
CREATE TABLE ast_config
(
    id serial NOT NULL,
    cat_metric int4 NOT NULL DEFAULT 0,
    var_metric int4 NOT NULL DEFAULT 0,
    filename varchar(128) NOT NULL DEFAULT ''::character varying,
    category varchar(128) NOT NULL DEFAULT 'default'::character varying,
    var_name varchar(128) NOT NULL DEFAULT ''::character varying,
    var_val varchar(128) NOT NULL DEFAULT ''::character varying,
    commented int2 NOT NULL DEFAULT 0,
    CONSTRAINT ast_config_id_pk PRIMARY KEY (id)
)
WITHOUT OIDS;
```

A brief explanation about the columns is required in order to understand how Asterisk takes the rows from the database and applies them to the configuration for the various modules you may load:

cat_metric
> The weight of the category within the file. A lower metric means it appears higher in the file (see the sidebar "A Word About Metrics").

var_metric
> The weight of an item within a category. A lower metric means it appears higher in the list. This is useful for things like codec order in *sip.conf* or *iax.conf* where you want `disallow=all` to appear first (metric of 0), followed by `allow=ulaw` (metric of 1), then by `allow=gsm` (metric of 2) (see the sidebar "A Word About Metrics").

filename
> The filename the module would normally read from the hard drive of your system (i.e., *musiconhold.conf*, *sip.conf*, *iax.conf*, etc.).

category
> The section name within the file, such as [general], but don't save to the database using the square brackets.

var_name
> The option on the left side of the equals sign (i.e., *disallow* is the `var_name` in `disallow=all`).

var_val
> The value to an option on the right side of the equals sign (i.e., *all* is the `var_val` in `disallow=all`).

commented
> Any value other than 0 will evaluate as if it were prefixed with a semicolon in the flat file (commented out).

A Word About Metrics

The metrics in static realtime are used to control the order that objects are read into memory. Think of the `cat_metric` and `var_metric` as the original line numbers in the flat file. A higher *cat_metric* is processed first (because Asterisk matches categories from bottom to top—this is why the order of users and peers can matter in *sip.conf* or *iax.conf*). A lower *var_metric* is processed first within the category because Asterisk will process the order of options top-down within the category (e.g., `disallow=all` should be set to a value lower than the `allow`'s value within the category to make sure it is processed first).

A simple file we can load from static realtime is the *musiconhold.conf* file. Let's start by moving this file to a temporary location:

```
# cd /etc/asterisk
# mv musiconhold.conf musiconhold.conf.old
```

In order for the classes to be removed from memory, we need to restart Asterisk. Then we can verify our classes are blank by running `moh show classes`:

```
*CLI> restart now
*CLI> moh show classes
*CLI>
```

So let's put the [default] class back into Asterisk, but now we'll load it from the database. Connect to PostgreSQL and execute the following INSERT statements:

```
INSERT INTO ast_config (filename,category,var_name,var_val)
VALUES ('musiconhold.conf','general','mode','files');
INSERT INTO ast_config (filename,category,var_name,var_val)
VALUES ('musiconhold.conf','general','directory','/var/lib/asterisk/moh');
```

You can verify your values have made it into the database by running a SELECT statement:

```
asterisk=# select filename,category,var_name,var_val from ast_config;
```

filename	category	var_name	var_val
musiconhold.conf	general	mode	files

```
musiconhold.conf | general        | directory   | /var/lib/asterisk/moh
(2 rows)
```

And now, there's just one last thing to modify in the *extconfig.conf* file in */etc/asterisk* directory to tell Asterisk to get the data for *musiconhold.conf* from the database. Add the following line to the end of the *extconfig.conf* file, then save it:

```
musiconhold.conf => odbc,asterisk,ast_config
```

Then connect to the Asterisk console and perform a reload:

```
*CLI> module reload
```

You can now verify that we have our music-on-hold classes loading from the database by running moh show classes:

```
*CLI> moh show classes
Class: general
        Mode: files
        Directory: /var/lib/asterisk/moh
```

And there you go; *musiconhold.conf* loaded from the database. You can perform the same steps in order to load other flat files from the database!

Dynamic Realtime

The dynamic realtime system is used to load objects that may change often: SIP/IAX2 users and peers, queues and their members, and voicemail. Since this information in the system may either be changing or new records are being added on a regular basis, we can utilize the power of the database to let us load this information on an as-needed basis.

All of realtime is configured in the */etc/asterisk/extconfig.conf* file, but dynamic realtime has well-defined configuration names such as sippeers. Defining something like SIP peers is done with the following format:

```
; extconfig.conf
sippeers => driver,database[,table]
```

The table name is optional, in which case Asterisk will use the predefined name (i.e., sippeers) as the table to look up the data. In our example, we'll be using the ast_sip peers table to store our SIP peer information.

 Remember that we have both SIP peers and SIP users; peers are endpoints we send calls to, and a user is something we receive calls from. A friend is shorthand that defines both.

So to configure Asterisk to load all SIP peers from a database using realtime, we would define something like:

```
; extconfig.conf
sippeers => odbc,asterisk,ast_sipfriends
```

To also load our SIP users from the database, define it like so:

```
sipusers => odbc,asterisk,ast_sipfriends
```

You may have noticed we used the same table for both the `sippeers` and `sipusers`. This is because there will be a type field (just as if you had defined the type in the *sip.conf* file) that will let us define a type of `user`, `peer`, or `friend`. When defining the table for SIP users and peers, we need at least the following:

```
+------+--------+-------+--------+-----+------------+----------+
|name  |host    |secret | ipaddr | port| regseconds | username |
+------+--------+-------+--------+-----+------------+----------+
|100   |dynamic |welcome|        |     |1096954152  |   1000   |
+------+--------+-------+--------+-----+------------+----------+
```

The `port`, `regseconds`, and `ipaddr` fields are required to let Asterisk store the registration information for the peer in order to know where to send the call. This is assuming the host is `dynamic`; however, if the peer is `static`, we would have to populate the `ipaddr` field ourselves. The `port` field is optional and would use the default standard port defined in the [general] section, and the `regseconds` would remain blank.) There are many more options for a SIP friend that we can define, such as the caller ID, and adding that information is as simple as adding the `callerid` column to the table. See the *sip.conf.sample* file for more options that can be defined for SIP friends.

Storing Call Detail Records

Call Detail Records (CDR) contain information about calls that have passed through your Asterisk system. They are discussed further in Chapter 13. This is a popular use of databases in Asterisk because CDR can be easier to manage if you store the records in a database (for example, you could keep track of many Asterisk systems in a single table).

Setting the systemname for Globally Unique IDs

CDRs consist of a unique identifier and several fields of information about the call (including source and destination channel, length of call, last application executed and so forth). In a clustered set of Asterisk boxes, it is theoretically possible to have duplication among unique identifiers since each Asterisk system considers only itself. To address this, we can automatically append a system identifier to the front of the unique ID by adding an option to */etc/asterisk/asterisk.conf*. For each of your boxes, set an identifier by adding something like:

```
[options]
systemname=toronto
```

Let's create a table in our database to store CDR. Log in to the PostgreSQL server with the *psql* application:

```
# psql -U asterisk -h localhost asterisk
Password:
```

And create the asterisk_cdr table:

```
asterisk=> CREATE TABLE asterisk_cdr
(
    id bigserial NOT NULL,
    calldate timestamptz,
    clid varchar(80),
    src varchar(80),
    dst varchar(80),
    dcontext varchar(80),
    channel varchar(80),
    dstchannel varchar(80),
    lastapp varchar(80),
    lastdata varchar(80),
    duration int8,
    billsec int8,
    disposition varchar(45),
    amaflags int8,
    accountcode varchar(20),
    uniqueid varchar(40),
    userfield varchar(255),
    CONSTRAINT asterisk_cdr_id_pk PRIMARY KEY (id)
)
WITHOUT OIDS;
```

You can verify the table was created by using the **\dt** command (describe tables):

```
asterisk=> \dt asterisk_cdr
             List of relations
 Schema |     Name     | Type  |  Owner
--------+--------------+-------+----------
 public | asterisk_cdr | table | asterisk
(1 row)
```

Next, configure Asterisk to store its CDR into the database. This is done in the */etc/asterisk/cdr_odbc.conf* file with the following configuration:

```
[global]
dsn=asterisk-connector
username=asterisk
password=welcome
loguniqueid=yes
table=asterisk_cdr
```

If Asterisk is already running, from the Asterisk CLI execute `module reload cdr_odbc.so`. You can also just type `reload`, to reload everything.

```
*CLI> reload
```

Verify the status of CDR by entering the following command and looking for `CDR registered backend: ODBC`:

```
*CLI> cdr status
CDR logging: enabled
CDR mode: simple
CDR registered backend: cdr-custom
CDR registered backend: cdr_manager
CDR registered backend: ODBC
```

Now, perform a call through your Asterisk box and verify you have data in the asterisk_cdr table. The easiest way to test a call is with the Asterisk CLI command `console dial` (assuming that you have a sound card and *chan_oss* installed). However, you can utilize any method at your disposal to place a test call:

```
*CLI> console dial 100@default
-- Executing [100@default:1] Playback("OSS/dsp", "tt-weasels") in new stack
-- <OSS/dsp> Playing 'tt-weasels' (language 'en')
```

Then connect to the database and perform a `SELECT` statement to verify you have data in the `asterisk_cdr` table. You could also do `SELECT * FROM asterisk_cdr;`, but that will return a lot more data:

```
# psql -U asterisk -h localhost asterisk
Password:

asterisk=> SELECT id,dst,channel,uniqueid,calldate FROM asterisk_cdr;
 id | dst | channel |      uniqueid        |        calldate
----+-----+---------+----------------------+------------------------
  1 | 100 | OSS/dsp | toronto-1171611019.0 | 2007-02-16 02:30:19-05
(1 rows)
```

Getting Funky with func_odbc: Hot-Desking

The `func_odbc` dialplan function is arguably the coolest and most powerful dialplan function in Asterisk. It allows you to create and use fairly simple dialplan functions that retrieve and use information from databases directly in the dialplan. There are all kinds of ways in which this might be used, such as managing users or allowing sharing of dynamic information within a clustered set of Asterisk machines.

What `func_odbc` allows you to do is define SQL queries to which you assign function names. In effect, you are creating custom functions that obtain their results by executing queries against a database. The *func_odbc.conf* file is where you specify the relationship between the function names you create and the SQL statements you wish them to perform. By referring to the named function in the dialplan, you can retrieve and update values in the database.

While using an external script to interact with a database (from which a flat file is created that Asterisk would read) has advantages (if the database went down, your system would continue to function and the script would simply not update any files until connectivity to the database was restored), a major disadvantage is that any changes you make to a user are not available until you run the update script. This is probably not a big issue on small systems, but on large systems, waiting for changes to take effect can cause issues, such as pausing a live call while a large file is loaded and parsed.

You can relieve some of this by utilizing a replicated database system. In the version of Asterisk following 1.4 (currently trunk), the syntax of the *func_odbc.conf* file changes slightly, but gives the ability to failover to another database system. This way you can cluster the database backend utilizing a master-master relationship (pgcluster; Slony-II), or a master-slave (Slony-I) replication system.

In order to get you into the right frame of mind for what follows, we want you to picture a Dagwood sandwich.[§]

Can you relay the total experience of such a thing by showing someone a picture of a tomato, or by waving a slice of cheese about? Not hardly. That is the conundrum we faced when trying to give a useful example of why func_odbc is so powerful. So, we decided to build the whole sandwich for you. It's quite a mouthful, but after a few bites of this, peanut butter and jelly is never going to be the same.

For our example, we decided to implement something that we think could have some practical uses. Let's picture a small company with a sales force of five people who have to share two desks. This is not as cruel as it seems, because these folks spend most of their time on the road, and they are each only in the office for at most one day each week.

Still, when they do get into the office, they'd like the system to know which desk they are sitting at, so that their calls can be directed there. Also, the boss wants to be able to track when they are in the office, and control calling privileges from those phones when no one is there.

This need is typically solved by what is called a *hot-desking* feature, so we have built one for you in order to show you the power of func_odbc.

Lets start with the easy stuff, and create two desktop phones in the *sip.conf* file.

```
; sip.conf
; HOT DESK USERS
[desk_1]
type=friend
host=dynamic
secret=my_special_secret
context=hotdesk
```

[§] And if you don't know what a Dagwood is, that's what Wikipedia is for. I am not that old.

```
qualify=yes

[desk_2]
type=friend
host=dynamic
secret=my_special_secret
context=hotdesk
qualify=yes

; END HOT DESK USERS
```

These two desk phones both enter the dialplan at the [hotdesk] context in *extensions.conf*. If you want to have these devices actually work, you will of course need to set the appropriate parameters in the devices themselves, but we've covered all that in Chapter 4.

That's all for *sip.conf*. We've got two slices of bread. Hardly a sandwich yet.

Now let's get the database part of it set up (we are assuming that you have an ODBC database created and working as outlined in the earlier parts of this chapter). First, connect to the database console like so:

```
# su - postgres
$ psql -U asterisk -h localhost asterisk
Password:
```

Then create the table with the following bit of code:

```
CREATE TABLE ast_hotdesk
(
  id serial NOT NULL,
  extension int8,
  first_name text,
  last_name text,
  cid_name text,
  cid_number varchar(10),
  pin int4,
  context text,
  status bool DEFAULT false,
  "location" text,
  CONSTRAINT ast_hotdesk_id_pk PRIMARY KEY (id)
)
WITHOUT OIDS;
```

After that, we populated the database with the following information (some of the values that you see actually would change only after the dialplan work is done, but we have it in here by way of example). At the PostgreSQL console, run the following commands:

```
asterisk=> INSERT INTO ast_hotdesk ('extension', 'first_name', 'last_name', 'cid_name',
'cid_number', 'pin', 'context', 'location') \
VALUES (1101, 'Leif', 'Madsen', 'Leif Madsen', '4165551101', '555', 'longdistance',
'desk_1');
```

Repeat the previous line and change the VALUES for all entries you wish to have in the database. You can view the data in the **ast_hotdesk** table by running a simple **SELECT** statement from the PostgreSQL console:

```
asterisk=> SELECT * FROM ast_hostdesk;
```

which would give you something like the following output:

```
| id | extension | first_name | last_name     | cid_name          | cid_number     | pin
|----+-----------+------------+---------------+-------------------+----------------+------
| 1 | 1101      | "Leif"     | "Madsen"      | "Leif Madsen"     | "4165551101"   | "555"
| 2 | 1102      | "Jim"      | "Van Meggelen"| "Jim Van Meggelen"| "4165551102"   | "556"
| 3 | 1103      | "Jared"    | "Smith"       | "Jared Smith"     | "4165551103"   | "557"
| 4 | 1104      | "Mark"     | "Spencer"     | "Mark Spencer"    | "4165551104"   | "558"
| 5 | 1105      | "Kevin"    | "Fleming"     | "Kevin Fleming"   | "4165551105"   | "559"

| context         | status  | location |$
+-----------------+---------+----------+
| "longdistance"  | "TRUE"  | "desk_1" |
| "longdistance"  | "FALSE" | ""       |
| "local"         | "FALSE" | ""       |
| "international" | "FALSE" | ""       |
| "local"         | "FALSE" | ""       |
```

We've got the condiments now, so let's get to our dialplan. This is where the magic is going to happen.

 Before you start typing, take note that we have placed all of the sample text that follows in appendix H, so while we encourage you to follow us along through the examples, you can also see what it all looks like as a whole, by checking the appendix (and by copying and pasting, if you have an electronic version of this book).

Somewhere in *extensions.conf* we are going to have to create the [hotdesk] context. To start, let's define a pattern-match extension that will allow the users to log in:

```
; extensions.conf
; Hot Desking Feature
[hotdesk]
; Hot Desk Login
exten => _110[1-5],1,NoOp()
exten => _110[1-5],n,Set(E=${EXTEN})
exten => _110[1-5],n,Verbose(1|Hot Desk Extension ${E} is changing status)
exten => _110[1-5],n,Verbose(1|Checking current status of extension ${E})
exten => _110[1-5],n,Set(${E}_STATUS=${HOTDESK_INFO(status,${E})})
exten => _110[1-5],n,Set(${E}_PIN=${HOTDESK_INFO(pin,${E})})
```

We're not done writing this extension yet, but let's pause for a moment and see where we're at so far.

When a sales agent sits down at a desk, they log in by dialing their own extension number. In this case we have allowed the 1101 through 1105 extensions to log in with our pattern match of _110[1-5]. You could just as easily make this less restrictive by using _11XX (allowing 1100 through 1199). This extension uses func_odbc to perform a lookup with the HOTDESK_INFO() dialplan function (which we will be creating shortly). This custom function (which we define in the *func_odbc.conf* file) performs an SQL statement and returns whatever is retrieved from the database.

We would define the new function HOTDESK_INFO() in *func_odbc.conf* like so:

```
[INFO]
prefix=HOTDESK
dsn=asterisk
read=SELECT ${ARG1} FROM ast_hotdesk WHERE extension = '${ARG2}'
```

That's a lot of stuff in just a few lines. Let's quickly cover them before we move on.

First of all, the prefix is optional. If you don't configure the prefix, then Asterisk adds "ODBC" to the name of the function (in this case INFO), which means this function would become ODBC_INFO(). This is not very descriptive of what the function is doing, so it can be helpful to assign a prefix that helps to relate your ODBC functions to the task they are performing. In this case we chose HOTDESK, which means that this custom function will be named HOTDESK_INFO.

The dsn attribute tells Asterisk which connection to use from *res_odbc.conf*. Since several database connections could be configured in *res_odbc.conf*, we specify which one to use here. In Figure 12-1, we show the relationship between the various file configurations and how they reference down the chain to connect to the database.

We then define our SQL statement with the read attribute. Dialplan functions have two different formats that they can be called with: one for retrieving information, and one for setting information. The read attribute is used when we call the HOTDESK_INFO() function with the retrieve format (and we could execute a separate SQL statement with the write attribute; we'll discuss the format for the write attribute a little bit later in this chapter).

Reading values from this function would take the format in the dialplan like so:

```
exten => s,n,Set(RETURNED_VALUE=${HOTDESK_INFO(status,1101)})
```

This would return the value located in the database within the status column where the extension column equals 1101. The status and 1101 we pass to the HOTDESK_INFO() function are then placed into the SQL statement we assigned to the read attribute, available as ${ARG1} and ${ARG2}, respectively. If we had passed a third option, this would have been available as ${ARG3}.

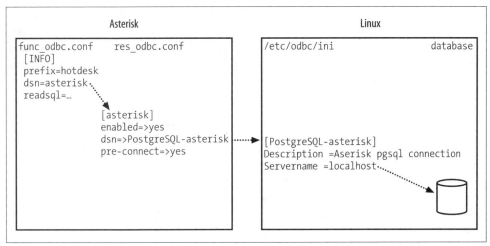

Figure 12-1. Relationships between func_odbc.conf, res_odbc.conf, /etc/odbc.ini (unixODBC), and the database connection

 Be sure that your data is unique enough that you only get a single row back. If more than one row is returned, Asterisk will see only the first row returned. With PostgreSQL, you could add a LIMIT 1 to the end of your SQL statement to limit a single row being returned, but this is not a good practice to rely on. A little further into this section we'll see how we can use the LIMIT and OFFSET PostgreSQL functions to loop through multiple rows of data!

After the SQL statement is executed, the value returned (if any) is assigned to the RETURNED_VALUE channel variable.

Using the ARRAY() function

In our example, we are utilizing two separate database calls and assigning those values to a pair of channel variables, (`${E}_STATUS` and `${E}_PIN`). This was done to simplify the example:

```
exten => _110[1-5],n,Set(${E}_STATUS=${HOTDESK_INFO(status,${E})})
exten => _110[1-5],n,Set(${E}_PIN=${HOTDESK_INFO(pin,${E})})
```

As an alternative, we could have returned multiple columns and saved them to separate variables utilizing the ARRAY() dialplan function. If we had defined our SQL statement in the *func_odbc.conf* file like so:

```
read=SELECT pin,status FROM ast_hotdesk WHERE extension = '${E}'
```

then we could use the ARRAY() function to save each column of information for the row to its own variable with a single call to the database:

```
exten => _110[1-5],n,Set(ARRAY(${E}_PIN,${E}_STATUS)=${HOTDESK_INFO(${E})})
```

So, in the first two lines of our following block of code we are passing the value status, and the value contained in the ${E} variable (e.g., 1101) to the HOTDESK_INFO () function. The two values are then replaced in the SQL statement with ${ARG1} and ${ARG2}, respectfully, the SQL statement is executed, and the value returned is assigned to the ${E}_STATUS channel variable.

OK, let's finish writing the pattern-match extension now:

```
exten => _110[1-5],n,Set(${E}_STATUS=${HOTDESK_INFO(status,${E})})
exten => _110[1-5],n,Set(${E}_PIN=${HOTDESK_INFO(pin,${E})})
exten => _110[1-5],n,GotoIf($[${ISNULL(${${E}_STATUS})}]?invalid_user,1)
; check if ${E}_STATUS is NULL
exten => _110[1-5],n,GotoIf($[${${E}_STATUS} = 1]?logout,1:login,1)
```

After assigning the value of the status column to the ${E}_STATUS variable (if you dialed extension 1101, then the variable name would be 1101_STATUS), we check if we received a value back from the database (error checking). We make use of the ISNULL() function to perform this check.

The last row in the block checks the status of the phone, and if currently logged in, will log off the agent. If not already logged in, it will go to extension login, priority 1 within the same context.[||]

 In the version following 1.4 (currently trunk) you can use the ${ODBCROWS} channel variable with statements executed by a readsql. We could have replaced the GotoIf() with something like:

```
exten => _110[1-5],n,GotoIf($[${ODBCROWS} < 0]?invalid_user,1)
```

The login extension runs some initial checks to verify the pin code entered by the agent. We allow him three tries to enter the correct pin, and if invalid, will send the call to the login_fail extension (which we will be writing later on).

```
exten => login,1,NoOp() ; set counter initial value
exten => login,n,Set(PIN_TRIES=0) ; set max number of login attempts
exten => login,n,Set(MAX_PIN_TRIES=3)
exten => login,n(get_pin),NoOp() ; increase pin try counter
exten => login,n,Set(PIN_TRIES=$[${PIN_TRIES} + 1])
exten => login,n,Read(PIN_ENTERED|enter-password|${LEN(${${E}_PIN})})
exten => login,n,GotoIf($[${PIN_ENTERED} = ${${E}_PIN}]?valid_login,1)
exten => login,n,Playback(invalid-pin)
exten => login,n,GotoIf($[${PIN_TRIES} <=${MAX_PIN_TRIES}]?get_pin:login_fail,1)
```

[||] Remember that in a traditional phone system all extensions must be numbers, but in Asterisk, extensions can have names as well. A possible advantage of using an extension that's not a number is that it will be much harder for a user to dial it from her phone and, thus, more secure. We're going to use several named extensions in this example. If you want to be absolutely sure that a malicious user cannot access those named extensions, simply use the trick that the AEL loader uses: start with a priority other than 1.

If the pin entered matches, we validate the login with the `valid_login` extension. First we utilize the `CHANNEL` variable to figure out which phone device we're calling from. The `CHANNEL` variable is usually populated with something such as: `SIP/desk_1-ab4034c`, so we make use of the `CUT()` function to first pull off the `SIP/` portion of the string and assign that to `LOCATION`. We then strip off the `-ab4034c` part of the string, discard it, and assign the remainder of `desk_1` to the `LOCATION` variable.

```
exten => valid_login,1,NoOp()
; CUT off the channel technology and assign to the LOCATION variable
exten => valid_login,n,Set(LOCATION=${CUT(CHANNEL,/,2)})
; CUT off the unique identifier and save the remainder to the LOCATION variable
exten => valid_login,n,Set(LOCATION=${CUT(LOCATION,-,1)})
```

We utilize yet another custom function, `HOTDESK_CHECK_PHONE_LOGINS()`, created in *func_odbc.conf* to check if any other users were previously logged in to this phone and had forgotten to log out. If the number of previously logged in users was greater than 0 (and should only ever be 1, but we check for more anyway and reset those, too), it runs the logic in the `logout_login` extension.

If no previous agents were logged in, we update the login status for this user with the `HOTDESK_STATUS()` function:

```
exten => valid_login,n,Set(ARRAY(USERS_LOGGED_IN)=${HOTDESK_CHECK_PHONE_
LOGINS(${LOCATION})})
exten => valid_login,n,GotoIf($[${USERS_LOGGED_IN} > 0]?logout_login,1)
exten => valid_login,n(set_login_status),NoOp()

; Set the status for the phone to '1' and where we're logged into
; NOTE: we need to escape the comma here because the Set() application has arguments
exten => valid_login,n,Set(HOTDESK_STATUS(${E})=1\,${LOCATION})
exten => valid_login,n,GotoIf($[${ODBCROWS} < 1]?error,1)
exten => valid_login,n,Playback(agent-loginok)
exten => valid_login,n,Hangup()
```

We create a write function in *func_odbc.conf* like so:

```
[STATUS]
prefix=HOTDESK
dsn=asterisk
write=UPDATE ast_hotdesk SET status = '${VAL1}', location = '${VAL2}' WHERE extension
= '${ARG1}'
```

The syntax is very similar to the **read** syntax discussed earlier in the chapter, but there are a few new things here, so let's discuss them before moving on.

The first thing you may have noticed is that we now have both `${VALx}` and `${ARGx}` variables in our SQL statement. These contain the values we pass to the function from the dialplan. In this case, we have two `VAL` variables, and a single `ARG` variable that were set from the dialplan via this statement:

```
Set(HOTDESK_STATUS(${E})=1\,${LOCATION})
```

Because the Set() dialplan application can also take arguments (you can set multiple variables and values by separating them with commas or pipes), you need to escape the comma with the backslash (\ ) so it is not processed by the expression parser for the Set() application, but rather parses it for the HOTDESK_STATUS() function.

Notice the syntax is slightly different from that of the read style function. This signals to Asterisk that you want to perform a write (this is the same syntax as other dialplan functions).

We are passing the value of the ${E} variable to the HOTDESK_STATUS() function, whose value is then accessible in the SQL statement within *func_odbc.conf* with the ${ARG1} variable. We then pass two values: 1 and ${LOCATION}. These are available to the SQL statement in the ${VAL1} and ${VAL2} variables, respectively.

As mentioned previously, if we had to log out one or more agents before logging in, we would check this with the logout_login extension. This dialplan logic will utilize the While() application to loop through the database and perform any database correction that may need to occur. More than likely this will execute only one loop, but it's a good example of how you might update or parse multiple rows in the database:

```
exten => logout_login,1,NoOp()
; set all logged in users on this device to logged out status
exten => logout_login,n,Set(ROW_COUNTER=0)
exten => logout_login,n,While($[${ROW_COUNTER} < ${USERS_LOGGED_IN}])
```

The ${USERS_LOGGED_IN} variable was set previously with the HOTDESK_CHECK_PHONE_LOG INS() function, which assigned a value of 1 or greater. We did this by counting the number of rows that were affected:

```
; func_odbc.conf
[CHECK_PHONE_LOGINS]
prefix=HOTDESK
dsn=asterisk
read=SELECT COUNT(status) FROM ast_hotdesk WHERE status = '1' AND location = '${ARG1}'
```

We then get the extension number of the user that is logged in with the HOTDESK_LOGGED_IN_USER() function. The LOCATION variable is populated with desk_1, which tells us which device we want to check on, and the ${ROW_COUNTER} contains which iteration of the loop we're on. These are both passed as arguments to the dialplan function. The result is then assigned to the WHO variable:

```
exten => logout_login,n,Set(WHO=${HOTDESK_LOGGED_IN_USER(${LOCATION},${ROW_COUNTER})})
```

The HOTDESK_LOGGED_IN_USER() function then pulls a specific row out of the database that corresponds with the iteration of the loops we are trying to process:

```
[LOGGED_IN_USER]
prefix=HOTDESK
dsn=asterisk
```

```
read=SELECT extension FROM ast_hotdesk WHERE status = '1'
AND location = '${ARG1}' ORDER BY id LIMIT '1' OFFSET '${ARG2}'
```

Now that we know what extension we want to update, we write to the HOTDESK_STATUS
() function, and assign a 0 to the status column where the extension number matches
the value in the ${WHO} variable (i.e., 1101). We then end the loop with EndWhile() and
return back to the valid_login extension at the set_login_status priority label (as discussed previously):

```
exten => logout_login,n,Set(HOTDESK_STATUS(${WHO})=0)           ; logout phone
exten => logout_login,n,Set(ROW_COUNTER=$[${ROW_COUNTER} + 1])
exten => logout_login,n,EndWhile()
exten => logout_login,n,Goto(valid_login,set_login_status)     ; return to logging in
```

The rest of the context should be fairly straightforward (if some of this doesn't make
sense, we suggest you go back and refresh your memory with Chapter 5 and Chapter 6). The one trick you may be unfamiliar with could be the usage of the ${ODB
CROWS} channel variable, which is set by the HOTDESK_STATUS() function. This tells us
how many rows were affected in the SQL UPDATE, which we assume to be 1. If the value
of ${ODBCROWS} is less than 1, then we assume an error and handle appropriately:

```
exten => logout,1,NoOp()
exten => logout,n,Set(HOTDESK_STATUS(${E})=0)
exten => logout,n,GotoIf($[${ODBCROWS} < 1]?error,1)
exten => logout,n,Playback(silence/1&agent-loggedoff)
exten => logout,n,Hangup()

exten => login_fail,1,NoOp()
exten => login_fail,n,Playback(silence/1&login-fail)
exten => login_fail,n,Hangup()

exten => error,1,NoOp()
exten => error,n,Playback(silence/1&connection-failed)
exten => error,n,Hangup()

exten => invalid_user,1,NoOp()
exten => invalid_user,n,Verbose(1|Hot Desk extension ${E} does not exist)
exten => invalid_user,n,Playback(silence/2&invalid)
exten => invalid_user,n,Hangup()
```

We also include the hotdesk_outbound context which will handle our outgoing calls
after we have logged the agent in to the system:

```
include => hotdesk_outbound
```

The hotdesk_outbound context utilizes many of the same principles and usage as previously discussed, so we won't approach it quite so thoroughly, but essentially the
[hotdesk_outbound] context will catch all dialed numbers from the desk phones. We first
set our LOCATION variable using the CHANNEL variable, then determine which extension
(agent) is logged in to the system and assign it to the WHO variable. If this variable is
NULL, then we reject the outgoing call. If not NULL, then we get the agent information
using the HOTDESK_INFO() function and assign it to several CHANNEL variables. This

includes the context to handle the call with, where we perform a `Goto()` to the context we have been assigned (which controls our outbound access).

If we try to dial a number that is not handled by our context (or one of the transitive contexts—i.e., international contains -> long distance, which also contains -> local), then the built-in extension i is executed which plays back a message stating the action cannot be performed, then hangs up the caller:

```
[hotdesk_outbound]
exten => _X.,1,NoOp()
exten => _X.,n,Set(LOCATION=${CUT(CHANNEL,/,2)})
exten => _X.,n,Set(LOCATION=${CUT(LOCATION,-,1)})
exten => _X.,n,Set(WHO=${HOTDESK_PHONE_STATUS(${LOCATION})})
exten => _X.,n,GotoIf($[${ISNULL(${WHO})}]?no_outgoing,1)
exten => _X.,n,Set(${WHO}_CID_NAME=${HOTDESK_INFO(cid_name,${WHO})})
exten => _X.,n,Set(${WHO}_CID_NUMBER=${HOTDESK_INFO(cid_number,${WHO})})
exten => _X.,n,Set(${WHO}_CONTEXT=${HOTDESK_INFO(context,${WHO})})
exten => _X.,n,Goto(${${WHO}_CONTEXT},${EXTEN},1)

[international]
exten => _011.,1,NoOp()
exten => _011.,n,Set(E=${EXTEN})
exten => _011.,n,Goto(outgoing,call,1)

exten => i,1,NoOp()
exten => i,n,Playback(silence/2&sorry-cant-let-you-do-that2)
exten => i,n,Hangup()

include => longdistance

[longdistance]
exten => _1NXXNXXXXXX,1,NoOp()
exten => _1NXXNXXXXXX,n,Set(E=${EXTEN})
exten => _1NXXNXXXXXX,n,Goto(outgoing,call,1)

exten => _NXXNXXXXXX,1,Goto(1${EXTEN},1)

exten => i,1,NoOp()
exten => i,n,Playback(silence/2&sorry-cant-let-you-do-that2)
exten => i,n,Hangup()

include => local

[local]
exten => _416NXXXXXX,1,NoOp()
exten => _416NXXXXXX,n,Set(E=${EXTEN})
exten => _416NXXXXXX,n,Goto(outgoing,call,1)

exten => i,1,NoOp()
exten => i,n,Playback(silence/2&sorry-cant-let-you-do-that2)
exten => i,n,Hangup()
```

If the call is allowed to be executed, then the call is sent to the [outgoing] context for call processing, where the caller ID name and number are set with the CALLERID()

function. The call is then placed via the SIP channel using the `service_provider` we created in the *sip.conf* file:

```
[outgoing]
exten => call,1,NoOp()
exten => call,n,Set(CALLERID(name)=${${WHO}_CID_NAME})
exten => call,n,Set(CALLERID(number)=${${WHO}_CID_NUMBER})
exten => call,n,Dial(SIP/service_provider/${E})
exten => call,n,Playback(silence/2&pls-try-call-later)
exten => call,n,Hangup()
```

Our `service_provider` might look something like this in *sip.conf*:

```
[service_provider]
type=friend
host=switch1.service_provider.net
username=my_username
fromuser=my_username
secret=welcome
context=incoming
canreinvite=no
disallow=all
allow=ulaw
```

And that's it! The complete dialplan utilized for the hot-desk feature is displayed in full in Appendix G.

How many things have you just thought of that you could apply *func_odbc* to? See why we're so excited about this feature as well?!

The func_odbc Backport

With Asterisk 1.4, you can use a backported version of *func_odbc* that uses a slightly different configuration format. This allows you to use multiple DSN connections to separate databases, and also to utilize the `${ODBCROWS}` channel variable for SQL `read` (`SELECT`) queries. To check out the backport of *func_odbc* and install it (which overwrites the existing *func_odbc.c* file):

```
# cd /usr/src/
# svn co http://svncommunity.digium.com/svn/func_odbc/1.4 ./func_odbc-1.4
# cp func_odbc-1.4/func_odbc.c ./asterisk-1.4/funcs
# cp: overwrite `./asterisk-1.4/funcs/func_odbc.c'? y
# cd asterisk-1.4
# make install
```

The version described in this chapter works with a stock Asterisk 1.4 system, but the backport (and Asterisk 1.6) will use the following syntax changes:

- `read` will become `readsql`
- `write` will become `writesql`
- `dsn` will become `readhandle` and `writehandle` (for separate read and write database locations)

- Multiple `readhandles` and `writehandles` can be listed, in order of preference, to perform failover when the primary handle cannot be contacted (limit of 5)
- `prefix` will remain unchanged

The current syntax in Asterisk 1.4 will continue to work in the backport, but will display a deprecation warning in the version to follow. Support will eventually be removed.

ODBC Voicemail

Asterisk contains the ability to store voicemail inside the database using the ODBC connector. This is useful in a clustered environment where you want to abstract the voicemail data from the local system so that multiple Asterisk boxes have access to the same data. Of course, you have to take into consideration that you are centralizing a part of Asterisk, and you need to take actions to protect that data, such as regular backups, and possibly clustering the database backend using replication. If you are using PostgreSQL, there are some good projects for doing this: PGcluster (*http:// pgfoundry.org/projects/pgcluster/*) and Slony-I (*http://gborg.postgresql.org/project/slo ny1/projdisplay.php*).

Asterisk stores the voicemail inside a Binary Large Object (BLOB). When retrieving the data, it pulls the information out of the BLOB and temporarily stores it on the hard drive while it is being played back to the user. Asterisk then removes the BLOB and records from the database when the user deletes the voicemail. Many databases, such as MySQL, contain native support for BLOBs, but PostgreSQL has a couple of extra steps required to utilize this functionality that we'll explore in this section. When you're done, you'll be able to record, play back, and delete voicemail data from the database just as if it were stored on the local hard drive.

 This section builds upon previous configuration sections in this chapter. If you have not already done so, be sure to follow the steps in the "Installing the Database" and "Installing and Configuring ODBC" sections before continuing. In the "Installing and Configuring ODBC" section, be sure you have enabled ODBC_STORAGE in the menuselect system under Voicemail Options.

Creating the Large Object Type

We have to tell PostgreSQL how to handle the large objects. This includes creating a trigger to clean up the data when we delete a record from the database that references a large object.

Connect to the database as the *asterisk* user from the console:

```
# psql -h localhost -U asterisk asterisk
Password:
```

At the PostgreSQL console, run the following script to create the large object type:

```
CREATE FUNCTION loin (cstring) RETURNS lo AS 'oidin' LANGUAGE internal IMMUTABLE STRICT;
CREATE FUNCTION loout (lo) RETURNS cstring AS 'oidout' LANGUAGE internal
IMMUTABLE STRICT;
CREATE FUNCTION lorecv (internal) RETURNS lo AS 'oidrecv' LANGUAGE internal
IMMUTABLE STRICT;
CREATE FUNCTION losend (lo) RETURNS bytea AS 'oidrecv' LANGUAGE internal
IMMUTABLE STRICT;

CREATE TYPE lo ( INPUT = loin, OUTPUT = loout, RECEIVE = lorecv, SEND = losend,
INTERNALLENGTH = 4, PASSEDBYVALUE );
CREATE CAST (lo AS oid) WITHOUT FUNCTION AS IMPLICIT;
CREATE CAST (oid AS lo) WITHOUT FUNCTION AS IMPLICIT;
```

We'll be making use of the PostgreSQL procedural language called pgSQL/PL to create
a function. This function will be called from a trigger that gets executed whenever we
modify or delete a record from the table used to store voicemail. This is so the data is
cleaned up and not left as an orphan in the database:

```
CREATE FUNCTION vm_lo_cleanup() RETURNS "trigger"
    AS $$
    declare
      msgcount INTEGER;
    begin
      -- raise notice 'Starting lo_cleanup function for large object with oid
        %',old.recording;
      -- If it is an update action but the BLOB (lo) field was not changed,
        dont do anything
      if (TG_OP = 'UPDATE') then
        if ((old.recording = new.recording) or (old.recording is NULL)) then
          raise notice 'Not cleaning up the large object table,
        as recording has not changed';
          return new;
        end if;
      end if;
      if (old.recording IS NOT NULL) then
        SELECT INTO msgcount COUNT(*) AS COUNT FROM voicemessages WHERE recording
        = old.recording;
        if (msgcount > 0) then
         raise notice 'Not deleting record from the large object table, as object is
          still referenced';
          return new;
        else
          perform lo_unlink(old.recording);
          if found then
            raise notice 'Cleaning up the large object table';
            return new;
          else
            raise exception 'Failed to cleanup the large object table';
            return old;
          end if;
        end if;
      else
       raise notice 'No need to cleanup the large object table, no recording on old row';
```

```
    return new;
  end if;
end$$
LANGUAGE plpgsql;
```

We're going to create a table called **voicemessages** where the voicemail information will be stored:

```
CREATE TABLE voicemessages
(
  uniqueid serial PRIMARY KEY,
  msgnum int4,
  dir varchar(80),
  context varchar(80),
  macrocontext varchar(80),
  callerid varchar(40),
  origtime varchar(40),
  duration varchar(20),
  mailboxuser varchar(80),
  mailboxcontext varchar(80),
  recording lo,
  label varchar(30),
  "read" bool DEFAULT false
);
```

And now we need to associate a trigger with our newly created table in order to perform cleanup whenever we make a change or deletion from the **voicemessages** table:

```
CREATE TRIGGER vm_cleanup AFTER DELETE OR UPDATE ON voicemessages FOR EACH ROW EXECUTE
PROCEDURE vm_lo_cleanup();
```

Configuring voicemail.conf for ODBC Storage

There isn't much to add to the *voicemail.conf* file to enable the ODBC voicemail storage. In fact, it's only three lines! Generally, you probably have multiple format types defined in the [**general**] section of *voicemail.conf*, however we need to set this to a single format. The wav49 format is a compressed WAV file format that should be playable on both Linux and Microsoft Windows desktops.

The **odbcstorage** option points at the name you defined in the *res_odbc.conf* file (if you've been following along in this chapter, then we called it *asterisk*). The **odbctable** option refers to the table where voicemail information should be stored. In the examples in this chapter we use the table named **voicemessages**:

```
[general]
format=wav49
odbcstorage=asterisk
odbctable=voicemessages
```

You may want to create a separate voicemail context, or you can utilize the default voicemail context:

```
[default]
1000 => 1000,J.P. Wiser
```

Now connect to your Asterisk console and unload then reload the *app_voicemail.so* module:

```
*CLI> module unload app_voicemail.so
   == Unregistered application 'VoiceMail'
   == Unregistered application 'VoiceMailMain'
   == Unregistered application 'MailboxExists'
   == Unregistered application 'VMAuthenticate'

*CLI> module load app_voicemail.so
 Loaded /usr/lib/asterisk/modules/app_voicemail.so => (Comedian Mail (Voicemail System))
   == Registered application 'VoiceMail'
   == Registered application 'VoiceMailMain'
   == Registered application 'MailboxExists'
   == Registered application 'VMAuthenticate'
   == Parsing '/etc/asterisk/voicemail.conf': Found
```

And verify that your new mailbox loaded successfully:

```
*CLI> voicemail show users for default
Context    Mbox  User                      Zone      NewMsg
default    1000  J.P. Wiser                              0
```

Testing ODBC Voicemail

Let's create some simple dialplan logic to leave and retrieve some voicemail from our test voicemail box. We can use the simple dialplan logic as follows:

```
[odbc_vm_test]
exten => 100,1,Voicemail(1000@default)     ; leave a voicemail
exten => 200,1,VoicemailMain(1000@default) ; retrieve a voicemail
```

Once you've updated your *extensions.conf* file, be sure to reload the dialplan:

```
*CLI> dialplan reload
```

You can either include the odbc_vm_test context into a context accessible by an existing user, or create a separate user to test with. If you wish to do the latter, you could define a new SIP user in *sip.conf* like so (this will work assuming the phone is on the local LAN):

```
[odbc_test_user]
type=friend
secret=supersecret
context=odbc_vm_test
host=dynamic
qualify=yes
disallow=all
allow=ulaw
allow=gsm
```

Don't forget to reload the SIP module:

```
*CLI> module reload chan_sip.so
```

And verify that the SIP user exists:

```
*CLI> sip show users like odbc_test_user
Username                Secret          Accountcode    Def.Context      ACL  NAT
odbc_test_user          supersecret                    odbc_vm_test     No   RFC3581
```

Then configure your phone or client with the username *odbc_test_user* and password *supersecret*, and then place a call to extension 100 to leave a voicemail. If successful, you should see something like:

```
-- Executing VoiceMail("SIP/odbc_test_user-10228cac", "1000@default") in new stack
-- Playing 'vm-intro' (language 'en')
-- Playing 'beep' (language 'en')
-- Recording the message
-- x=0, open writing: /var/spool/asterisk/voicemail/default/1000/tmp/dlZunm format:
   wav49, 0x101f6534
-- User ended message by pressing #
-- Playing 'auth-thankyou' (language 'en')
== Parsing '/var/spool/asterisk/voicemail/default/1000/INBOX/msg0000.txt': Found
```

We can now make use of the psql application again to make sure the recording really did make it into the database:

```
# psql -h localhost -U asterisk asterisk
Password:
```

Then run a SELECT statement to verify that you have some data in the voicemessages table:

```
localhost=# SELECT id,dir,callerid,mailboxcontext,recording FROM voicemessages;
id | dir                                        | callerid    | mailboxcontext | recording
---+--------------------------------------------+-------------+----------------+-------
 1 | /var/spool/asterisk/voicemail/default/1000/INBOX | +18005551212 | default        | 47395
(1 row)
```

If the recording was placed in the database, we should get a row back. You'll notice that the recording column contains a number (which will most certainly be different from that listed here), which is really the object ID of the large object stored in a system table. Let's verify that the large object exists in this system table with the lo_list command:

```
localhost=# \lo_list
   Large objects
  ID    | Description
--------+-------------
 47395  |
(1 row)
```

What we're verifying is that the object ID in the voicemessages table matches that listed in the large object system table. We can also pull the data out of the database and store it to the hard drive so we can play the file back to make sure our message was saved correctly:

```
localhost=# \lo_export 47395 /tmp/voicemail-47395.wav
lo_export
```

Then verify the audio with your favorite audio application, such as the **play** application:

```
# play /tmp/voicemail-47395.wav

Input Filename : /tmp/voicemail-47395.wav
Sample Size    : 8-bits
Sample Encoding: wav
Channels       : 1
Sample Rate    : 8000

Time: 00:06.22 [00:00.00] of 00:00.00 (  0.0%) Output Buffer: 298.36K

Done.
```

And now that we've confirmed everything was stored in the database correctly, we can try listening to it via the **VoicemailMain()** application by dialing extension 200:

```
*CLI>
    -- Executing VoiceMailMain("SIP/odbc_test_user-10228cac", "1000@default") in new stack
    -- Playing 'vm-password' (language 'en')
    -- Playing 'vm-youhave' (language 'en')
    -- Playing 'digits/1' (language 'en')
    -- Playing 'vm-INBOX' (language 'en')
    -- Playing 'vm-message' (language 'en')
    -- Playing 'vm-onefor' (language 'en')
    -- Playing 'vm-INBOX' (language 'en')
    -- Playing 'vm-messages' (language 'en')
    -- Playing 'vm-opts' (language 'en')
    -- Playing 'vm-first' (language 'en')
    -- Playing 'vm-message' (language 'en')
    == Parsing '/var/spool/asterisk/voicemail/default/1000/INBOX/msg0000.txt': Found
```

Conclusion

In this chapter, we learned about several areas where Asterisk can integrate with a relational database. This is useful for systems where you need to start scaling by clustering multiple Asterisk boxes working with the same centralized information, or when you want to start building external applications to modify information without requiring a reload of the system (i.e., not requiring the modification of flatfiles).

## CHAPTER 13

Managing Your Asterisk System

It won't be covered in the book. The source code has to be useful for something,
after all.

—Larry Wall

While there is a cornucopia of creative things that you are going to want to do with your spanking-new Asterisk system, there are also some basic, unglamorous, dare we say, boring things that need to be discussed.

Call Detail Recording

Without even being told, Asterisk assumes that you want to store CDR information.[*]

By default, Asterisk will create a CSV file and place it in the folder */var/log/asterisk/cdr-csv/*.[†] To the naked eye, this file looks like a bit of a mess. If, however, you separate each line according to the commas, you will find that each line contains information about a particular call, and that the commas separate the following values:

accountcode
> Assigned if configured for the channel in the channel configuration file (i.e., *sip.conf*). The account code is assigned on a per-channel basis. You can also change this value from the dialplan by setting CDR(accountcode).

src
> Received Caller ID (string, 80 characters).

---

[*] If you are wondering why such an obviously simple thing seems to be such an achievement, the reason is simply that many traditional PBXes do not have this capability built in. With those systems, you have to purchase some sort of third-party appliance even just to capture the raw call data. Asterisk simply stores it. No drama. No cost. No kidding.

[†] A Comma Separated Values (CSV) file is a common method of formatting database-type information in a text file. You can open CSV files with a text editor, but most spreadsheet and database programs will also read them and properly parse them into rows and columns.

dst
> Destination extension.

dcontext
> Destination context.

clid
> Caller ID with text (80 characters).

channel
> Channel used (80 characters).

dstchannel
> Destination channel, if appropriate (80 characters).

lastapp
> Last application, if appropriate (80 characters).

lastdata
> Last application data (arguments, 80 characters).

start
> Start of call (date/time).

answer
> Answer of call (date/time).

end
> End of call (date/time).

duration
> Total time in system, in seconds (integer), from dial to hangup.

billsec
> Total time call is up, in seconds (integer), from answer to hangup.

disposition
> What happened to the call (ANSWERED, NO ANSWER, BUSY).

amaflags
> What flags to use (DOCUMENTATION, BILL, IGNORE, etc.), specified on a per-channel basis, like *accountcode*. AMA flags stand for Automated Message Accounting flags, which are somewhat standard (supposedly) in the industry.

userfield
> A user-defined field, maximum 255 characters.

Storing CDRs in a Database

CDRs can also be stored in a database. Asterisk currently supports SQLite, PostgreSQL, MySQL, and unixODBC, but we will cover only ODBC in this book (see Chapter 12). Many people prefer to store their CDRs in a database, so that queries can be run to help with billing and resource management.

Managing Logs

Asterisk activity generates events that will cause the creation of an entry in either the main system logs, or in Asterisk's own logfiles. On a busy system (or a system that is experiencing a severe problem), these logfiles can grow very large, very quickly. If debugging is turned on, the processes involved in writing to these logfiles can begin to have an effect on system performance. By default, Asterisk will simply add to the files until the hard drive is full. Fortunately, Linux provides utilities to handle the rotation of logfiles (so that no single file becomes too large), and also the deletion of older logfiles (which will prevent the system from getting clogged with logfiles).

The *logrotate* utility is normally run once per day by the operating system. Unfortunately, since there is no script installed to instruct *logrotate* on how to handle Asterisk, its logfiles will grow unchecked until a rotate script is added to handle them. In order to make that happen, we need to set up parameters for Asterisk in a file in the */etc/logrotate.d* directory. This file will need to rotate the current logfile, and then send Asterisk instructions to rotate its own logger (causing it to stop using the now old logfile, and generate a new file).

Create a new file */etc/logrotate.d/asterisk* and place the following lines in it:

```
/var/log/asterisk/* /var/log/asterisk/cdr-csv {
missingok
sharedscripts
monthly
rotate 12
postrotate
    asterisk -rx "logger rotate" > /dev/null 2> /dev/null
endscript
}
```

This file tells the *logrotate* utility to rotate the Asterisk logs every month, save 12 months worth of logs, and then tell Asterisk that the logfiles have been rotated (which will cause Asterisk to create new logfiles and begin writing to them). We selected these values arbitrarily. Feel free to adjust them to suit your needs.

Running Asterisk As a Non-root User

By default, Asterisk runs as the *root* user, and while we don't have any hard data, our own experiences lead us to conclude that the vast majority of Asterisk systems are run in this default state. From a security perspective, this represents an unacceptable risk–strangely, one which most of us seem willing to take.

Running Asterisk as non-*root* is not terribly hard to achieve, but it requires a few extra steps, and debugging it can be frustrating if you do not understand how Linux permissions work. However, from a security perspective it is well worth the effort.

We're going to run Asterisk as the user *asterisk*, so we need to create that user on our system first. The following commands will be run as *root*. We'll tell you when to switch and use the *asterisk* user that we're about to create:

```
# adduser -c "Asterisk PBX" asterisk
# passwd asterisk
```

Now that you've created the *asterisk* user, let's switch to that user, with which we'll perform the rest of the commands. Once we **su** to the *asterisk* user,[‡] we can download a copy of Asterisk via SVN, FTP or WGET, and then compile and install. We're going to grab a copy of Asterisk from the SVN repository in the following example.

 1.4.5 is the current release version at the time of this writing, but it won't be by the time you read this, so check the Asterisk web site for the latest version. In other words, don't just type 1.4.5 whenever you see us refer to it. Find out what is current and use that instead.

```
# su - asterisk
$ svn co http://svn.digium.com/svn/asterisk/tags/1.4.5 asterisk-1.4.5
$ cd asterisk-1.4.5
$ ./configure --prefix=$HOME/asterisk-bin --sysconfdir=$HOME/asterisk-bin
  --localstatedir=$HOME/asterisk-bin
$ make menuselect
$ make install
```

When running the *./configure* script with the `--prefix` flag, we're telling the system to install the binary components into our `$HOME`[§] directory under the subdirectory called *asterisk-bin*. The `--sysconfdir` flag tells the system where to place the configuration files, and `--localstatedir` tells the system where to install additional files, such as sounds. The key here is that since we are downloading, compiling, and installing as the user *asterisk*, everything that only gets created will be assigned to that user, and have the permissions granted to that user.

We can now install the sample files as well into the *$HOME/asterisk-bin/asterisk* directory:

```
$ make samples
```

Test starting up Asterisk with the following command:

```
$ ./asterisk-bin/sbin/asterisk -cvvv
```

Normally, Asterisk needs to be run as a service. During installation, the `make config` command will install the init scripts. Unfortunately, this will not work when you are logged in as the user *asterisk*, because only the *root* user has the authority to make

---

‡ **su** historically means super-user, but nowadays it could also mean switch-user or substitute-user. The - in the command tells **su** to use the environment for that user (for example to use the `PATH` for that user)

§ `$HOME` is a system variable that defines the path to the home directory for the current user, i.e., */home/asterisk*.

changes to system startup commands. It would appear that what we need to do is log in as *root*, navigate to the */home/asterisk/asterisk-1.4.5* folder, and run the `make config` command again (now with the authority to really make it happen). Problem solved, right?

Yes, but not quite. If you run the `service asterisk start` command, you will find that it complains that it cannot find *asterisk*. Know why? Because the init script figures the *asterisk* executable got installed in */usr/sbin*, where it would be if we had installed *asterisk* as *root*. So, we need to tell the init script where to find *asterisk* and the *safe_asterisk* script, like this:

```
# ln -s /home/asterisk/asterisk-bin/sbin/asterisk /usr/sbin/asterisk
# ln -s /home/asterisk/asterisk-bin/sbin/safe_asterisk /usr/sbin/safe_asterisk
```

Since the init script utilizes the *safe_asterisk* script, and by default wants to start Asterisk as the root user, we have to modify the *safe_asterisk* script telling it to run Asterisk as our non-root user. So open up the *safe_asterisk* script with your favorite text editor and look for the `ASTARGS` variable (around line 78). Then add **-U asterisk** between the quotes like so:

```
#
# Don't fork when running "safely"
#
ASTARGS="-U asterisk"
```

Go ahead and start Asterisk by running `service asterisk start` and verify Asterisk is running as the `asterisk` user using the `ps` command:

```
# service asterisk start
# ps aux | grep asterisk

503     30659 0.0 1.8  26036 8692 pts/2   Sl   15:07   0:00
/home/asterisk/asterisk-bin/sbin/asterisk -U asterisk -vvvg -c
```

The 503 is actually our *asterisk* user, which we verify by looking at the */etc/passwd* file:

```
# cat /etc/passwd

asterisk:x:503:503:Asterisk PBX:/home/asterisk:/bin/bash
```

Reboot the system to ensure that everything comes up as required. Keep in mind that a lot of things that you do with Asterisk might assume that you are running as *root*, so keep an eye out for errors that relate to a lack of permission. Your Asterisk process may think it is the superuser, but we have clipped its wings somewhat.

Why go through the trouble? The advantage of this is simply that if any security vulnerability in Asterisk[||] allows someone to access the box through the Asterisk account,

---

[||] If you walk up to any system that's running Asterisk, hook a keyboard and screen up to it, and press Alt-F9; you will be connected to the Asterisk CLI. Press ! and hit Return, and you will have a shell. If Asterisk is running as *root*, you now own that system.

he will be limited to system activities allowed by that account. When Asterisk is run as *root*, a security compromise gives the intruder full control of your system.

Customizing System Prompts

In keeping with the seemingly limitless flexibility of Asterisk, you can also modify the system prompts. This is very simple to explain, but generally difficult to do well.

With more than 300 system prompts in the main distribution, and an additional 600 in the *asterisk-sounds* add-on, if you're contemplating customizing all of them you'd better have either a lot of money or a lot of time on your hands.

An audio engineer is also recommended to ensure that the recordings are normalized to −3 dB and that all prompts start and end at a zero-crossing point (with just the right amount of silence prepended and appended).

The Voice

If you are interested in The Voice of Asterisk, she is Allison Smith, and she can deliver customized recordings for you to use on your own system.

This is a powerful concept, as very few PBXes allow you to use the same voice in your custom recordings as is used by the system prompts.

Allison is the voice of the system prompts in both English and Spanish.

There are also prompts available in French. Montreal's own June Wallack is the voice of the French Asterisk prompts (and also does prompts in flawless English,[#] should you want the same voice for both languages).

To find out how to get your own voice prompts, visit the Digium web site, *http://www.digium.com/products/voice*.

Once you have the recordings, the actual implementation is easy—simply replace the files in */var/lib/asterisk/sounds* with the ones you have created.

Alternatively, you can opt to record your own prompts and place them in a folder of your choosing. When you refer to sound files with the `Playback()` or `Background()` applications, you can refer to the full pathname of the sound file, or to any subdirectory of */var/lib/asterisk/sounds/*.

Note that the default sounds that come with Asterisk are delivered in format. We would not normally recommend storing them in this format (unless you have a lot of channels

---

[#] We were going to say "accentless English", but then we'd have to apologize to folks from the British Isles, Australia, South Africa, and who knows where else. We are not experts in languages, dialects, and such, but our ears tell us that there is a type of accent that in North America is common for professional voices. This is the accent that is common to the Pacific coast from San Diego to Seattle, and most of English-speaking Canada as well. Both June and Allison deliver English prompts in this accent, and we think it sounds great.

that will be entering the system using the GSM codec). Sure, you save some hard drive space, but the extra load on your CPU when it has to transcode all these files (not to mention the lower overall quality of the sound) makes the use of GSM undesirable, to our thinking. Use uncompressed files (such as *.wav*, *.ulaw* or *.alaw*) and your CPU will not have to work as hard. As an added bonus, your prompts will sound better.

Sound Recording from the Dialplan

Surprisingly, one of the easiest ways to get respectable-quality recordings is not through a PC with fancy editing software, but rather through a telephone set. There are many reasons for this, but the most important is that the telephone will tend to filter out background noise (such as white noise caused by HVAC equipment) and will record at a consistent audio level.

This little addition to your dialplan will allow you to easily create recordings, which will be placed in your system's */tmp/* folder (from there, you can rename them and move them wherever you'd like):

```
exten => _66XX,1,Wait(2)
exten => _66XX,n,Record(/tmp/prompt${EXTEN:2}:wav)
exten => _66XX,n,Wait(1)
exten => _66XX,n,Playback(/tmp/prompt${EXTEN:2})
exten => _66XX,n,Wait(2)
exten => _66XX,n,Hangup()
```

This snippet will allow you to dial from 6600 to 6699, and it will record prompts in the */tmp/* folder using the names *prompt00.wav* to *prompt99.wav*. After you complete recording (by pressing the # key), it will play your prompt back to you and hang up.

Be sure to move your prompts out of the */tmp/* dir to the Asterisk sounds directory. To keep the dialplan readable, rename your *prompt<XX>* files to something more meaningful. For example:

```
mv /tmp/prompt00.wav /var/lib/asterisk/sounds/custom/welcome-message.wav
```

Music on Hold

Any popular PBX system offers the ability to supply a source of music to be played for callers while on hold. Asterisk allows for a lot of creativity in this regard.

Nowadays, everyone is familiar with the MP3 music format, and there is a lot of interest in using MP3s as a music-on-hold source. The concept sure seems like a good idea, but there are a few things that we think should be given some consideration:

- MP3 files are extremely complex, and require a substantial amount of CPU to decode. If you have a lot of channels pulling music from the system (for example, people sometimes like to listen to music through their phone, or a call center may have several callers on hold), the load on the CPU caused by all of the transcoding of the stored MP3 files could place too much demand on a machine that is otherwise suitable to the performance needs of the system.

- Current-generation hard drives hold a lot of data, so there may not be any reason to worry about cutting down hard drive use. Compressed audio makes sense from a distribution standpoint (an MP3 is a much smaller download than the equivalent in .wav format), but once on your system, do we really care how much space they take up?

- MP3 files don't usually come with the right sort of licensing. ;-)

Taking all of this into consideration, we recommend that you convert your music sources into the native format of the various codecs you may be supporting. For example, if you support μlaw for your internal phones, and G.729 on your VoIP circuits, you will want to store your music in both formats so that Asterisk will not have to perform transcoding to play music to calls on those channels.

Free Music

A lot of people do not realize that playing music on hold requires a special license. This is true even if you play music from CDs that you own, or from the radio. To ensure that there is no ambiguity, we recommend avoiding the whole matter and using only music that does not come encumbered with the kind of licenses that the music industry seems to prefer.

There are many web sites where you can go to get music that is licensed in a manner that is suitable for music on hold. Two that we have found are *http://en.wikipedia.org/wiki/Wikipedia:Sound/list* and *http://www.opsound.org/*.

Both offer a sizeable collection of music that can be easily downloaded. Note that this may not all be of professional quality, so listen to all of it before you commit it to your music-on-hold collection.[*]

We often use public domain music (or Creative Commons licensed music) on our systems. Creative Commons music often comes in ogg-Vorbis format (which is conceptually similar to MP3, but not compatible). In order to play *.ogg* or *.mp3* files on our Asterisk system, we are going to convert them to a format that Asterisk can easily handle. This requires the following steps:

1. We need to make sure that SoX, the Sound eXchange utility, is installed. If not, run the following command to install it:

---

[*] Seed a search with the term "Creative Commons music" to find more freely usable music.

```
$ yum install sox
```

2. Download the music that you have chosen to a working folder on your system (*/tmp* is probably a suitable location). As an example, the following command downloaded some nice piano music by Pachelbel for us:

```
$ wget http://upload.wikimedia.org/wikipedia/commons/6/62/Pachelbel%27s_Canon.ogg
```

3. Now we have to convert the song from ogg-Vorbis format to a format more suitable to Asterisk:

```
$ sox Pachelbel\'s_Canon.ogg -r 8000 -c 1 -s -w moh1.wav resample -ql
```

 You may also need to adjust the amplitude with the -v option.

We've now taken our source file, converted it to a .wav file suitable to Asterisk,[†] and saved the resulting file as *moh1.wav*.

4. Almost done now. We just need to create a folder for the permanent home of the new files (*/tmp* is certainly no place for them):

```
$ mkdir /var/lib/asterisk/mohwav
```

and then move them there:

```
$ mv *.wav /var/lib/asterisk/mohwav
```

5. Since we have placed our music files in a different folder from that where Asterisk installs its sample music, we will need to change the configuration file to reflect this. Edit your */etc/asterisk/musiconhold.conf* file with one that contains the following:

```
[default]
mode=files
directory=/var/lib/asterisk/mohwav
random=yes
```

As for what to play, that will depend on what image you want to project to your callers. Regardless of your choice, you should keep some things in mind:

- People don't actually want to be on hold, so they are not usually planning to be there for long. This means that there is not much point in providing them with a mind-expanding musical experience. If things go as they hope, they won't be there long enough to get into it.

---

† Note that we could have used any format that was compatible with Asterisk; we've just chosen .wav for this example because it is easy for the CPU to transcode into μlaw/alaw/slin on the fly, yet remains easy to work with in other environments.

- The fidelity on a phone system does not allow for accurate reproduction of tones. Heavy bass generally sounds terrible, and high frequencies will typically just end up as noise. Keep the music simple, and it is more likely to sound good.
- Musical tastes are as varied as people, and while it might be nice to try and cover a wide range of styles, music that is too eclectic is more likely to annoy than enlighten.

Classical music addresses all of the above criteria, and it is easy to obtain. It also sounds classy (go figure!), so it is a pretty safe choice, although we'll admit it doesn't usually score any points in the hipness department.

Asterisk includes three songs with the source code download that are licensed for use with Asterisk. These songs are intended as samples. Since there are only three of them, people who call you regularly will quickly tire of them. We have a recurring nightmare in which the worldwide success of Asterisk means that the human race is forced to listen to the same three songs as music on hold. That is why we wrote this section for you.

Randomizing Music on Hold

In a traditional PBX, music on hold usually comes from a single source. Everyone that is hearing music is hearing the exact same thing at the exact same time, and even when no one is on hold, the music is still playing. On Asterisk, the music is not playing until a need for it arises, and each caller gets her own music source. If Asterisk were to simply start playing songs in the order it found them, each call placed on hold would always hear the same song starting from the beginning. In order to simulate traditional music on hold, Asterisk can (and normally should) be set to play the music in a random fashion. This means that it will select which file to play at random. If you have enough different songs in your music-on-hold directory, you will minimize the chance that someone who calls frequently will have to listen to the same songs all of the time.

Conclusion

This chapter could easily become a book (and possibly will one day). We have chosen a few topics to cover that we think will provide value to most readers, but there are certainly many more topics that can be discussed. As with so many things in this book, we have merely scratched the surface.

## CHAPTER 14

Potpourri

The first 90 percent of the task takes 90 percent of the time, and the last 10 percent of the task takes the other 90 percent of the time.

—The Ninety:Ten Rule

The toughest part of writing this book was not finding things to write about, but rather deciding what we would not be able to write about. Now that we've covered the basics, you are ready to be told the truth: we have not taught you anywhere near all that there is to know about Asterisk.

Now please understand, this is not because we didn't want to give you our very best; it's merely because Asterisk is, well, limitless (or so we believe).

In this chapter, we want to give you a taste of some of the wonders Asterisk holds in store for you. Nearly every section in this chapter could become a book in itself (and they *will* become books, if Asterisk succeeds in the way we think it is going to).

Festival

Festival is a popular open source text-to-speech engine. The basic premise of using Festival with Asterisk is that your dialplan can pass a body of text to Festival, which will then "speak" the text to the caller. Probably the most obvious use for Festival would be to have it read your email to you when you are on the road.*

---

* Probably the coolest use of Festival is in Simon Ditner's ZoIP, a port of the famous Zork game to a fully speech-enabled engine running on Asterisk (ZoIP also uses Sphinx, which we will not be covering in this book). We're going to have to come up with a new kind of name for this sort of thing. It's not a video game, since there is no screen, so do we need to call these audio games? Regardless, check it out at *http://www.zoip.org*.

Getting Festival Set Up and Ready for Asterisk

There are currently two ways to use Festival with Asterisk. The first (and easiest) method—without having to patch and recompile Festival—is to add the following text to Festival's configuration file (*festival.scm*, usually located in */etc/* or */usr/share/festival/*):

```
(define (tts_textasterisk string mode)
"(tts_textasterisk STRING MODE)

Apply tts to STRING. This function is specifically designed for use in
server mode so a single function call may synthesize the string. This
function name may be added to the server safe functions."
(let ((wholeutt (utt.synth (eval (list 'Utterance 'Text string)))))
(utt.wave.resample wholeutt 8000)
(utt.wave.rescale wholeutt 5)
(utt.send.wave.client wholeutt)))
```

You may place this text anywhere in the file, as long as it is not between any other parentheses.

The second (and more traditional) way is to compile Festival with an Asterisk-specific patch (located in the *contrib/* directory of the Asterisk source).

Information on both of these methods is contained in the *README.festival* file, located in the *contrib/* directory of the Asterisk source.

For either method, you'll need to modify the Festival access list in the *festival.scm* file. Simply search for the word "localhost" and replace it with the fully qualified domain name of your server.

Both of these methods set up Festival to be able to correctly communicate with Asterisk. After setting up Festival, you should start the Festival server. You can then call the `Festival()` application from within your dialplan.

Configuring Asterisk for Festival

The Asterisk configuration file that deals with Festival is aptly called *festival.conf*. Inside this file, you specify the hostname and port of your Festival server, as well some settings for the caching of Festival speech. For most installations (if you're going to run Festival on your Asterisk server), the defaults will work just fine.

Starting the Festival Server

To start the Festival server for debugging purposes, simply run `festival` with the `--server` argument, like this:

```
[root@asterisk ~]# festival --server
```

Once you're sure that the Festival server is running and not rejecting your connections, you can start Festival by typing:

```
[root@asterisk ~]# festival_server 2>&1 >/dev/null &
```

Calling Festival from the Dialplan

Now that Festival is configured and the Festival server is started, let's call it from within a simple dialplan:

```
exten => 123,1,Answer()
exten => 123,2,Festival(Asterisk and Festival are working together)
```

 You should always call the Answer() application before calling Festival(), to ensure that a channel is established.

As Asterisk connects to Festival, you should see output like this in the terminal where you started the Festival server:

```
[root@asterisk ~]# festival --server
server    Sun May  1 18:38:51 2005 : Festival server started on port 1314
client(1) Sun May  1 18:39:20 2005 : accepted from asterisk.localdomain
client(1) Sun May  1 18:39:21 2005 : disconnected
```

If you see output like the following, it means you didn't add the host to the access list in *festival.scm*:

```
[root@asterisk ~]# festival --server
server    Sun May  1 18:30:52 2005 : Festival server started on port 1314
client(1) Sun May  1 18:32:32 2005 : rejected from asterisk.localdomain not
in access list
```

Yet Another Way to Use Festival with Asterisk

Some people in the Asterisk community have reported success with passing text to Festival's *text2wave* utility and then having Asterisk play back the resulting *.wav* file. For example, you might do something like this:

```
exten => 124,1,Answer()
exten => 124,2,System(echo "This is a test of Festival" | /usr/bin/text2wave
-scale 1.5 -F 8000 -o /tmp/festival.wav)
exten => 124,3,Playback(/tmp/festival)
exten => 124,4,System(rm /tmp/festival.wav)
exten => 124,5,Hangup()
```

This method also allows you to call other text-to-speech engines, such as the popular speech engine from Cepstral *http://www.cepstral.com*, which is an inexpensive commercial derivative of Festival with very good-sounding voices. For this example, we'll assume that Cepstral is installed in */usr/local/cepstral/*:

```
exten => 125,1,Answer()
exten => 125,2,System(/usr/local/cepstral/bin/swift -o /tmp/swift.wav
```

```
"This is a test of Cepstral")
exten => 125,3,Playback(/tmp/swift)
exten => 125,4,System(rm /tmp/swift.wav)
exten => 125,5,Hangup()
```

Call Files

Call files allow you to create calls through the Linux shell. These powerful events are triggered by depositing a *.call* file in the directory */var/spool/asterisk/outgoing/*. The actual name of the file does not matter, but it's good form to give the file a meaningful name and to end the filename with *.call*.

When a call file appears in the outgoing folder, Asterisk will almost immediatelyact on the instructions contained therein.[†]

Call files are formatted in the following manner. First, we define where we want to call:

```
Channel: channel
```

We can control how long to wait for a call to be answered (the default is 45 seconds), how long to wait between call retries, and the maximum number of retries. If MaxRetries is omitted, the call will be attempted only once:

```
WaitTime: number
RetryTime: number
MaxRetries: number
```

If the call is answered, we specify where to connect it here:

```
Context: context-name
Extension: ext
Priority: priority
```

Alternatively, we can specify a single application and pass arguments to it:

```
Application: Playback()
Data: hello-world
```

Next, we set the Caller ID of the outgoing call:

```
CallerID: Asterisk 800-555-1212
```

Then we set channel variables, as follows:

```
SetVar: john=Zap/1/5551212
SetVar: sally=SIP/1000
```

and add a CDR account code:

```
Account: documentation
```

[†] We're talking within milliseconds here. Don't believe us? Try it for yourself!

When you create a call file, do *not* do so from the spool directory. Asterisk monitors the spool aggressively and will try to grab your file before you've even finished writing it! Create call files in some other folder, make a copy in the same folder, and then mv the copy into the spool directory. Note that we said mv, not cp. This is important, because the way that Linux copies files means that the file appears in the destination folder before it is completely there. Contrast that with a mv operation, which will not allow the file to appear in the destination folder until the move operation is complete. If you copy, there is a very good chance that Asterisk will read the file before it is all there, which will cause unexpected results.

DUNDi

If there were any concerns that Mark Spencer was in danger of running out of good ideas, Distributed Universal Number Discovery (DUNDi) ought to lay such thoughts to rest. DUNDi is poised to be as revolutionary as Asterisk. The DUNDi web site (*http://www.dundi.com*) says it best: "DUNDi™ is a peer to peer system for locating Internet gateways to telephony services. Unlike traditional centralized services (such as the remarkably simple and concise *ENUM* standard; *http://www.faqs.org/rfc/rfc2916.txt*), DUNDi is fully distributed with no centralized authority whatsoever." DUNDi is somewhat of a routing protocol for VoIP.

How Does DUNDi Work?

Think of DUNDi as a large phone book that allows you to ask peers if they know of an alternative VoIP route to an extension number or PSTN telephone number.

For example, assume that you are connected to the *DUNDi-test* network (a free and open network that terminates calls to traditional PSTN numbers). You ask your friend Bob if he knows how to reach 1-212-555-1212, a number for which you have no direct access. Bob replies, "I don't know how to reach that number, but let me ask my peer Sally."

Bob asks Sally if she knows how to reach the requested number, and she responds with, "You can reach that number at IAX2/dundi:*very_long_password@hostname/extension*." Bob then stores the address in his database and passes on to you the information about how to reach 1-800-555-1212 via VoIP, allowing you an alternative method of reaching the same destination through a different network.

Because Bob has stored the information he found, he'll be able to provide it to any peers who later request the same number from him, so the lookup won't have to go any further. This helps reduce the load on the network and increases response times for numbers that are looked up often. (However, it should be noted that DUNDi creates a rotating key and, thus, stored information is valid for a limited period of time.)

DUNDi performs lookups dynamically, either with a `switch =>` statement in your *extensions.conf* file or with the use of the `DUNDiLookup()` application. DUNDi is available only in Asterisk version 1.2 or higher.

You can use the DUNDi protocol in a private network as well. Say you're the Asterisk administrator of a very large enterprise installation, and you wish to simplify the administration of extension numbers. You could use DUNDi in this situation, allowing multiple Asterisk boxes (presumably located at each of the company's locations and peered with one another) to perform dynamic lookups for the VoIP addresses of extensions on the network.

Configuring Asterisk for Use with DUNDi

There are three files that need to be configured for DUNDi: *dundi.conf*, *extensions.conf*, and *iax.conf*.[‡] The *dundi.conf* file controls the authentication of peers whom we allow to perform lookups through our system. This file also manages the list of peers to whom we might submit our own lookup requests. Since it is possible to run several different networks on the same box, it is necessary to define a different section for each peer, and then configure the networks in which that peer is allowed to perform lookups. Additionally, we need to define which peers we wish to use to perform lookups.

The General Peering Agreement

The General Peering Agreement (GPA) is a legally binding license agreement that is designed to prevent abuse of the DUNDi protocol. Before connecting to the *DUNDi-test* group, you are required to sign a GPA. The GPA is used to protect the members of the group and to create a "trust" between the members. It is a requirement of the *DUNDi-test* group that your complete and accurate contact information be configured in *dundi.conf*, so that members of your peer group can contact you. The GPA can be found in the *doc/* subdirectory of the Asterisk source.

General configuration

The [general] section of *dundi.conf* contains parameters relating to the overall operation of the DUNDi client and server:

```
; DUNDi configuration file
;
[general]
;
department=IT
organization= toronto.example.com
```

‡ The *dundi.conf* and *extensions.conf* files must be configured. We have chosen to configure *iax.conf* for our address advertisement on the network, but DUNDi is protocol-agnostic—thus *sip.conf*, *h323.conf*, or *mgcp.conf* could be used instead.

```
locality=Toronto
stateprov=ON
country=CA
email=support@toronto.example.com
phone=+19055551212
;
; Specify bind address and port number.  Default is 4520
;bindaddr=0.0.0.0
port=4520
entityid=FF:FF:FF:FF:FF:FF
ttl=32
autokill=yes
;secretpath=dundi
```

The entity identifier defined by `entityid` should generally be the Media Access Control (MAC) address of an interface in the machine. The entity ID defaults to the first Ethernet address of the server, but you can override this with `entityid`, as long as it is set to the MAC address of *something* you own. The MAC address of the primary external interface is recommended. This is the address that other peers will use to identify you.

The Time To Live (`ttl`) field defines how many peers away we wish to receive replies from and is used to break loops. Each time a request is passed on down the line because the requested number is not known, the value in the TTL field is decreased by one, much like the TTL field of an ICMP packet. The TTL field also defines the maximum number of seconds we are willing to wait for a reply.

When you request a number lookup, an initial query (called a `DPDISCOVER`) is sent to your peers requesting that number. If you do not receive an acknowledgment (`ACK`) of your query (`DPDISCOVER`) within 2,000 ms (enough time for a single transmission only) and `autokill` is equal to `yes`, Asterisk will send a `CANCEL` to the peers. (Note that an acknowledgment is not necessarily a reply to the query; it is just an acknowledgment that the peer has received the request.) The purpose of `autokill` is to keep the lookup from stalling due to hosts with high latency. In addition to the `yes` and `no` options, you may also specify the number of milliseconds to wait.

The *pbx_dundi* module creates a rotating key and stores it in the local Asterisk database (AstDB). The key name `secret` is stored in the `dundi` family. The value of the key can be viewed with the `database show` command at the Asterisk console. The database family can be overridden with the `secretpath` option.

Creating mapping contexts

The *dundi.conf* file defines DUNDi contexts that are mapped to dialplan contexts in your *extensions.conf* file. DUNDi contexts are a way of defining distinct and separate directory service groups. The contexts in the mapping section point to contexts in the *extensions.conf* file, which control the numbers that you advertise. When you create a peer, you need to define which mapping contexts you will allow this peer to search. You do this with the `permit` statement (each peer may contain multiple `permit` statements). Mapping contexts are related to dialplan contexts in the sense that they are a security boundary for your peers.

Phone numbers must be advertised in the following format:

```
<country_code><area_code><prefix><number>
```

For example, a complete North American number could be advertised as 14165551212.

All DUNDi mapping contexts take the form of:

```
dundi_context => local_context,weight,technology,destination[,options]]
```

The following configuration creates a DUNDi mapping context that we will use to advertise our local phone numbers to the *DUNDi-test* group. Note that this should all appear on one line:

```
dundi-test => dundi-local,0,IAX2,dundi:${SECRET}@toronto.example.com/
${NUMBER}, nounsolicited,nocomunsolicit,nopartial
```

In this example, the mapping context is **dundi-test**, which points to the **dundi-local** context within *extensions.conf* (providing a listing of phone numbers to reply to). Numbers that resolve to the PBX should be advertised with a *weight* of zero (directly connected). Numbers higher than 0 indicate an increased number of hops or paths to reach the final destination. This is useful when multiple replies for the same lookup are received at the end that initially requested the number; a *weight* with a lower number will be the preferred path.

If we can reply to a lookup, our response will contain the method by which the other end can connect to the system. This includes the technology to use (such as IAX2, SIP, H323, and so on), the username and password with which to authenticate, which host to send the authentication to, and finally the extension number.

Asterisk provides some shortcuts to allow us to create a "template" with which we can build our responses. The following channel variables can be used to construct the template:

${SECRET}
Replaced with the password stored in the local AstDB

${NUMBER}
The number being requested

${IPADDR}
The IP address to connect to

It is generally safest to statically configure the hostname, rather than making use of the ${IPADDR} variable. The ${IPADDR} variable will sometimes reply with an address in the private IP space, which is unreachable from the Internet.

Defining DUNDi peers

DUNDi peers are defined in the *dundi.conf* file. Peers are identified by the unique layer-two MAC address of an interface on the remote system. The *dundi.conf* file is where we define what context to search for peers requesting a lookup and which peers we want to use when doing a lookup for a particular network:

```
[00:00:00:00:00:00] ; Remote Office
model = symmetric
host = montreal.example.com
inkey = montreal
outkey = toronto
include = dundi-test
permit = dundi-test
qualify = yes
dynamic=yes
```

The remote peer's identifier (MAC address) is enclosed in square brackets ([ ]). The inkey and outkey are the public/private key pairs that we use for authentication. Key pairs are generated with the *astgenkey* script, located in the *./asterisk/contrib/scripts/* source directory. Be sure to use the -n flag so that you don't have to initialize passwords every time you start Asterisk:

```
# cd /var/lib/asterisk/keys
# /usr/src/asterisk/contrib/scripts/astgenkey -n toronto
```

The resulting keys, *toronto.pub* and *toronto.key*, will be placed in your */var/lib/asterisk/keys/* directory. The *toronto.pub* file is the public key, which you should post to a web server so that it is easily accessible for anyone with whom you wish to peer. When you peer, you can give your peers the HTTP-accessible public key, which they can then place in their */var/lib/asterisk/keys/* directories.

After you have downloaded the keys, you must reload the *res_crypto.so* and *pbx_dundi.so* modules in Asterisk:

```
*CLI> module reload res_crypto.so
   -- Reloading module 'res_crypto.so' (Cryptographic Digital Signatures)
   -- Loaded PRIVATE key 'toronto'
   -- Loaded PUBLIC key 'toronto'

*CLI> module reload pbx_dundi.so
   -- Reloading module 'pbx_dundi.so' (Distributed Universal Number
      Discovery
(DUNDi))

   == Parsing '/etc/asterisk/dundi.conf': Found
```

Then, create the dundi user in the *iax.conf* file to allow connections into your Asterisk system. When a call is authenticated, the extension number being requested is passed to the dundi-local context in the *extensions.conf* file, where the call is then handled by Asterisk.

Allowing remote connections

Here is the user definition for the `dundi` user:

```
[dundi]
type=user
dbsecret=dundi/secret
context=dundi-local
disallow=all
allow=ulaw
allow=g726
```

Instead of using a static password, Asterisk regenerates passwords every 3,600 seconds (1 hour). The value is stored in */dundi/secret* of the Asterisk database and advertised using the `${SECRET}` variable defined within the mapping context lines in *dundi.conf*. You can see the current keys for all peers, including your local public and private keys, by performing a `show keys` at the Asterisk CLI.

The `context` entry `dundi-local` is where authorized callers are sent in *extensions.conf*. From there, we can manipulate the call just as we would in the dialplan of any other incoming connection.

Configuring the dialplan

The *extensions.conf* file handles what numbers you advertise and what you do with the calls that connect to them. The `dundi-local` context performs double duty:

- It controls the numbers we advertise, referenced by the `dundi` mapping context in *dundi.conf*.

- It controls what to do with the call, referenced by the `dundi` user in *iax.conf*.

You have the power of dialplan pattern matching to advertise ranges of numbers and to control the incoming calls. In the following dialplan, we are only advertising the number +1-416-555-1212, but pattern matching could just as easily have been employed to advertise a range of numbers or extensions:

```
[dundi-local]
exten => 14165551212,1,NoOp(dundi-local: Number advertisement and incoming)
exten => 14165551212,n,Answer()
exten => 14165551212,n(call),Dial(SIP/1000)
exten => 14165551212,n,Voicemail(u1000)
exten => 14165551212,n,Hangup()
exten => 14165551212,n(call)+101,Voicemail(b1000)
exten => 14165551212,n,Hangup()
```

Alternative Voicemail Storage Methods

Asterisk's normal way of storing voicemail is to simply record the message in a file, which is placed on the local hard drive under the */var/spool/asterisk/voicemail* tree. While this works well enough for simple PBX deployments, there are more advanced ways of doing this that can be very useful in larger, distributed networks, or environments where tighter integration with external applications is desired.

Storing Voicemail in an IMAP Server

The ability to store voice messages in the same location as regular email is something that the telecom industry has been promising for a long time. They called it Unified Messaging, and while most PBXes now offer some sort of unified messaging, it is typically very expensive to license and implement.

Naturally, Asterisk cuts through all the silliness and just allows you to have your voicemailbox integrated into an IMAP environment. There are several advantages to storing your voicemail on an IMAP server. When you listen to a voicemail on your phone, the message is set to the *read* state on the IMAP server. This means that your email client will also note that it has been read. By the same token, if you listen to the message from your email client, the voicemail will turn off the message notification light on any phones that are assigned to that mailbox. Deleting a message from one place will cause it to be deleted from every place. So once deleted, the message is truly gone. This is Unified Messaging, the holy grail of voicemail to email integration, but Asterisk humbly decides not to make a big deal of it.

IMAP integration is still new functionality, so there are a few things that need to be added in order to get it to function. First off, Asterisk needs to have an IMAP client installed so that it can communicate with the IMAP server. Pretty much any IMAP server works (even Exchange Server), and the authors have personally tested IMAP voicemail support with both the Courier-IMAP and Dovecot IMAP servers. The IMAP server may be on the same physical machine as the Asterisk installation, or it may be on the other side of the globe. To be able to access the IMAP server, Asterisk requires an IMAP client library. This library is the University of Washington's free IMAP client, named *c-client*. To install the *c-client* you simply need to navigate to your */usr/src* directory and run the following commands:

```
# wget ftp://ftp.cac.washington.edu/mail/imap.tar.Z
```

This downloads the source code. Extract it with:

```
# tar zxvf imap.tar.Z
```

 You'll want to pay special attention to the name of the directory that is created by this command, as the directory name will probably change again by the time you read this. During the production of this book, the directory name has changed four times. The last time we checked, it was named */usr/src/imap-2006h*.

Navigate into the resulting folder and run:

```
# make lrh IP6=4
```

This will compile everything Asterisk needs to make use of the IMAP client libraries.[§]

Now we have to recompile Asterisk with the IMAP capabilities. We'll need to navigate to the location of our Asterisk source files (such as */usr/src/asterisk*), and run the following command:

```
# /configure --with-imap=/usr/src/imap-2006h
```

The we need to rerun `make menuconfig` to incorporate IMAP storage into the compile. Under Voicemail Build Options select the `IMAP_STORAGE` parameter, and then press x to save and exit. This ensures that when we compile Asterisk, it will build the IMAP module as well. Obviously, the next step then is to recompile and reinstall Asterisk. A simple way to do this is, in your terminal, to run:

```
# make && make install
```

OK, so we've got the module compiled and installed. Now it's time to configure it.

In the */etc/asterisk* folder, we'll need to add a few lines to the *voicemail.conf* file, in the [general] section:

```
imapserver=localhost
imapport=143
expungeonhangup=yes
authuser=vmail
authpassword=vmailsecret
imapfolder=Voicemail
```

Since Dovecot is available in the CentOS package repository, installing a small IMAP server to handle your virtual (voicemail) users on your Asterisk box is simple:

```
# yum install dovecot
```

Now make sure that IMAP support is enabled in /etc/dovecot.conf by uncommenting the protocols line so that it appears as follows:

```
protocols = imap imaps
```

After you've enabled IMAP support, create the user account that will be storing the mail:

```
# groupadd vmail
# useradd vmail -g vmail -s /bin/true -c "asterisk voicemail user" -p vmailsecret
  -d /var/spool/asterisk/imap-voicemail vmail
# chown -R vmail.vmail /var/spool/asterisk/imap-voicemail
```

Now restart Dovecot and Asterisk, and you should be good to go.

```
# service dovecot restart
# service asterisk restart
```

[§] The 1rh option tells the compiler that this is a Linux Red Hat system. The IP6=4 option tells the compiler that we don't want to compile in support for IPV6. Read the *Makefile* for other options. For RHEL 5 or CentOS 5, you should use 1r5 instead of 1rh.

Congratulations! You've successfully installed basic IMAP voicemail support with Asterisk! This is just the tip of the iceberg, though. With IMAP voicemail storage, it is easy to implement shared (e.g., departmental) voicemail using shared IMAP folders. Many companies already have departmental email, so having a shared voicemail box is a very natural and logical progression of the technology. With IMAP voicemail storage, each employee can manage several voicemail boxes without becoming confused as to whether a particular voicemail message is for them personally or for a department to which they belong. There is nothing unusual to configure from Asterisk's point of view; you simply call the VoiceMail() application with the desired mailbox and context, and make sure that the department employees have the shared IMAP folder included in their email client's folder list.

Finally, you may want to use per-mailbox authorization (i.e., each voicemail box authenticates as a specific user) instead of a global Asterisk IMAP user. Asterisk supports this through the imapuser and imappassword options in the individual voicemail box definition entries:

```
[imapvoicemail]
100 => 1234,Sue's Mailbox,,,imapuser=sue@example.tld|imapsecret=suesimapsecret
101 => 5555,Bob's Mailbox,,,imapuser=bob@example.tld|imapsecret=bobsimapsecret
```

In this particular example, if a message is left in IMAP mailbox 100 in the imapvoicemail context, Asterisk will authenticate to the IMAP server as sue@example.tld, using suesimapsecret as the password. Similarly, bob@example.tld/bobsimapsecret will be used to authenticate if a message is left in mailbox 101 of the same voicemail context.

Storing Voicemail in an ODBC Database

In case you missed it, you can also store voicemail in a database via the ODBC connector. See Chapter 12 for details!

Asterisk and Jabber (XMPP)

The name Jabber is actually the original name for the IETF XMPP protocols (RFC 3920-3923). Since Jabber is by far a better name than XMPP, the original name has stuck. This protocol was originally designed to be a decentralized, nonproprietary, open-standards messaging and presence framework. It supports offline message delivery and encryption, and has grown to include voice messaging, which Asterisk supports.

It is interesting to note that in the beginning, Jabber was seen as a competitor to the SIMPLE protocol, which is SIP-based. XMPP is designed as a more general protocol, and is of course XML-based.

Asterisk can be configured to utilize XMPP in several capacities. It may utilize XMPP as a presence framework (e.g., Extension 205 is away or on the phone), or, through the voice messaging framework JINGLE, support full voice communications with other services such as Google Talk.

Unlike other messaging networks such as MSN and Yahoo!, XMPP is decentralized. Anyone may have his own Jabber server and run any number of services on his server. You send messages in much the same way that you send email: the Jabber server you use contacts the Jabber server of the other person, and a direct connection is established. If the other person is not online, the message is stored, and when she log in to her Jabber server, any stored messages are delivered. With encryption (the XMPP protocol supports TLS), it is no wonder why many businesses are starting to implement their internal messaging networks using this amazing protocol, and Asterisk is able to seamlessly integrate into this communications network.

Conclusion

That's pretty much all that this chapter is going to teach you, but it's nowhere near all there is to learn. Hopefully, you are starting to get an idea of how big this Asterisk thing really is.

In the next chapter, we're going to try and predict the future of telecom, and we'll discuss how (and why) we believe that Asterisk is well positioned to play a starring role.

## CHAPTER 15

Asterisk: The Future of Telephony

First they ignore you, then they laugh at you, then they fight you, then you win.

—Mahatma Gandhi

We have arrived at the final chapter of this book. We've covered a lot, but we hope that you now realize that we have barely begun to scratch the surface of this phenomenon called Asterisk. To wrap things up, we want to spend some time exploring what we might see from Asterisk and open source telephony in the near future.

While prognostication is always a thankless task, we are confident in asserting that open source communications engines such as Asterisk herald a shift in thinking that will transform the telecommunications industry. In this chapter, we will discuss some of our reasons for this belief.

The Problems with Traditional Telephony

Although Alexander Graham Bell is most famously remembered as the father of the telephone, the reality is that during the latter half of the 1800s, dozens of minds were working toward the goal of carrying voice over telegraph lines. These people were mostly business-minded folks, looking to create a product through which they might make their fortunes.

We have come to think of traditional telephone companies as monopolies, but this was not true in their early days. The early history of telephone service took place in a very competitive environment, with new companies springing up all over the world, often with little or no respect for the patents they might be violating. Many famous monopolies got their start through the waging (and winning) of patent wars.

It's interesting to contrast the history of the telephone with the history of Linux and the Internet. While the telephone was created as a commercial exercise, and the telecom industry was forged through lawsuits and corporate takeovers, Linux and the Internet arose out of the academic community, which has always valued the sharing of knowledge over profit.

The cultural differences are obvious. Telecommunications technologies tend to be closed, confusing, and expensive, while networking technologies are generally open, well-documented, and competitive.

Closed Thinking

If one compares the culture of the telecommunications industry to that of the Internet, it is sometimes difficult to believe the two are related. The Internet was designed by enthusiasts, whereas contributing to the development of the PSTN is impossible for any individual to contemplate. This is an exclusive club; membership is not open to just anyone.[*]

The International Telecommunication Union (ITU) clearly exhibits this type of closed thinking. If you want access to its knowledge, you have to be prepared to pay for it. Membership requires proof of your qualifications, and you will be expected to pay thousands of dollars to gain access to its library of publications.

Although the ITU is the United Nations's sanctioned body responsible for international telecommunications, many of the VoIP protocols (SIP, MGCP, RTP, STUN) come not from the hallowed halls of the ITU, but rather from the IETF (which publishes all of its standards free to all, and allows anyone to submit an Internet Draft for consideration).

Open protocols such as SIP may have a tactical advantage over ITU protocols, such as H.323, due to the ease with which one can obtain them. Although H.323 is widely deployed by carriers as a VoIP protocol in the backbone, it is much more difficult to find H.323-based endpoints; newer products are far more likely to support SIP.

The success of the IETF's open approach has not gone unnoticed by the mighty ITU. It has recently become possible to download up to three documents free of charge from the ITU web site.[†] Openness is clearly on its minds. Recent statements by the ITU suggest that there is a desire to achieve "Greater participation in ITU by civil society and the academic world." Mr. Houlin Zhao, the ITU's Director of the Telecommunication Standardization Bureau (TSB), believes that "ITU should take some steps to encourage this."[‡]

The roadmap to achieving this openness is unclear, but the ITU is beginning to realize the inevitable.

---

[*] Contrast this with the IETF's membership page, which states: "The IETF is not a membership organization (no cards, no dues, no secret handshakes :-)... It is open to any interested individual... Welcome to the IETF." Talk about community!

[†] Considering the thousands of documents available, and the fact that each document generally contains references to dozens more, the value of this free information is difficult to judge.

[‡] *http://www.itu.int/ITU-T/tsb-director/itut-wsis/files/wg-wsis-Zhao-rev1.pdf*

As for Asterisk, it embraces both the past and the future: H.323 support is available, although the community has for the most part shunned H.323 in favor of the IETF protocol SIP and the darling of the Asterisk community, IAX.

Limited Standards Compliancy

One of the oddest things about all of the standards that exist in the world of legacy telecommunications is the various manufacturers' seeming inability to implement them consistently. Each manufacturer desires a total monopoly, so the concept of interoperability tends to take a back seat to being first to market with a creative new idea.

The ISDN protocols are a classic example of this. Deployment of ISDN was (and in many ways still is) a painful and expensive proposition, as each manufacturer decided to implement it in a slightly different way. ISDN could very well have helped to usher in a massive public data network, 10 years before the Internet. Unfortunately, due to its cost, complexity, and compatibility issues, ISDN never delivered much more than voice, with the occasional video or data connection for those willing to pay. ISDN is quite common (especially in Europe, and in North America in larger PBX implementations), but it is not delivering anywhere near the capabilities that were envisioned for it.

As VoIP becomes more and more ubiquitous, the need for ISDN will disappear.

Slow Release Cycles

It can take months, or sometimes years, for the big guys to admit to a trend, let alone release a product that is compatible with it. It seems that before a new technology can be embraced, it must be analyzed to death, and then it must pass successfully through various layers of bureaucracy before it is even scheduled into the development cycle. Months or even years must pass before any useful product can be expected. When those products are finally released, they are often based on hardware that is obsolete; they also tend to be expensive and to offer no more than a minimal feature set.

These slow release cycles simply don't work in today's world of business communications. On the Internet, new ideas can take root in a matter of weeks and become viable in extremely short periods of time. Since every other technology must adapt to these changes, so too must telecommunications.

Open source development is inherently better able to adapt to rapid technological change, which gives it an enormous competitive advantage.

The spectacular crash of the telecom industry may have been caused in large part by an inability to change. Perhaps that continued inability is why recovery has been so slow. Now, there is no choice: change, or cease to be. Community-driven technologies such as Asterisk will see to that.

Refusing to Let Go of the Past and Embrace the Future

Traditional telecommunications companies have lost touch with their customers. While the concept of adding functionality beyond the basic telephone is well understood, the idea that the user should be the one defining this functionality is not.

Nowadays, people have nearly limitless flexibility in every other form of communication. They simply cannot understand why telecommunications cannot be delivered as flexibly as the industry has been promising for so many years. The concept of flexibility is not familiar to the telecom industry, and very well might not be until open source products such as Asterisk begin to transform the fundamental nature of the industry. This is a revolution similar to the one Linux and the Internet willingly started more than 10 years ago (and IBM unwittingly started with the PC, 15 years before that). What is this revolution? The commoditization of telephony hardware and software, enabling a proliferation of tailor-made telecommunications systems.

Paradigm Shift

In "Paradigm Shift" (*http://tim.oreilly.com/articles/paradigmshift_0504.html*), Tim O'Reilly talks about a paradigm shift that is occurring in the way technology (both hardware and software) is delivered.[§] O'Reilly identifies three trends: *the commoditization of software*, *network-enabled collaboration*, and *software customizability (software as a service)*. These three concepts provide evidence to suggest that open source telephony is an idea whose time has come.

The Promise of Open Source Telephony

> *Every good work of software starts by scratching a developer's personal itch.*
>
> —Eric S. Raymond, *The Cathedral and the Bazaar*

In his book *The Cathedral and the Bazaar* (O'Reilly), Eric S. Raymond explains that "Given enough eyeballs, all bugs are shallow." The reason open source software development produces such consistent quality is simple: crap can't hide.

The Itch That Asterisk Scratches

In this era of custom database and web site development, people are not only tired of hearing that their telephone system "can't do that," they quite frankly just don't believe it. The creative needs of the customers, coupled with the limitations of the technology, have spawned a type of creativity born of necessity: telecom engineers are like

---

§ Much of the following section is merely our interpretation of O'Reilly's article. To get the full gist of these ideas, the full read is highly recommended.

---

contestants in an episode of *Junkyard Wars*, trying to create functional devices out of a pile of mismatched components.

The development methodology of a proprietary telephone system dictates that it will have a huge number of features, and that the number of features will in large part determine the price. Manufacturers will tell you that their products give you hundreds of features, but if you only need five of them, who cares? Worse, if there's one missing feature you really can't do without, the value of that system will be diluted by the fact that it can't completely address your needs.

The fact that a customer might only need 5 out of 500 features is ignored, and that customer's desire to have 5 unavailable features that address the needs of his business is dismissed as unreasonable.[||] Until flexibility becomes standard, telecom will remain stuck in the last century—all the VoIP in the world notwithstanding.

Asterisk addresses that problem directly and solves it in a way that few other telecom systems can. This is extremely disruptive technology, in large part because it is based on concepts that have been proven time and time again: "the closed-source world cannot win an evolutionary arms race with open-source communities that can put orders of magnitude more skilled time into a problem."[#]

Open Architecture

One of the stumbling blocks of the traditional telecommunications industry has been its apparent refusal to cooperate with itself. The big telecommunications giants have all been around for over a hundred years. The concept of closed, proprietary systems is so ingrained in their culture that even their attempts at standards compliancy are tainted by their desire to get the jump on the competition, by adding that one feature that no one else supports. For an example of this thinking, one simply has to look at the VoIP products being offered by the telecom industry today. While they claim standards compliance, the thought that you would actually expect to be able to connect a Cisco phone to a Nortel switch, or that an Avaya voicemail system could be integrated via IP to a Siemens PBX, is not one that bears discussing.

In the computer industry, things are different. Twenty years ago, if you bought an IBM server, you needed an IBM network and IBM terminals to talk to it. Now, that IBM server is likely to interconnect to Dell terminals though a Cisco network (and run Linux, of all things). Anyone can easily think of thousands of variations on this theme. If any

---

[||] From the perspective of the closed-source industry, this attitude is understandable. In his book *The Mythical Man-Month: Essays on Software Engineering* (Addison-Wesley), Fred Brooks opined that "the complexity and communication costs of a project rise with the square of the number of developers, while work done only rises linearly." Without a community-based development methodology, it is very difficult to deliver products that at best are little more than incremental improvements over their predecessors, and at worst are merely collections of patches.

[#] Eric S. Raymond, *The Cathedral and the Bazaar*.

one of these companies were to suggest that we could only use their products with whatever they told us, they would be laughed out of business.

The telecommunications industry is facing the same changes, but it's in no hurry to accept them. Asterisk, on the other hand, is in a big hurry to not only accept change, but embrace it.

Cisco, Nortel, Avaya, and Polycom IP phones (to name just a few) have all been successfully connected to Asterisk systems. There is no other PBX in the world today that can make this claim. None.

Openness is the power of Asterisk.

Standards Compliance

In the past few years, it has become clear that standards evolve at such a rapid pace that to keep up with them requires an ability to quickly respond to emerging technology trends. Asterisk, by virtue of being an open source, community-driven development effort, is uniquely suited to the kind of rapid development that standards compliance demands.

Asterisk does not focus on cost-benefit analysis or market research. It evolves in response to whatever the community finds exciting—or necessary.

Lightning-Fast Response to New Technologies

After Mark Spencer attended his first SIP Interoperability Test (SIPIT) event, he had a rudimentary but working SIP stack for Asterisk coded within a few days. This was before SIP had emerged as the protocol of choice in the VoIP world, but he saw its value and momentum and ensured that Asterisk would be ready.

This kind of foresight and flexibility is typical in an open source development community (and very unusual in a large corporation).

Passionate Community

The *Asterisk-users* list receives many email messages per day. More than 10,000 people are subscribed to it. This kind of community support is unheard of in the world of proprietary telecommunications, while in the open source world it is commonplace.

The very first AstriCon event was expected to attract 100 participants. Nearly 500 showed up (far more wanted to but couldn't attend). This kind of community support virtually guarantees the success of an open source effort.

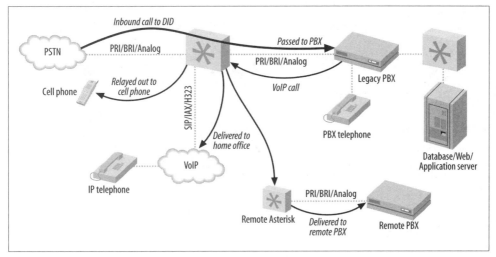

Figure 15-1. Asterisk as a PBX gateway

Some Things That Are Now Possible

So what sorts of things can be built using Asterisk? Let's look at some of the things we've come up with.

Legacy PBX migration gateway

Asterisk can be used as a fantastic bridge between an old PBX and the future. You can place it in front of the PBX as a gateway (and migrate users off the PBX as needs dictate), or you can put it behind the PBX as a peripheral application server. You can even do both at the same time, as shown in Figure 15-1.

Here are some of the options you can implement:

Keep your old PBX, but evolve to IP
> Companies that have spent vast sums of money in the past few years buying proprietary PBX equipment want a way out of proprietary jail, but they can't stomach the thought of throwing away all of their otherwise functioning equipment. No problem—Asterisk can solve all kinds of dilemmas, from replacing a voicemail system to providing a way to add IP-based users beyond the nominal capacity of the system.

Find-me-follow-me
> Provide the PBX a list of numbers where you can be reached, and it will ring them all whenever a call to your DID (Direct Inward Dialing, a.k.a. phone number) arrives. Figure 15-2 illustrates this technology.

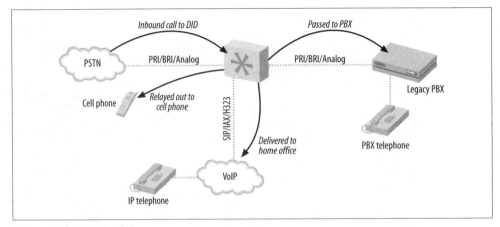

Figure 15-2. Find-me-follow-me

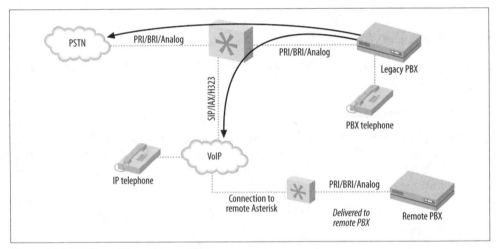

Figure 15-3. VoIP-enabling a legacy PBX

VoIP calling

If a legacy telephony connection from an Asterisk PBX to an old PBX can be established, Asterisk can provide access to VoIP services, while the old PBX continues to connect to the outside world as it always has. As a gateway, Asterisk simply needs to emulate the functions of the PSTN, and the old PBX won't know that anything has changed. Figure 15-3 shows how you can use Asterisk to VoIP-enable a legacy PBX.

Low-barrier IVR

Many people confuse the term "Interactive Voice Response," or IVR, with the Automated Attendant (AA). Since the Automated Attendant was the very first thing IVR was used for, this is understandable. Nevertheless, to the telecom industry, the term IVR represents far more than an AA. An AA generally does little more than present a way for callers to be transferred to extensions, and it is built into most proprietary voicemail systems—but IVR can be so much more.

IVR systems are generally very expensive not only to purchase, but also to configure. A custom IVR system will usually require connectivity to an external database or application. Asterisk is arguably the perfect IVR, as it embraces the concepts of connectivity to databases and applications at its deepest level.

Here are a few examples of relatively simple IVRs that an Asterisk system could be used to create:

Weather reporting
> Using the Internet, you can obtain text-based weather reports from around the world in a myriad of ways. Capturing these reports and running them through a purpose-built parser (Perl would probably eat this up) would allow the information to be available to the dialplan. Asterisk's sound library already contains all of the required prompts, so it would not be an onerous task to produce an interactive menu to play current forecasts for anywhere in the world.

Math programs
> Ed Guy (the architect of Pulver's FWD network) did a presentation at AstriCon 2004 in which he talked about a little math program he'd cooked up for his daughter to use. The program took him no more than an hour to write. What it did was present her with a number of math questions, the answers to which she keyed into the telephone. When all the questions were tabulated, the system presented her with her score. This extremely simple Asterisk application would cost tens of thousands of dollars to implement on any closed PBX platform, assuming it could be done at all. See Chapter 9 for further details. As is so often the case, things that are simple for Asterisk would be either impossible or massively expensive with any other IVR system.

Distributed IVR
> The cost of a proprietary IVR system is such that when a company with many small retail locations wants to provide IVR, it is forced to transfer callers to a central server to process the transactions. With Asterisk, it becomes possible to distribute the application to each node and, thus, handle the requests locally. Literally thousands of little Asterisk systems deployed at retail locations across the world could serve up IVR functionality in a way that would be impossible to achieve with any other system. No more long-distance transfers to a central IVR server, no more huge trunking facility dedicated to the task—more power with less expense.

These are three rather simple examples of the potential of Asterisk.

Conference rooms

This little gem is going to end up being one of the killer functions of Asterisk. In the Asterisk community, everyone finds themselves using conference rooms more and more, for purposes such as these:

- Small companies need an easy way for business partners to get together for a chat.
- Sales teams have a meeting once per week where everyone can dial in from wherever they are.
- Development teams designate a common place and time to update one another on progress.

Home automation

Asterisk is still too much of an über-geek's tool to be able to serve in the average home, but with no more than average Linux and Asterisk skills, the following things become plausible:

Monitoring the kids
> Parents who want to check up on the babysitter (or the kids home alone) could dial an extension context protected by a password. Once authenticated, a two-way audio connection would be created to all the IP phones in the house, allowing Mom and Dad to listen for trouble. Creepy? Yes. But an interesting concept nonetheless.

Locking down your phones
> Going out for the night? Don't want the babysitter tying up the phone? No problem! A simple tweak to the dialplan, and the only calls that can be made are to 911, your cell phone, and the pizza parlor. Any other call attempt will get the recording "We are paying you to babysit our kids, not make personal calls."
>
> Pretty evil, huh?

Controlling the alarm system
> You get a call while on vacation that your Mom wants to borrow some cooking utensils. She forgot her key and is standing in front of the house shivering. Piece of cake; a call to your Asterisk system, a quick digit string into the context you created for the purpose, and your alarm system is instructed to disable the alarm for 15 minutes. Mom better get her stuff and get out quick, though, or the cops'll be showing up!

Managing teenagers' calls
> How about allocating a specific phone-time limit to your teenagers? To use the phone, they have to enter their access codes. They can earn extra minutes by doing chores, scoring all As, dumping that annoying bum with the bad haircut—you get the idea. Once they've used up their minutes... click... you get your phone back.
>
> Incoming calls can be managed as well, via Caller ID. "Donny, this is Suzy's father. She is no longer interested in seeing you, as she has decided to raise her standards a bit. Also, you should consider getting a haircut."

The Future of Asterisk

We've come to love the Internet, both because it is so rich in content and inexpensive, and, perhaps more importantly, because it allows us to define how we communicate. As its ability to carry richer forms of media advances, we'll find ourselves using it more and more. Once Internet voice delivers quality that rivals (or betters) the capabilities of the PSTN, the phone company had better look for another line of business. The PSTN will cease to exist and become little more than one more communications protocol that the Internet happily carries for us. As with most of the rest of the Internet, open source technologies will lead this transformation.

Speech Processing

The dream of having our technical inventions talk to us is older than the telephone itself. Each new advance in technology spurs a new wave of eager experimentation. Generally, results never quite meet expectations, possibly because as soon as a machine says something that *sounds* intelligent, most people assume that it *is* intelligent.

People who program and maintain computers realize their limitations and, thus, tend to allow for their weaknesses. Everybody else just expects their computers and software to work. The amount of thinking a user must do to interact with a computer is often inversely proportional to the amount of thinking the design team did. Simple interfaces belie complex design decisions.

The challenge, therefore, is to design a system that has anticipated the most common desires of its users, and that can also adroitly handle unexpected challenges.

Festival

The Festival text-to-speech server can transform text into spoken words. While this is a whole lot of fun to play with, there are many challenges to overcome.

For Asterisk, an obvious value of text-to-speech might be the ability to have your telephone system read your emails back to you. Of course, if you've noticed the somewhat poor grammar, punctuation, and spelling typically found in email messages these days, you can perhaps appreciate the challenges this poses.

One cannot help but wonder if the emergence of text-to-speech will inspire a new generation of people dedicated to proper writing. Seeing spelling and punctuation errors on the screen is frustrating enough; having to hear a computer speak such things will require a level of Zazen that few possess.

Speech recognition

If text-to-speech is rocket science, speech recognition is science fiction.

Speech recognition can actually work very well, but unfortunately this is generally true only if you provide it with the right conditions—and the right conditions are not those found on a telephone network. Even a perfect PSTN connection is considered to be at the lowest acceptable limit for accurate speech recognition. Add in compressed and lossy VoIP connections, or a cell phone, and you will discover far more limitations than uses.

Asterisk now has an entire speech API so that outside companies (or even open source projects) can tie their speech recognition engines into Asterisk. One company that has done this is LumenVox. By using its speech recognition engine along with Asterisk, you can make voice-driven menus and IVR systems in record time! For more information, see *http://www.lumenvox.com*.

High-Fidelity Voice

As we gain access to more and more bandwidth, it becomes less and less easy to understand why we still use low-fidelity codecs. Many people do not realize that Skype uses a higher fidelity than a telephone; it's a large part of the reason why Skype has a reputation for sounding so good.

If you were ever to phone CNN, wouldn't you love to hear James Earl Jones's mellifluous voice saying "This is CNN," instead of some tinny electronic recording? And if you think Allison Smith* sounds good through the phone, you should hear her in person!

In the future, we will expect, and get, high-fidelity voice though our communications equipment.

Beginning in Asterisk 1.4, there is limited support for the G.722 codec. As more and more hardware vendors start building support for high-fidelity voice into their VoIP hardware, you'll see more support in Asterisk for making better-than PSTN quality calls.

Video

While most of this book focuses on audio, video is also supported in many ways within Asterisk. Video support is not complete, however. The problem is not so much one of functionality as it is one of bandwidth and processing power. More significantly, it is not yet important enough to the community to merit the attention it needs.

---

* Allison Smith is The Voice of Asterisk; it is her voice in all of the system prompts. To have Allison produce your own prompt, simply visit *http://thevoice.digium.com*.

The challenge of video-conferencing

The concept of video-conferencing has been around since the invention of the cathode ray tube. The telecom industry has been promising a video-conferencing device in every home for decades.

As with so many other communications technologies, if you have video-conferencing in your house, you are probably running it over the Internet, with a simple, inexpensive webcam. Still, it seems that people see video-conferencing as a bit gimmicky. Yes, you can see the person you're talking to, but there's something missing.

Why we love video-conferencing

Video-conferencing promises a richer communications experience than the telephone. Rather than hearing a disembodied voice, the nuances of speech that come from eye-to-eye communication are possible.

Why video-conferencing may never totally replace voice

There are some challenges to overcome, though, and not all of them are technical.

Consider this: using a plain telephone, people working from their home offices can have business conversations, unshowered, in their underwear, feet on the desk, coffee in hand. A similar video conversation would require half an hour of grooming to prepare for, and couldn't happen in the kitchen, on the patio, or... well, you get the idea.

Also, the promise of eye-to-eye communication over video will never happen as long as the focal points of the participants are not in line with the cameras. If you look at the camera, your audience will see you looking at them, but you won't see them. If you look at your screen to see whom you are talking to, the camera will show you looking down at something—not at your audience. That looks impersonal. Perhaps if a video-phone could be designed like a TelePromptR, where the camera was behind the screen, it wouldn't feel so unnatural. As it stands, there's something psychological that's missing. Video ends up being a gimmick.

Wireless

Since Asterisk is fully VoIP-enabled, wireless is all part of the package.

Wi-Fi

Wi-Fi is going to be the office mobility solution for VoIP phones. This technology is already quite mature. The biggest hurdle is the cost of handsets, which can be expected to improve as competitive pressure from around the world drives down prices.

Wi-MAX

Since we are so bravely predicting so many things, it's not hard to predict that Wi-MAX spells the beginning of the end for traditional cellular telephone networks.

With wireless Internet access within the reach of most communities, what value will there be in expensive cellular service?

Unified Messaging

This is a term that has been hyped by the telecom industry for years, but adoption has been far slower than predicted.

Unified Messaging is the concept of tying voice and text-messaging systems into one. With Asterisk, the two don't need to be artificially combined, as Asterisk already treats them the same way.

Just by examining the terms, *unified* and *messaging*, we can see that the integration of email and voicemail must be merely the beginning; Unified Messaging needs to do a lot more than just that if it is to deserve its name.

Perhaps we need to define "messaging" as communication that does not occur in real time. In other words, when you send a message, you expect that the reply may take moments, minutes, hours, or even days to arrive. You compose what you wish to say, and your audience is expected to compose a reply.

Contrast this with conversing, which happens in real time. When you talk to someone on a telephone connection, you expect no more than a few seconds' delay before the response arrives.

In 2002, Tim O'Reilly delivered a speech titled "Watching the Alpha Geeks: OS X and the Next Big Thing" (*http://www.macdevcenter.com/pub/a/mac/2002/05/14/ oreilly_wwdc_keynote.html*), in which he talked about someone piping IRC through a text-to-speech engine. One could imagine doing the reverse as well, allowing us to join an IRC or instant messaging chat over our Wi-Fi phone, our Asterisk PBX providing the speech-to-text-to-speech translations.

Peering

As monopoly networks such as the PSTN give way to community-based networks like the Internet, there will be a period of time where it is necessary to interconnect the two. While the traditional providers would prefer that the existing model be carried into the new paradigm, it is increasingly likely that telephone calls will become little more than another application the Internet happily carries.

But a challenge remains: how to manage the telephone numbering plan with which we are all familiar and comfortable?

E.164

The ITU defined a numbering plan in its E.164 specification. If you've used a telephone to make a call across the PSTN, you can confidently state that you are familiar with the concept of E.164 numbering. Prior to the advent of publicly available VoIP, nobody cared about E.164 except the telephone companies—nobody needed to.

Now that calls are hopping from PSTN to Internet to who-knows-what, some consideration must be given to E.164.

ENUM

In response to this challenge, the IETF has sponsored the Telephone Number Mapping (ENUM) working group, the purpose of which is to map E.164 numbers into the Domain Name System (DNS).

While the concept of ENUM is sound, it requires cooperation from the telecom industry to achieve success. However, cooperation is not what the telecom industry is famous for and, thus, far ENUM has foundered.

e164.org

The folks at *e164.org* are trying to contribute to the success of ENUM. You can log on to this site, register your phone number, and inform the system of alternative methods of communicating with you. This means that someone who knows your phone number can connect a VoIP call to you, as the *e164.org* DNS zone will provide the IP addressing and protocol information needed to connect to your location.

As more and more people publish VoIP connectivity information, fewer and fewer calls will be connected through the PSTN.

DUNDi

Distributed Universal Number Discovery (DUNDi) is an open routing protocol designed to maintain dynamic telecom routing tables between compatible systems (see Chapter 14 for more information). While Asterisk is currently the only PBX to support DUNDi, the openness of the standard ensures that anyone can implement it.

DUNDi has huge potential, but it is very much in its infancy. This is the one to watch.

Challenges

As is true with any worthwhile thing, Asterisk will face challenges. Let's take a glance at what some of them may be.

Too much change, too few standards

These days, the Internet is changing so fast, and offers so much diverse content, that it is impossible for even the most attentive geek to keep on top of it all. While this is as it should be, it also means that an enormous amount of technology churn is an inevitable part of keeping any communications system current.

VoIP spam

Yes, it's coming. There will always be people who believe they have the right to inconvenience and harass others in their pursuit of money. Efforts are under way to try and address this, but only time will tell how efficacious they will be.

Fear, uncertainty, and doubt

The industry is making the transition from ignorance to laughter. If Gandhi is correct, we can expect the fight to begin soon.

As their revenue streams become increasingly threatened by open source telephony, the traditional industry players are certain to mount a fear campaign, in hopes of undermining the revolution.

Bottleneck engineering

There is a rumor making the rounds that the major network providers will begin to artificially cripple VoIP traffic by tagging and prioritizing the traffic of their premium VoIP services and, worse, detecting and bumping any VoIP traffic generated by services not approved by them.

Some of this is already taking place, with service providers blocking traffic of certain types through their networks, ostensibly due to some public service being rendered (such as blocking popular file-sharing services to protect us from piracy). In the United States, the FCC has taken a clear stand on the matter and fined companies that engage in such practices. In the rest of the world, regulatory bodies are not always as accepting of VoIP.

What seems clear is that the community and the network will find ways around blockages, just as they always have.

Regulatory wars

The recently departed Chairman of the United States Federal Communications Commission, Michael Powell, delivered a gift that may well have altered the path of the VoIP revolution. Rather than attempting to regulate VoIP as a telecom service, he has championed the concept that VoIP represents an entirely new way of communicating and requires its own regulatory space in which to evolve.

VoIP will become regulated, but not everywhere as a telephony service. Some of the regulations that may be created include:

Presence information for emergency services

One of the characteristics of a traditional PSTN circuit is that it is always in the same location. This is very helpful to emergency services, as they can pinpoint the location of a caller by identifying the address of the circuit from which the call was placed. The proliferation of cell phones has made this much more difficult to achieve, since a cell phone does not have a known address. A cell phone can be plugged into any network and can register to any server. If the phone does not identify its physical location, an emergency call from it will provide no clue as to the where the caller is. VoIP creates similar challenges.

Call monitoring for law enforcement agencies

Law enforcement agencies have always been able to obtain wiretaps on traditional circuit-switched telephone lines. While regulations are being enacted that are designed to achieve the same end on the network, the technical challenge of delivering this functionality will probably never be completely solved. People value their privacy, and the more governments want to stifle it, the more effort will be put toward maintaining it.

Anti-monopolistic practices

These practices are already being seen in the U.S., with fines being levied against network providers who attempt to filter traffic based on content.

When it comes to regulation, Asterisk is both a saint and a devil: a saint because it feeds the poor, and a devil because it empowers the phrackers and spammers like nothing ever has. The regulation of open source telephony may in part be determined by how well the community regulates itself. Concepts such as DUNDi, which incorporate anti-spam processes, are an excellent start. On the other hand, concepts such as Caller ID-spoofing are ripe with opportunities for abuse.

Quality of service

Due to the best-effort reality of the TCP/IP-based Internet, it is not yet known how well increasing realtime VoIP traffic will affect overall network performance. Currently, there is so much excess bandwidth in the backbone that best-effort delivery is generally quite good indeed. Still, it has been proven time and time again that whenever we are provided with more bandwidth, we figure out a way to use it up. The 1 MB DSL connection undreamt of five years ago is now barely adequate.

Perhaps a corollary of Moore's Law[†] will apply to network bandwidth. QoS may become moot, due to the network's ability to deliver adequate performance without any special processing. Organizations that require higher levels of reliability may elect to pay a premium for a higher grade of service. Perhaps the era of paying by the minute for long-distance connections will give way to paying by the millisecond for guaranteed low latency, or by the percentage point for reduced packet loss. Premium services will

---

[†] Gordon Moore wrote a paper in 1965 that predicted the doubling of transistors on a processor every few years.

offer the five-nines reliability[‡] the traditional telecom companies have always touted as their advantage over VoIP.

Complexity

Open systems require new approaches toward solution design. Just because the hardware and software are cheap doesn't mean the solution will be. Asterisk does not come out of the box ready to run; it has to be designed and built, and then maintained. While the base software is free, and the hardware costs will be based on commodity pricing, it is fair to say that the configuration costs for a highly customized system will be a sizeable part of the solution costs—in many cases, because of its high degree of complexity and configurability, more than would be expected with a traditional PBX.

The rule of thumb is generally considered to be something like this: if it can be done in the dialplan, the system design will be roughly the same as for any similarly featured traditional PBX. Beyond that, only experience will allow one to accurately estimate the time required to build a system.

There is much to learn.

Opportunities

Open source telephony creates limitless opportunities. Here are some of the more compelling ones.

Tailor-made private telecommunications networks

Some people would tell you that price is the key, but we believe that the real reason Asterisk will succeed is because it is now possible to build a telephone system as one would a web site: with complete, total customization of each and every facet of the system. Customers have wanted this for years. Only Asterisk can deliver.

Low barrier to entry

Anyone can contribute to the future of communicating. It is now possible for someone with an old $200 PC to develop a communications system that has intelligence to rival the most expensive proprietary systems. Granted, the hardware would not be production-ready, but there is no reason the software couldn't be. This is one of the reasons why closed systems will have a hard time competing. The sheer number of people who have access to the required equipment is impossible to equal in a closed shop.

[‡] This term refers to 99.999%, which is touted as the reliability of traditional telecom networks. Achieving five nines requires that service interruptions for an entire year total no more than 5 minutes and 15 seconds. Many people believe that VoIP will need to achieve this level of reliability before it can be expected to fully replace the PSTN. Many other people believe that the PSTN doesn't even come close to five-nines reliability. We believe that this could have been an excellent term to describe high reliability, but marketing departments abuse it far too frequently.

Hosted solutions of similar complexity to corporate web sites

The design of a PBX was always a kind of art form, but before Asterisk, the art lay in finding creative ways to overcome the limitations of the technology. With limitless technology, those same creative skills can now be properly applied to the task of completely answering the needs of the customer. Open source telephony engines such as Asterisk will enable this. Telecom designers will dance for joy, as their considerable creative skills will now actually serve the needs of their customers, rather than be focused on managing kludge.

Proper integration of communications technologies

Ultimately, the promise of open source comes to nothing if it cannot fulfill the need people have to solve problems. The closed industries lost sight of the customer and tried to fit the customer to the product.

Open source telephony brings voice communications in line with other information technologies. It is finally possible to properly begin the task of integrating email, voice, video, and anything else we might conceive of over flexible transport networks (whether wired or wireless), in response to the needs of the user, not the whims of monopolies.

Welcome to the future of telecom!

VoIP Channels

VoIP channels in Asterisk represent connections to the protocols they support. Each protocol you wish to use requires a configuration file, containing general parameters defining how your system handles the protocol as well as specific parameters for each channel (or device) you will want to reference in your dialplan. In this appendix, we'll take an in-depth look at the IAX and SIP configuration files.

IAX

The IAX configuration file (*iax.conf*) contains all of the configuration information Asterisk needs to create and manage IAX protocol channels. The sections in the file are separated by headings, which are formed by a word framed in square brackets ([ ]). The name in the brackets will be the name of the channel, with one notable exception: the [general] section, which is not a channel, is the area where global protocol parameters are defined.

This section examines the various general and channel-specific settings for *iax.conf*. We will define each parameter and then give an example of its use. Certain options may have several valid arguments. These arguments are listed beside the option, separated with the pipe symbol (|). For example, bandwidth=low|medium|high means that the bandwidth option accepts one of the values low, medium, or high as its argument.

You can insert comments anywhere in the *iax.conf* file, by preceding the comment text with the semicolon character (;). Everything to the right of the semicolon will be ignored. Feel free to use comments liberally.

General IAX Settings

The first non-comment line in your *iax.conf* file must be the heading [general]. The parameters in this section will apply to all connections using this protocol, unless defined differently in a specific channel's definition. Since some of these settings can be defined on a per-channel basis, we have identified settings that are always global with the tag "(global)" and those that can optionally be configured for individual channels with the tag "(channel)." If you define a channel parameter under the [general] section, you do not need to define it in each channel; its value becomes the default. Keep in mind that setting a parameter in the [general] section does not prevent you from setting it differently for specific channels; it merely makes this setting the default. Also keep in mind that not defining these parameters may, in some cases, cause a system default to be used instead.

Here are the parameters that you can configure:

accountcode (*channel*)

> The account code can be defined on a per-user basis. If defined, this account code will be assigned to a call record whenever no specific user account code is set. The accountcode name configured will be used as the *filename.csv* in the */var/log/asterisk/cdr-csv/* directory to store Call Detail Records (CDRs) for the user/peer/friend:
>
> ```
> accountcode=iax-username
> ```

adsi (*channel*)

> Support for ADSI (Analog Display Services Interface) can be enabled if you have ADSI-compatible CPE equipment:
>
> ```
> adsi=yes|no
> ```

allow *and* disallow (*channel*)

> Specific codecs can be allowed or disallowed, limiting codec use to those preferred by the system designer. allow and disallow can also be defined on a per-channel basis. Keep in mind that allow statements in the [general] section will carry over to each of the channels, unless you reset with a disallow=all. Codec negotiation is attempted in the order in which the codecs are defined. Best practice suggests that you define disallow=all, followed by explicit allow statements for each codec you wish to use. If nothing is defined, allow=all is assumed:
>
> ```
> disallow=all
> allow=ulaw
> allow=gsm
> allow=ilbc
> ```

amaflags (*channel*)

> Automatic Message Accounting (AMA) is defined in the Telcordia Family of Documents listed under FR-AMA-1. These documents specify standard mechanisms for generation and transmission of CDRs. You can specify one of four AMA flags to apply to all IAX connections:
>
> ```
> amaflags=default|omit|billing|documentation
> ```

authdebug (*global*)

> You can minimize the amount of authorization debugging by disabling it with authdebug=no. Authorization debugging is enabled by default if not explicitly disabled:

```
authdebug=no
```

autokill (*global*)

To minimize the danger of stalling when a host is unreachable, you can set
`autokill` to `yes` to specify that any new connection should be torn down if an `ACK`
is not received within 2,000 ms. (This is obviously not advised for hosts with high
latency.) Alternatively, you can replace `yes` with the number of milliseconds to wait
before considering a peer unreachable. `autokill` configures the wait for all IAX2
peers, but you can configure it differently for individual peers with the use of the
`qualify` command:

```
autokill=1500
```

bandwidth (*channel*)

`bandwidth` is a shortcut that may help you get around using `disallow=all` and multiple `allow` statements to specify which codecs to use. The valid options are:

high

> Allows all codecs (G.723.1, GSM, law, alaw, G.726, ADPCM, slinear,
> LPC10, G.729, Speex, iLBC)

medium

> Allows all codecs except slinear, law, and alaw

low

> Allows all medium codecs except G.726 and ADPCM

```
bandwidth=low|medium|high
```

bindport and bindaddr (*global*)

These optional parameters allow you to control the IP interface and port on which
you wish to accept IAX connections. If omitted, the port will be set to 4569, and
all IP addresses in your Asterisk system will accept incoming IAX connections. If
multiple bind addresses are configured, only the defined interface will accept IAX
connections.[*] The address 0.0.0.0 tells Asterisk to listen on all interfaces:

```
bindport=4569
bindaddr=192.168.0.1
```

codecpriority (*channel*)

The `codecpriority` option controls which end of an inbound call leg will have
priority over the negotiation of codecs. If set in the [general] section, the selected
options will be inherited by all user entries in the channel configuration file; however, they can be defined in the individual user entries for more granular control.
If set in both the [general] and user sections, the user entry will override the entry

---

[*] Currently, Asterisk will only work with a single **bindaddr** option. If you wish to listen to more than one address,
you'll need to use 0.0.0.0. Note that Asterisk will work in a multihomed environment, but not with multi-
address interfaces. Asterisk will use the system's routing table to select which interface it sends the packet
out on and it will use the primary address on that interface as the source.

that is configured in the [general] section. If this parameter is not configured, the value defaults to host.

Valid options include:

caller
> The inbound caller has priority over the host.

host
> The host has priority over the inbound caller.

disabled
> Codec preferences are not considered; this is the default behavior before the implementation of codec preferences.

reqonly
> Codec preferences are ignored, and the call is accepted only if the requested codec is available:

```
codecpriority=caller|host|disabled|reqonly
```

delayreject (*global*)
> If an incorrect password is received on an IAX channel, this will delay the sending of the REGREQ or AUTHREP reject messages, which will help to secure against brute-force password attacks. The delay time is 1,000 ms:

```
delayreject=yes|no
```

forcejitterbuffer (*channel*)
> Since Asterisk attempts to bridge channels (endpoints) directly together, the endpoints are normally allowed to perform jitter buffering themselves. However, if the endpoints have a poor jitter buffer implementation, you may wish to force Asterisk to perform jitter buffering no matter what. You can force jitter buffering to be performed with forcejitterbuffer=yes:

```
forcejitterbuffer=yes
```

iaxthreads and iaxmaxthreads (*global*)
> The iaxthreads setting specifies the number of IAX helper threads that are created on startup to handle IAX communications.

> The iaxmaxthreads setting specifies the maximum number of additional IAX helper threads that may be spawned to handle higher IAX traffic levels:

```
iaxthreads=10
iaxmaxthreads=100
```

jitterbuffer (*channel*)
> *Jitter* refers to the varying latency between packets. When packets are sent from an end device, they are sent at a constant rate with very little latency variation. However, as the packets traverse the Internet, the latency between the packets may become varied; thus, they may arrive at the destination at different times, and possibly even out of order.

The jitter buffer is, in a sense, a staging area where the packets can be reordered and delivered in a regulated stream. Without a jitter buffer, the user may perceive anomalies in the stream, experienced as static, strange sound effects, garbled words, or, in severe cases, missed words or syllables.

The jitter buffer affects only data received from the far end. Any data you transmit will not be affected by your jitter buffer, as the far end will be responsible for the de-jittering of its incoming connections.

The jitter buffer is enabled with the use of `jitterbuffer=yes`:

```
jitterbuffer=yes|no
```

language (*channel*)
This sets the language flag to whatever you define. The global default language is English. The language that is set is sent by the channel as an information element. It is also used by applications such as `SayNumber()` that have different files for different languages. Keep in mind that languages other than English are not explicitly installed on the system, and it is up to you to configure the system to ensure that the language you specify is handled properly:

```
language=en
```

`mailboxdetail` (*global*)
If `mailboxdetail` is set to `yes`, the new/old message count is sent to the user, instead of a simple statement of whether new and old messages exist. `mailboxdetail` can also be set on a per-peer basis:

```
mailboxdetail=yes
```

`maxjitterbuffer` (*channel*)
This parameter is used to set the maximum size of the jitter buffer, in milliseconds. Be sure not to set `maxjitterbuffer` too high, or you will needlessly increase your latency:

```
maxjitterbuffer=500
```

`maxjitterinterps` (*channel*)
The maximum number of interpolation frames the jitter buffer should return in a row. Since some clients do not send CNG/DTX frames to indicate silence, the jitter buffer will assume silence has begun after returning this many interpolations. This prevents interpolating throughout a long silence:

```
maxjitterinterps=10
```

`maxregexpire` and `minregexpire` (*channel*)
Specifies the maximum and minimum time intervals for registration expiration, in seconds:

```
maxregexpire=180
minregexpire=60
```

mohinterpret (*channel*)

This option specifies a preference for which music-on-hold class this channel should listen to when put on hold if the music class has not been set on the channel with Set(CHANNEL(musicclass)=*whatever*) in the dialplan, and the peer channel putting this all on hold did not suggest a music class.

If this option is set to **passthrough**, then the hold message will always be passed through as signaling instead of generating hold music locally.

This option may be specified globally, or on a per-user or per-peer basis:

```
mohinterpret=default
```

mohsuggest (*channel*)

This option specifies which music-on-hold class (as defined in *musiconhold.conf*) to suggest to the peer channel when this channel places the peer on hold. It may be specified globally or on a per-user or per-peer basis:

```
mohsuggest=default
```

nochecksums (*global*)

If set, Asterisk will disable the calculation of UDP checksums and no longer check UDP checksums on systems supporting this feature:

```
nochecksums=yes
```

regcontext (*channel*)

By specifying the context that contains the actions to perform, you can configure Asterisk to perform a number of actions when a peer registers to your server. This option works in conjunction with **regexten**, which specifies the extension to execute. If no **regexten** is configured, the peer name is used as the extension. Asterisk will dynamically create and destroy a NoOp at priority 1 for the extension. All actions to be performed upon registration should start at priority 2. More than one **regexten** may be supplied, if separated by an **&**. **regcontext** can be set on a per-peer basis or globally:

```
regcontext=registered-phones
```

regexten (*channel*)

The **regexten** option is used in conjunction with **regcontext** to specify the extension to be executed within the configured context. If **regexten** is not explicitly configured, the peer name is used as the extension to match:

```
regexten=myphone
```

resyncthreshold (*channel*)

The resynchronize threshold is used to resynchronize the jitter buffer if a significant change is detected over a few frames, assuming that the change was caused by a timestamp mixup. The resynchronization threshold is defined as the measured jitter plus the **resyncthreshold** value, defined in milliseconds:

```
resyncthreshold=1000
```

rtautoclear (*global*)

This specifies whether or not Asterisk should auto-expire friends created on the fly on the same schedule as if they had just registered. If set to **yes**, when the registration expires, the friend will vanish from the configuration until requested again. If set to an integer, friends expire within that number of seconds instead of the normal registration interval:

```
rtautoclear=yes|no|seconds
```

rtcachefriends (*global*)

If **rtcachefriends** is turned on, Asterisk will cache friends that come from the realtime engine, just as if they had come from *iax.conf*. This often helps with items such as message-waiting indications on realtime peers:

```
rtcachefriends=yes|no
```

rtignoreregexpire (*global*)

If **rtignoreregexpire** is set to **yes**, and a realtime peer's registration has expired (based on its registration interval), then Asterisk will continue to use the IP address and port stored in the database:

```
rtignoreregexpire=yes|no
```

rtupdate (*global*)

If set to **yes** Asterisk will update the IP address, origination port, and registration period of a peer upon registration. Defaults to **yes**:

```
rtupdate=yes|no
```

tos (*global*)

Asterisk can set the Type of Service (TOS) bits in the IP header to help improve performance on routers that respect TOS bits in their routing calculations. The following values are valid: CS0, CS1, CS2, CS3, CS4, CS5, CS6, CS7, AF11, AF12, AF13, AF21, AF22, AF23, AF31, AF32, AF33, AF41, AF42, AF43, and ef (expedited forwarding). You may also use a numeric value for the TOS bits.

For more information, see the *doc/ip-tos.txt* file in the Asterisk source directory.

trunk (*channel*)

IAX2 trunking enables Asterisk to send media (as mini-frames) from multiple channels using a single header. The reduction in overhead makes the IAX2 protocol more efficient when sending multiple streams to the same endpoint (usually another Asterisk server):

```
trunk=yes|no
```

trunkfreq (*channel*)

trunkfreq is used to control how frequently you send trunk messages, in milliseconds. Trunk messages are sent in conjunction with the **trunk=yes** command:

```
trunkfreq=20
```

`trunktimestamps` (*channel*)

Specifies whether or not Asterisk should send timestamps for the individual sub-frames within trunk frames. There is a small bandwidth penalty for sending these timestamps (less than 1 kbps/call), but they ensure that frame timestamps get sent end-to-end properly. If both ends of all your trunks go directly to TDM *and* your `trunkfreq` equals the frame length for your codecs, you can probably suppress these. The receiver must also support this feature, although it does not also need to have it enabled:

```
trunktimestamps=yes|no
```

Retrieving Dialplan Information from a Remote Asterisk Box

Asterisk can retrieve dialplan information from another Asterisk box with the use of a `switch =>` statement. When this occurs, the Asterisk IAX channel driver must wait for a reply from the remote box before it can continue with other IAX-related processes. This is especially troubling when you have multiple `switch` statements nested throughout multiple boxes: if a `switch` statement has to traverse several boxes, there could be an appreciable delay before a result is returned.

When the global `iaxcompat` option is set to `yes`, Asterisk will spawn a separate thread when the `switch` lookup is being performed. The use of this thread allows the main IAX channel driver to continue on with other processes while the thread waits for the reply. A small performance hit is incurred with this option:

```
iaxcompat=yes|no
```

Registering to Other Servers with register Statements

The register switch (`register =>`) is used to register your Asterisk box to a remote server —this lets the remote end know where you are, in case you are configured with a dynamic IP address. Note that `register` statements are used only when the remote end has you configured as a peer, and when `host=dynamic`.

The basic format for a `register` statement is:

```
register => username:password@remote-host
```

The *password* is optional (if not configured on the remote system).

Alternatively, you can specify an RSA key by framing the appropriate RSA key name[†] in square brackets ([ ]):

```
register => username:[rsa-key-name]@remote-host
```

[†] Asterisk RSA keys are typically located in */var/lib/asterisk/keys/*. You can generate your own keys using the *astkeygen* script.

By default, `register` requests will be sent via port 4569. You can direct them to a different port by explicitly specifying it, as follows:

```
register => username:password@remote-host:1234
```

IAX Channel Definitions

With the general settings defined, we can now define our channels. Defining a guest channel is recommended whenever you want to accept anonymous IAX calls. This is a very common way for folks in the Asterisk community to contact one another. Before you decide that this is not for you, keep in mind that anyone whom you want to be able to connect to you via IAX (without you specifically configuring an account for them) will need to connect as a guest. This account, in effect, becomes your "IAX phone number." Your guest channel definition should look something like this:

```
[guest]
type=user
context=incoming
callerid="Incoming IAX Guest"
```

 No doubt the spammers will find a way to harass these addresses, but in the short term this has not proven to be a problem. In the long term, we'll probably use DUNDi (see Chapter 14 for more information).

If you wish to accept calls from the Free World Dialup network, Asterisk comes with a predefined security key that ensures that anonymous connections cannot spoof an incoming Free World Dialup call. You'll want to set up an `iaxfwd` channel:

```
[iaxfwd]
type=user
context=incoming
auth=rsa
inkeys=freeworlddialup
```

If you have resources advertised on a DUNDi network, the associated user must be defined in *iax.conf*:

```
[dundi]
type=user
dbsecret=dundi/secret
context=dundi-incoming
```

If you have IAX-based devices (such as an IAXy), or IAX-based users at a remote node, you may want to provide them with their own user definition with which to connect to the system.

Let's say you have a user on a remote node for whom you want to define an IAX user. We'll call this hypothetical user `sushi`. The user definition might look something like this:

```
[sushi]
type=user
context=local_users
auth=md5,plaintext,rsa
secret=wasabi
transfer=no
jitterbuffer=yes
callerid="Happy Tempura" <(800) 555-1234>
accountcode=seaweed
deny=0.0.0.0/0.0.0.0
permit=192.168.1.100/255.255.255.0
language=en
```

IAX Authentication

IAX provides authentication mechanisms to allow for a reasonable level of security between endpoints. This does not mean that the audio information cannot be captured and decoded, but it does mean that you can more carefully control who is allowed to make connections to your system. Three levels of security are supported on IAX channels. The `auth` option defines which authentication method to use on the channel: `plaintext`, `md5`, or `rsa`.

`plaintext`, in IAX, offers very little security. While it will prevent connection to the channel unless a valid password is supplied, the fact that the password is stored in *iax.conf* in plain text and is transmitted and received as plain text makes this a very insecure authentication method.

`md5` improves the security on the network connection; however, both ends still require a plain-text `secret` in the *iax.conf* file. Here's how it works: Box A requests a connection with Box B, which in turn replies with an authorization request including a randomly generated number. Box A then generates an MD5 hash using the value supplied in the `secret` field of *iax.conf* and the random number from Box B. The hash is returned in the authorization reply, and Box B compares it to the hash it generated locally. If the hashes match, authorization is granted.

The `rsa` method provides the most security. Before using RSA authentication, each end must create a public and private key pair through the *astgenkey* script, typically located in */usr/src/asterisk/contrib/scripts/*. The public key must then be given to the far end. Each end of the circuit must include the public key of the far end in its channel definition, using the `inkeys` and `outkey` parameters.

RSA keys are stored in */var/lib/asterisk/keys/*. Public keys are named *name*.`pub`; private keys are named *name*.`key`. Private keys must be encrypted with 3DES.

Incoming calls from this user will arrive in the context `local_users` and will ask the system to accept the Caller ID `Happy Tempura <(800) 555-1234>`. The system will be willing to accept MD5, plain-text, or RSA authentication from this user, so long as the password `wasabi` is provided and the call comes from the IP address `192.168.1.100`. All calls related to this channel will be assigned the account code `seaweed`. Because we've

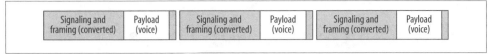

Figure A-1. Trunking disabled

set `transfer` to `no`, the media path for this channel will always pass through Asterisk; it cannot be redirected to another IAX node.

If you yourself are a remote node, and you need to connect into a remote node as a user, you would define that main node as your peer:

```
[sashimi_platter]
type=peer
username=sushi
secret=wasabi
host=192.168.1.101
qualify=yes
trunk=yes
```

A `peer` is called from the dialplan by using `Dial()` with the name contained in square brackets. If you need to authenticate to the peer with a username, you can set that username and secret in the `username` and `secret` fields.

> Remember, an incoming call from a `user` specified in *iax.conf* must authenticate using the name specified in square brackets. When Asterisk itself is calling to an outside `peer` however, you can use the `username` setting to set the authentication username.

`host` is specified using either IP dotted notation or a fully qualified domain name (FQDN). You can determine the latency between you and the remote host, and whether the peer is alive, with `qualify=yes`. To minimize the amount of overhead for multiple calls going to the same peer, you can `trunk` them.

Trunking is unique to IAX and is designed to take advantage of the fact that two large sites may have multiple, simultaneous VoIP connections between them. IAX trunking reduces overhead by loading the audio from several concurrent calls into each signaling packet.[‡] You can enable trunking for a channel with `trunk=yes` in *iax.conf*.

Figure A-1 shows a channel with trunking disabled, and Figure A-2 shows a channel with trunking enabled.

Channel-specific parameters

Now, let's take a look at the channel-specific parameters:

[‡] You can think of IAX trunking as carpooling for VoIP packets. This becomes very useful in many situations, as the IP overhead (UDP headers, IP headers, and so forth) is often bigger than the audio payload it carries. If you have several concurrent calls between two Asterisk boxes, you definitely want to turn on IAX trunking!

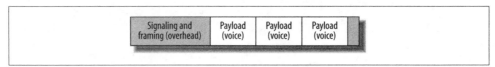

Figure A-2. Trunking enabled

callerid

> You can set a suggested Caller ID string for a user or peer with `callerid`. If you define a Caller ID field for a user, any calls that come in on that channel will have that Caller ID assigned to them, regardless of what the far end sends to you. If you define Caller ID for a *peer*, you are requesting that the far end use that to identify you (although you have no way of ensuring it will do so). If you want incoming users to be able to define their own Caller IDs (i.e., for guests), make sure you do not set the `callerid` field:
>
> ```
> callerid=John Smith <(800) 555-1234>
> ```

defaultip

> The `defaultip` setting complements `host=dynamic`. If a host has not yet registered with your server, you'll attempt to send messages to the default IP address configured here:
>
> ```
> defaultip=192.168.1.101
> ```

inkeys

> You can use the `inkeys` option to authenticate a user with the use of an RSA key. To associate more than one RSA key with a user channel definition, separate the key names with a colon (`:`). Any one of those keys will be sufficient to validate a connection. The "inkey" is the public key you distribute to your users:
>
> ```
> inkeys=server_one:server_two
> ```

mailbox

> If you associate a mailbox with a peer within the channel definition, voicemail will send a message waiting indication (MWI) to the nodes on the end of that channel. If the mailbox number is in a voicemail context other than `default`, you can specify it as *mailbox* @ *context*. To associate multiple mailboxes with a single peer, use multiple `mailbox` statements:
>
> ```
> mailbox=1000@internal
> ```

outkey

> You can use the `outkey` option to authenticate a peer with the use of an RSA key. Only one RSA key may be used for outgoing authentication. The "outkey" is not distributed; it is your private key:
>
> ```
> outkey=private_key
> ```

qualify

You can set **qualify** to **yes**, **no**, or a time in milliseconds. If you set **qualify=yes**, **PING** messages will be sent periodically to the remote peers to determine whether they are available and what the latency between replies is. The peers will respond with **PONG** messages. A peer will be determined unreachable if no reply is received within 2,000 ms (to change this default, instead set **qualify** to the number of milliseconds to wait for the reply).

qualifyfreqok and qualifyfreqnotok

These two settings are used to determine how frequently Asterisk should ping a peer when **qualify** is set. The **qualifyfreqok** setting determines how often to ping the peer when it's in an OK state, and **qualifyfreqnotok** determines how often to ping the peer when it's not in an OK state.

qualifysmoothing

You can set **qualifysmoothing** to **yes** or **no**. If enabled, Asterisk will take the average of the last two qualify times. This helps eliminate having peers marked as **LAGGED**, especially on a lossy network.

sendani

The SS7 PSTN network uses Automatic Number Identification (ANI) to identify a caller, and Caller ID is what is delivered to the user. The Caller ID is generated from the ANI, so it's easy to confuse the two. Blocking Caller ID sets a privacy flag on the ANI, but the backbone network still knows where the call is coming from:

    sendani=yes

> ANI has been around for a while. Its original purpose was to deliver the billing number of the originating party on a long-distance call to the terminating office. Unlike Caller ID, ANI does not require SS7, as it can be transmitted using DTMF. Also, ANI cannot be blocked.

transfer

You can set **transfer** to **yes**, **no**, or **mediaonly**. If set to **yes**, Asterisk will transfer the call away from itself if it can, in order to make the packet path shorter between the two endpoints. (This obviously won't work if Asterisk needs to transcode or translate between protocols, or if network conditions don't allow the two endpoints to talk directly to each other.) If it is set to **no**, Asterisk will not try to transfer the call away from itself.

If set to **mediaonly**, Asterisk will attempt to transfer the media stream so that it goes directly between the two endpoints, but the call signaling (call setup and teardown messages) will still go through Asterisk. This is useful because it ensures that call detail records are still accurate, even though the media is no longer flowing through the Asterisk box.

SIP

Just as with IAX, the SIP configuration file (*sip.conf*) contains configuration information for SIP channels. The headings for the channel definitions are formed by a word framed in square brackets ([ ])—again, with the exception of the [general] section, where we define global SIP parameters. Don't forget to use comments generously in your *sip.conf* file. Precede the comment text with a semicolon; everything to the right will be ignored.

General SIP Parameters

The following options are to be used within the [general] section of *sip.conf*:

allowexternalinvites
> If set to no, this setting disables INVITE and REFER messages to non-local domains. See the domain setting.
>
>   allowexternalinvites=yes|no

allowguest
> If set to no, this disallows guest SIP connections. The default is to allow guest connections. SIP normally requires authentication, but you can accept calls from users who do not support authentication (i.e., do not have a secret field defined). Certain SIP appliances (such as the Cisco Call Manager v4.1) do not support authentication, so they will not be able to connect if you set allowguest=no:
>
>   allowguest=no|yes

allowoverlap
> If set to no, overlap dialing is disabled:
>
>   allowoverlap=no|yes

allowsubscribe
> Specifies whether or not to allow external devices to subscribe to extension status (as set in the hint priority). Defaults to yes:
>
>   allowsubscribe=yes|no

allowtransfers
> If set to no, transfers are disabled for all SIP calls, unless specifically enabled on a per-user or per-peer basis:
>
>   allowtransfers=no|yes

alwaysauthreject
> If this option is enabled, whenever Asterisk rejects an INVITE or REGISTER, it will always reject it with a 401 Unauthorized message instead of letting the caller know whether there was a matching user or peer for his request:
>
>   alwaysauthreject=no|yes

autodomain

Set this option to **yes** to have Asterisk add the local hostname and local IP addresses to the domain list:

```
autodomain=yes|no
```

bindaddr *and* **bindport**

These optional parameters allow you to control the IP interface and port on which you wish to accept SIP connections. If omitted, the port will be set to 5060, and all IP addresses in your Asterisk system will accept incoming SIP connections. If multiple bind addresses are configured, only those interfaces will listen for connections. The address 0.0.0.0 tells Asterisk to listen on all interfaces:

```
bindaddr=0.0.0.0
bindport=5060
```

buggymwi

This setting allows Asterisk to send a message-waiting indication to certain Cisco SIP phones with firmware that doesn't fully support the message-waiting Internet RFC. Enable this option to avoid getting error messages when sending MWI messages on phones with this bug:

```
buggymwi=no|yes
```

callevents

Set this to **yes** when you want SIP to generate Manager events. This will be important if you have external programs that use the Asterisk Manager interface, such as the Flash Operator Panel:

```
callevents=yes
```

checkmwi

This option specifies the default amount of time, in seconds, between mailbox checks for peers:

```
checkmwi=30
```

compactheaders

You can set **compactheaders** to **yes** or **no**. If it's set to **yes**, the SIP headers will use a compact format, which may be required if the size of the SIP header is larger than the maximum transmission unit (MTU) of your IP headers, causing the IP packet to be fragmented. Do not use this option unless you know what you are doing:

```
compactheaders=yes|no
```

defaultexpiry

This sets the default SIP registration expiration time, in seconds, for incoming and outgoing registrations. A client will normally define this value when it initially registers, so the default value you set here will be used only if the client does not specify a timeout when it registers. If you are registering to another user agent server (UAS), this is the registration timeout that it will send to the far end:

```
defaultexpiry=300
```

directrtpsetup

This setting configures the direct RTP setup between two endpoints without the need for RE-INVITEs.

```
directrtpsetup=yes|no
```

 As of the time that this book was written, **directrtpsetup** was still considered experimental, and as such should not be enabled unless you fully understand the consequences. This option will not work for video calls and cases where the called party sends RTP payloads and FMTP headers in the **200 OK** response that do not match the caller's INVITE request.

domain

Sets the default domain for this Asterisk server. If configured, Asterisk will allow **INVITE** and **REFER** messages only to nonlocal domains. You can use the CLI command **sip show domains** to list the local domains:

```
domain=example.com
```

dumphistory

You can set **dumphistory** to **yes** or **no** to enable or disable the printing of the SIP history report at the end of the SIP dialog. The SIP history is printed to the DEBUG logging channel:

```
dumphistory=yes|no
```

externhost

externhost takes a fully qualified domain name as its argument. If Asterisk is behind NAT, the SIP header will normally use the private IP address assigned to the server. If you set this option, Asterisk will perform periodic DNS lookups on the hostname and replace the private IP address with the IP address returned from the DNS lookup:

```
externhost=my.hostname.tld
```

 The use of **externhost** is not recommended in production systems, because if the IP address of the server changes, the wrong IP address will be set in the SIP headers until the next lookup is performed. The use of **externip** is recommended instead.

externip

externip takes an IP address as its argument. If Asterisk is behind NAT, the SIP header will normally use the private IP address assigned to the server. The remote server will not know how to route back to this address; thus, it must be replaced with a valid, routeable address:

>   externip=*216.239.39.104*

externrefresh

>   If `externhost` is used, `externrefresh` configures how long, in seconds, should pass
>   between DNS lookups:
>
>   externrefresh=*30*

g726nonstandard

>   This parameter can be set when dealing with peers that incorrectly use the wrong
>   encoding for the G.726 codec. This setting tells Asterisk to use AAL2 packing order
>   instead of RFC3551 packing order if the peer negotiates G726-32 audio. Ordina-
>   rily, that would be contrary to the RFC3551 specification, as the peer should be
>   negotiating AAL2-G726-32 instead. You may need to set this option if you're using
>   a Sipura or Grandstream device:
>
>   g726nonstandard=yes

ignoreregexpire (*global*)

>   If `ignoreregexpire` is set to `yes`, Asterisk could do one of two things, for:
>
>   *Non-realtime peers*
>
>   >   When their registration expires, the information will *not* be removed from
>   >   memory or the Asterisk database. If you attempt to place a call to the peer, the
>   >   existing information will be used in spite of it having expired.
>
>   *Realtime peers*
>
>   >   When the peer is retrieved from realtime storage, the registration information
>   >   will be used regardless of whether it has expired or not; if it expires while the
>   >   realtime peer is still in memory (due to caching or other reasons), the infor-
>   >   mation will not be removed from realtime storage:
>   >
>   >   ignoreregexpire=yes|no

jbenable

>   Enables the use of an RTP jitter buffer on the receiving side of a SIP channel. De-
>   faults to `no`. An enabled jitter buffer will be used *only* if the sending side can create
>   and the receiving side cannot accept jitter. The SIP channel can accept jitter; thus
>   a jitter buffer on the receiving side will be used only if it is forced and enabled:
>
>   jbenable=yes|no

jbforce

>   Forces the use of the RTP jitter buffer on the receiving side of a SIP channel. Defaults
>   to `no`:
>
>   jbforce=yes|no

jbimpl

>   This setting is used to specify which jitter buffer implementation to use, the
>   `fixed` jitter buffer or the `adaptive` jitter buffer. If the `fixed` jitter buffer is used, it
>   will always be the size defined by `jbmaxsize`. If the `adaptive` jitter buffer is specified,

then the jitter buffer will vary in size up to the maximum size specified by `jbmax size`. This setting defaults to `fixed`:

```
jbimpl=fixed|adaptive
```

jblog

Specifies whether or not to enable jitter buffer frame logging. Defaults to `no`:

```
jblog=yes|no
```

jbmaxsize

Sets the maximum length of the jitter buffer, in milliseconds:

```
jbmaxsize=200
```

jbresyncthreshold

Jump in the frame timestamps over which the jitter buffer is resynchronized. This is useful to improve the quality of the voice, with big jumps in/broken timestamps that are usually sent from exotic devices and programs. Defaults to `1000`:

```
jbresyncthreshold=1000
```

limitonpeers

This setting tells Asterisk to apply call limits to peers only. This will improve call limits and status notification for devices set to `type=friend` because the peer limit will be checked, and not create a separate limit for the user and peer portions of a friend:

```
limitonpeers=yes|no
```

localnet

`localnet` is used to tell Asterisk which IP addresses are considered local, so that the address in the SIP header can be translated to that specified by `externip` or the IP address can be looked up with `externhost`. The IP addresses should be specified in CIDR notation:

```
localnet=192.168.1.0/24
localnet=172.16.0.0/16
```

matchexterniplocally

Specifies that Asterisk should substitute the `externip` or `externhost` setting only if it matches your `localnet` setting. Unless you have some sort of strange network setup you will not need to enable this:

```
matchexterniplocally=yes|no
```

maxexpiry

This sets the maximum amount of time, in seconds, until a peer's registration expires:

```
maxexpiry=3600
```

minexpiry

This sets the minimum amount of time, in seconds, allowed for a registration or subscription:

```
maxexpiry=3600
```

notifymimetype

This takes as its argument a string specifying the MIME type used for the message-waiting indication (MWI) in the SIP NOTIFY message. The most common setting for this field is text/plain, although it can be customized if need be:

```
notifymimetype=text/plain
```

notifyringing

Specifies whether Asterisk should notify subscriptions on RINGING state:

```
notifyringing=yes|no
```

notifyhold

Specifies whether Asterisk should notify subscriptions on HOLD state:

```
notifyhold=yes|no
```

pedantic

You can set pedantic to yes or no. Setting it to yes enables slow pedantic checking for phones that require it, such as the Pingtel, and enables more strict SIP RFC compliancy. In an effort to improve performance, SIP RFC compliance is not normally strictly adhered to:

```
pedantic=yes
```

realm

This option sets the realm for digest authentication. Set realm to your fully qualified domain name, which must be globally unique:

```
realm=mybox.example.com
```

recordhistory

You can set recordhistory to yes or no to enable or disable SIP history recording for all channels:

```
recordhistory=yes|no
```

registerattempts

Specifies how many times Asterisk will attempt its outbound registrations before giving up. This setting defaults to 0, which means that Asterisk will retry indefinitely:

```
registerattempts=0
```

registertimeout

Specifies how often Asterisk should attempt to re-register to other devices:

```
registertimeout=30
```

relaxdtmf

You can set `relaxdtmf` to **yes** or **no**. Setting it to **yes** will relax the DTMF detection handling. Use this if Asterisk is having a difficult time determining the DTMF on the SIP channel. Note that this may cause "talkoff," where Asterisk incorrectly detects DTMF when it should not:

```
relaxdtmf=yes|no
```

rtautoclear (*global*)

This specifies whether or not Asterisk should auto-expire friends created on the fly on the same schedule as if they had just registered. If set to **yes**, when the registration expires, the friend will vanish from the configuration until requested again. If set to an integer, friends expire within that number of seconds instead of the normal registration interval:

```
rtautoclear=yes|no|seconds
```

rtcachefriends (*global*)

If `rtcachefriends` is turned on, Asterisk will cache friends that come from the realtime engine, just as if they had come from *sip.conf*. This often helps with items such as message-waiting indications on realtime peers:

```
rtcachefriends=yes|no
```

rtsavesysname (*global*)

Specifies whether or not Asterisk should save the systemname in the realtime database at the time of registration:

```
rtsavesysname=yes|no
```

rtupdate (*global*)

If set to **yes** Asterisk will update the IP address, origination port, and registration period of a peer upon registration. Defaults to **yes**:

```
rtupdate=yes|no
```

sipdebug

Specifies whether or not Asterisk should turn on SIP debugging from the time that Asterisk loads the SIP channel driver:

```
sipdebug=yes|no
```

sendrpid

Specifies whether or not Asterisk should send a Remote-Party-ID header:

```
sendrpid=yes|no
```

srvlookup

DNS SRV records are a way of setting up a logical, resolvable address where you can be reached. This allows calls to be forwarded to different locations without the need to change the logical address. By using SRV records, you gain many of the advantages of DNS, whereas disabling them removes the ability to place SIP calls based on domain names.

 Currently, the support for SRV records in Asterisk is somewhat lacking. If multiple SRV records are returned, Asterisk will use only the first record.

Using DNS SRV record lookups is highly recommended. To enable them, set srvlookup=yes in the [general] section of *sip.conf*:

```
srvlookup=yes
```

t1min

This is the minimum round-trip time for messages to monitored hosts in milliseconds. Defaults to 100 milliseconds:

```
t1min=100
```

subscribecontext

Limits SUBSCRIBE requests to the specified context. This is useful if you want to limit subscriptions to internal extensions, for example.

This option may also be set on a per-user or per-peer basis:

```
subscribecontext=internal
```

t38pt_udptl

Setting t38pt_udptl to yes enables T.38 fax (UDPTL) passthrough on SIP-to-SIP calls, provided both parties have T.38 support. This setting must be enabled in the general section for all devices to work. You can then disable it on a per-device basis:

```
t38pt_udptl=yes|no
```

 T.38 fax passthrough works only in SIP-to-SIP calls, without any local or agent channel being used. Asterisk cannot currently originate or terminate T.38 fax calls; it can only passthrough UDPTL from one device to another.

tos_sip, tos_audio, *and* tos_video

Asterisk can set the TOS bits in the IP header to help improve performance on routers that respect TOS bits in their routing calculations. The tos_sip, tos_audio, and tos_video settings control the TOS bits for the SIP messages, the RTP audio, and RTP video respectively. Valid: CS0, CS1, CS2, CS3, CS4, CS5, CS6, CS7, AF11, AF12, AF13, AF21, AF22, AF23, AF31, AF32, AF33, AF41, AF42, AF43, and ef (expedited forwarding). You may also use a numeric value for the TOS bits.

For more information, see the *doc/ip-tos.txt* file in the Asterisk source directory.

trustrpid

Specifies whether or not Asterisk should trust the value in the Remote-Party-ID header:

```
trustrpid=yes|no
```

useragent

useragent takes as its argument a string specifying the value for the useragent field in the SIP header. The default value is **asterisk**:

```
useragent=Asterisk PBX v1.4
```

usereqphone

The usereqphone option tells Asterisk to add ;user=phone to SIP URIs that contain a valid phone number:

```
usereqphone
```

videosupport (*both*)

You can set videosupport to yes or no. You can turn it off on a per-peer basis if general video support is enabled, but you can't enable it for one peer only without enabling it in the general section:

```
videosupport=yes|no
```

vmexten

This option sets the dialplan extension to reach the voicemailbox, and will be sent in the Message-Account section of the MWI NOTIFY message. Set this if your SIP device supports the Message-Account setting. This option defaults to **asterisk**:

```
vmexten=8500
```

SIP Channel Definitions

Now that we've covered the global SIP parameters, we will discuss the channel-specific parameters. These parameters can be defined for a user, a peer, or both (as noted in parentheses):

accountcode (*both*)

The account code can be defined on a per-user basis. If defined, this account code will be assigned to a call record whenever no specific user account code is set. The accountcode name configured will be used as the *<filename>.csv* in the */var/log/asterisk/cdr-csv/* directory to store CDRs for the user/peer/friend:

```
accountcode=iax-username
```

allow *and* **disallow** (*both*)

Specific codecs can be allowed or disallowed, limiting codec use to those preferred by the system designer. allow and disallow can also be defined on a per-channel basis. Keep in mind that allow statements in the [general] section will carry over to each of the channels, unless you reset with a disallow=all. Codec negotiation is attempted in the order in which the codecs are defined. Best practice suggests that you define `disallow=all`, followed by explicit `allow` statements for each codec you wish to use. If nothing is defined, `allow=all` is assumed:

```
disallow=all
allow=ulaw
allow=gsm
allow=ilbc
```

amaflags *(both)*

Automatic Message Accounting (AMA) is defined in the Telcordia Family of Documents listed under FR-AMA-1. These documents specify standard mechanisms for generation and transmission of CDRs. You can specify one of four AMA flags (`default`, `omit`, `billing`, or `documentation`) to apply to all SIP connections:

```
amaflags=documentation
```

callerid *(both)*

You can set a suggested Caller ID string for a user or peer with `callerid`. If you define a Caller ID field for a user, any calls that come in on that channel will have that Caller ID assigned to them, regardless of what the far end sends to you. If Caller ID is defined for a peer, you are requesting that the far end use that to identify you (keep in mind, however, that you have no way to ensure that it will do so). If you want incoming callers to be able to define their own Caller IDs (i.e., for guests), make sure you do not set the `callerid` field:

```
callerid=John Smith <(800) 555-1234>
```

callgroup *and* pickupgroup *(both)*

You can use the `callgroup` parameter to assign a channel definition to one or more groups, and you can use the `pickupgroup` option in conjunction with this parameter to allow a ringing phone to be answered from another extension. The `pickupgroup` option is used to control which callgroups a channel may pick up—a channel is given authority to answer another ringing channel if it is assigned to the same `pickupgroup` as the ringing channel's callgroup. By default, remote ringing extensions can be answered with *8 (this is configurable in the *features.conf* file):

```
callgroup=1,3-5
pickupgroup=1,3-5
```

callingpres *(both)*

Sets Caller ID presentation for this user/peer. This setting takes one of the following options:

allowed_not_screened

Presentation allowed, not screened

allowed_passed_screen

Presentation allowed, passed screen

`allowed_failed_screen`
Presentation allowed, failed screen

`allowed`
Presentation allowed, network number

`prohib_not_screened`
Presentation prohibited, not screened

`prohib_passed_screen`
Presentation prohibited, passed screen

`prohib_failed_screen`
Presentation prohibited, failed screen

`prohib`
Presentation prohibited, network number

`unavailable`
Number unavailable

`=yes|no`

canreinvite (*both*)

The SIP protocol tries to connect endpoints directly. However, Asterisk must remain in the transmission path between the endpoints if it is required to detect DTMF (for more information, see Chapter 4):

`canreinvite=no`

context (both)

A context is assigned to a channel definition to direct incoming calls into the matching context in *extensions.conf*, where call handling is performed (see Chapters Chapter 4 and Chapter 5). Any channel connecting to an Asterisk machine has to have a context defined into which it will arrive. The context is essential for any user channel definition; if you do not define a context, incoming calls will be directed to the **default** context:

`context=incoming`

 You should be aware of an unusual scenario that will require a context definition for a peer. When a call comes through the SIP channel, it first tries to find a matching user definition (based on the user name in square brackets and the secret). If it can't find any matching users, it then looks for matching peers, based on the IP address that the call is coming from. Since peers don't normally have contexts, this will cause such a call to arrive in the **default** context. While this will work, the **default** context shouldn't really be used to handle incoming calls. The solution is to define a context, on a per-peer basis, for any peers that might match on incoming calls. To experiment with this, you can call your Free World Dialup number; the call will come right back to you.

defaultip (*peer*)

The `defaultip` setting complements `host=dynamic`. If a host has not yet registered with your server, you'll attempt to send messages to the default IP address configured here:

```
defaultip=192.168.1.101
```

deny (*both*)

Specific IP addresses and ranges can be controlled with the `deny` option. To restrict access from a range of IP addresses, use a subnet mask—for example, `deny=192.168.1.0/255.255.255.0`. You can also deny all addresses with `deny=0.0.0.0/0.0.0.0` and then allow only certain addresses with the `permit` command. Be aware of the security implications of this setting (see also `permit`):

```
deny=0.0.0.0/0.0.0.0
```

disallow (*both*)

See `allow`.

dtmfmode (*both*)

You can set `dtmfmode` to `inband`, `rfc2833`, or `info`. DTMF digits can be sent either in band (as part of the audio stream), or out of band (as signaling information), using the RFC 2833 or INFO methods. The `inband` method works reliably only when using an uncompressed codec such as G.711, μlaw, or alaw. The recommended method is to use `rfc2833`; however, some devices—such as those by Grandstream—support the `info` method:

```
dtmfmode=rfc2833
```

 In Asterisk 1.4, Variable Length DTMF was introduced in order to allow Asterisk to correctly signal to the far end the duration of a key press on the phone connected to the incoming channel (per IETF RFC 2833). Older Asterisk systems do not understand the variable-length parameter. In older Asterisk systems, DTMF delivered via RFC 2833 may not be correctly interpreted, leading to strange effects in sessions such as voicemail. If you want to have the older (pre-1.4) behaviour of the `rfc2833` setting, you must add the `rfc2833compensate=yes` option to the peer in *sip.conf* that defines communication with your pre-1.4 Asterisk system.

fromdomain (*peer*)

This allows you to set the domain in the `From:` field of the SIP header. It may be required by some providers for authentication:

```
fromdomain=my.hostname.tld
```

fromuser (*peer*)

This allows you to set the username with which to authenticate. The name contained within the square brackets of the channel definition is usually used, but this can be overridden with the `fromuser` option. This allows a channel definition to be referenced with a name other than that used to authenticate:

```
fromuser=john_smith
```

host (*peer*)

This configures the host to which this peer is to connect. Use a fully qualified domain name:

```
host=remote.hostname.tld
```

incominglimit (*both*)

This option limits the total number of simultaneous calls for a peer or user. It sets the max number of simultaneous outgoing calls for a peer, or the max number of incoming calls for a user.

```
incominglimit=3
```

insecure (*both*)

When an `INVITE` is received from a remote location, Asterisk attempts to authenticate the string of characters before the @ sign on the `INVITE` line received in the SIP header with the name of a channel definition in *sip.conf*. If the remote end is a user agent, it will authenticate based on a **user** definition. However, if the remote end is a SIP proxy service, it will authenticate on the **peer** entry. When calls come from a provider such as Free World Dialup, which acts as a proxy for the true remote end who is calling you, that provider cannot authenticate the call on behalf of the endpoint. Since it would be impractical to have an authentication configured for every FWD user, and since FWD cannot respond to a 407 Proxy Authentication Required response, there must be an alternate way to allow calls from these callers.

If you set `insecure=invite`, you'll determine which peer to match on by comparing the IP address or hostname and port number to those provided in the **Contact** field of the SIP header with the **host** and **port** options in *sip.conf*. If a match is found, authentication will not be required on the initial `INVITE`, and the call will be allowed.

If you have multiple endpoints behind a NAT device, you need to enable `insecure=port` to match against only the IP address. To not require authentication on the incoming `INVITE` for the peer, set `insecure=invite,port`:

```
insecure=invite
```

language (*both*)

This sets the language flag to whatever you define. The global default language is English. The language that is set is sent by the channel as an information element. It is also used by applications such as `SayNumber()` that have different files for different languages. Keep in mind that languages other than English are not explicitly installed on the system, and it is up to you to configure the system to ensure that the language you specify is handled properly:

```
language=en
```

mailbox (*peer*)

If you associate a `mailbox` with a peer within the channel definition, voicemail will send a MWI to the nodes on the end of that channel. If the mailbox number is in a voicemail context other than `default`, you can specify it as *mailbox @ context*. To associate multiple mailboxes with a single peer, use multiple `mailbox` statements:

```
mailbox=1000@internal
```

maxcallbitrate (*both*)

Sets the maximum bitrate for an individual call from this user or to this peer. Defaults to 384 Kb/s:

```
maxcallbitrate=384
```

md5secret (*both*)

If you do not wish to have plain-text secrets in your *sip.conf* files, you can use `md5secret` to configure the MD5 hash that can be used for authentication. To generate the MD5 hash from the Linux console, use the following command:

```
# echo -n "username:realm:secret" | md5sum
```

Be sure to use the -n flag, or `echo` will add a \n to the end of the string; the line feed will then be calculated into the MD5 hash, creating the incorrect hash. The *realm*, if not specified with the `realm` option (discussed in the list of general SIP parameters), defaults to `asterisk`. If both an `md5secret` and a `secret` are specified in the same channel definition, the `secret` will be ignored:

```
md5secret=0bcbe762982374c276fb01af6d272dca
```

mohinterpret (*channel*)

This option specifies a preference for which MoH class this channel should listen to when put on hold if the music class has not been set on the channel with `Set (CHANNEL(musicclass)=whatever)` in the dialplan, and the peer channel putting this one on hold did not suggest a music class.

This option may be specified globally, or on a per-user or per-peer basis:

```
mohinterpret=default
```

mohsuggest (*channel*)

This option specifies which music-on-hold class (as defined in *musiconhold.conf*) to suggest to the peer channel when this channel places the peer on hold. It may be specified globally or on a per-user or per-peer basis:

```
mohsuggest=default
```

musicclass (*both*)

This option sets the default music-on-hold class:

```
musicclass=classical
```

nat (*both*)

You can set nat to yes, no, or never. If you set it to yes, Asterisk ignores the IP address in the SIP and SDP headers and responds to the address and port in the IP header. The never option is for devices that cannot handle rport in the SIP header, such as the Uniden UIP200:

```
nat=yes|no|never
```

permit (*both*)

See deny.

pickupgroup (*both*)

See callgroup.

port (*peer*)

You can use this to define the port on which to listen for SIP signaling, if you want to listen on a nonstandard port. (The default port for SIP signaling is 5060.)

```
port=5060
```

progressinband (*both*)

You can set progressinband to yes, no, or never, to configure whether or not to generate in-band ringing. Normally, Asterisk will send the progress of a call via a few methods, such as 183 Session Progress, 180 Ringing, 486 Busy, and so on. If you set progressinband=yes, Asterisk will indicate the call progress in band by generating tones:

```
progressinband=yes|no|never
```

promiscredir (*both*)

You can set promiscredir to yes or no. Normally, when you perform call forwarding on a phone, Asterisk will use the Local channel (for example, local/18005551212@peer). If you set promiscredir=yes, Asterisk will use the SIP channel instead, which enables you to forward the calls to remote boxes:

```
promiscredir=yes|no
```

 Note that if Asterisk performs a redirect to itself when promiscredir=yes, the system will receive an INVITE with the same Caller ID and detect a loop to itself. SIP does not have the ability to perform a hairpin call, so the channel will then be destroyed.

qualify (*peer*)

You can set qualify to yes, no, or a time in milliseconds. If you set qualify=yes, NOTIFY messages will be sent periodically to the remote peers to determine whether they are available and what the latency between replies is. A peer is determined unreachable if no reply is received within 2,000 ms (to change this default, instead set qualify to the number of milliseconds to wait for the reply). Use this option in conjunction with nat=yes to keep the path through the NAT device alive:

```
qualify=yes|no|seconds
```

regcontext (*peer*)

By specifying the context that contains the actions to perform, you can configure Asterisk to perform a number of actions when a peer registers to your server. This option works in conjunction with **regexten** by specifying the extension to execute. If no **regexten** is configured, the peer name is used as the extension. Asterisk will dynamically create and destroy a **NoOp** at priority 1 for the extension. All actions to be performed upon registration should start at priority 2. More than one **regexten** may be supplied, if separated by an **&**. **regcontext** can be set on a per-peer basis or globally:

```
regcontext=peer_registrations
```

regexten (*peer*)

The **regexten** option is used in conjunction with **regcontext** to specify the extension that is executed within the configured context. If **regexten** is not explicitly configured, the peer name is used as the extension to match:

```
regexten=1000
```

rtpholdtimeout (*peer*)

This takes as its argument an integer, specified in seconds. It terminates a call if no RTP data is received while on hold within the time specified. The value of **rtphold timeout** must be greater than that of **rtptimeout** (see also **rtptimeout**):

```
rtpholdtimeout=120
```

rtpkeepalive (*peer*)

Specifies how often Asterisk should send keepalives in the RTP stream, in seconds. Defaults to zero, which means Asterisk won't send any RTP keepalives:

```
rtpkeepalive=45
```

rtptimeout (*peer*)

This takes as its argument an integer, specified in seconds. It terminates a call if no RTP data is received within the time specified:

```
rtptimeout=60
```

secret (*both*)

This sets the password to use for authentication:

```
secret=welcome
```

setvar (*both*)

This sets a channel variable, which will be available when a channel to the peer or user is created and will be destroyed when the call is hung up. For example, to set the channel variable **foo** with a value of **bar**, use:

```
setvar=foo=bar
```

username (*peer*)

The username field allows you to attempt contact with a peer before it has registered with you. At registration, a SIP device tells Asterisk which SIP URI to use to contact it. The username is used in conjunction with defaultip to create the SIP URI in the SIP INVITE header. This might be useful following a reboot, in order to place a call. The endpoints will not attempt to register with the server until their registration timeouts expire, so you will not know their locations. For nondynamic hosts, you will require the username to be specified, as it is used to construct the authorization username:

```
username=john_smith
```

Application Reference

Applications are the core functionality of the dialplan. Generally these all will operate on the channel, whereas functions, described in Appendix F, merely return values that can be used by applications. There are a few applications that still simply return values, but these will probably be deprecated in a future version and replaced with dialplan functions.

There are a few things to keep in mind about applications. First, when they exit, they will either terminate normally or abnormally. Abnormal termination almost always occurs when an application detects that the channel has hung up (or if it doesn't, the dialplan will detect that shortly thereafter). An application may also exit abnormally when it wishes to indicate to the dialplan that some condition has not been satisfied and that it should force a hangup. In all other cases, an application will exit normally, which indicates that processing should continue at the next priority in the dialplan.

In many cases, if you wish to override the application's wish to cause a hangup, you may wrap the application in a `TryExec()`.

In many places throughout this reference, you will see what's described as a *label*. This is shorthand for describing a location in the dialplan, whether it is simply a *priority*; an *extension* and a *priority*; or a *context*, an *extension*, and a *priority*. Note that if a *text label* is defined for a particular priority, the *priority* may be replaced with that *text label* in any of those cases. See the `GotoIf()` application for more information and an example.

 You will find many of the examples in this appendix to contain numbered priorities, which is not the preferred method of writing dialplans. We prefer the use of the 'n' priority for all priority numbers except 1 (which is required), but we have decided to utilize them in order to make some of the examples more clear.

AddQueueMember() — Dynamically adds queue members to the specified call queue

`AddQueueMember(queuename[,interface[,penalty,[option,[membername]]]])`

Dynamically adds the specified *interface* to an existing queue named *queuename*, as specified in *queues.conf*. If specified, *penalty* sets the penalty for queues to use this member. Members with a lower penalty are called before members with a higher penalty.

The AddQueueMember() application sets a channel variable named AQMSTATUS upon completion. The AQMSTATUS variable will be set to one of the following values:

```
ADDED
MEMBERALREADY
NOSUCHQUEUE
```

Calling AddQueueMember() without an *interface* argument will use the interface that the caller is currently using.

If the *option* argument is set to j, Asterisk cannot add the *interface* to the specified queue, and there exists an n+101 priority (where n is the number of the current priority), the call will jump to that priority.

The *membername* argument may be set to the name of the queue member. Consequently, this name will show up in the entries of the *queue_log* as well as Asterisk Manager Interface events, making it easier to identify the agent for reporting purposes:

```
; add SIP/3000 to the techsupport queue, with a penalty of 1
exten => 123,1,AddQueueMember(techsupport,SIP/3000,1)
```

See Also

Queue(), RemoveQueueMember(), PauseQueueMember(), UnpauseQueueMember(), AgentLogin() , *queues.conf*

ADSIProg()
Loads an ADSI script into an ADSI-capable phone

ADSIProg(*script*)

Programs an Analog Display Services Interface (ADSI) phone with the given *script*. If none is specified, the default script, *asterisk.adsi*, is used. The path for the *script* is relative to the Asterisk configuration directory (usually */etc/asterisk/*). You may also provide the full path to the script.

To get the CPE ID and other information from your ADSI-capable phone, use the GetCPEID() application:

```
; program the ADSI phone with the telcordia-1.adsi script
exten => 123,1,ADSIProg(telcordia-1.adsi)
```

See Also

GetCPEID(), *adsi.conf*

AgentCallbackLogin()

```
AgentCallbackLogin([AgentNumber][,[options][,[exten]@context]])
```

Allows a call agent identified by *AgentNumber* to log in to the call queue system, to be called back when a call comes in for that agent.

When a call comes in for the agent, Asterisk calls the specified *exten* (with an optional *context*).

The *options* argument may contain the letter s, which causes the login to be silent:

```
; silently log in as agent number 42, and have Asterisk
; call extension 123 in the internal context
; when a call comes in for this agent
exten => 123,1,AgentCallbackLogin(42,s,123@internal)
```

This application is deprecated, and the functionality has been replaced with AEL dialplan logic located in the *doc/queues-with-callback-members.txt* file within the Asterisk source.

See Also

Queue(), AgentLogin(), AddQueueMember(), RemoveQueueMember(), PauseQueueMember(), UnpauseQueueMember(), AGENT, *agents.conf*, *queues.conf*

AgentLogin()

```
AgentLogin([AgentNumber][,options])
```

Logs the current caller in to the call queue system as a call agent (optionally identified by *AgentNumber*). While logged in, the agent can receive calls and will hear a beep on the line when a new call comes in. The agent can hang up the current call by pressing the asterisk (∗) key. If *AgentNumber* is not specified, the caller will be prompted to enter her agent number. Agents are defined in *agents.conf*.

The *options* argument may contain the letter s, which causes the login to be silent:

```
; silently log in the caller as agent number 42, as defined in agents.conf
exten => 123,1,AgentLogin(42,s)
```

See Also

Queue(), AddQueueMember(), RemoveQueueMember(), PauseQueueMember(), UnpauseQueue-Member(), AGENT, *agents.conf*, *queues.conf*

AgentMonitorOutgoing()

```
AgentMonitorOutgoing([options])
```

Records all outbound calls made by a call agent.

This application tries to figure out the ID of the agent who is placing an outgoing call based on a comparison of the Caller ID of the current interface and the global variable set by the AgentCallbackLogin() application. As such, it should be used only in conjunction with (and after!) the AgentCallbackLogin() application. It uses the monitoring functions in the chan_agent module instead of the Monitor() application to record the calls. This means that call recording must be configured correctly in the *agents.conf* file.

By default, recorded calls are saved to the */var/spool/asterisk/monitor/* directory. This may be overridden by changing the savecallsin parameter in *agents.conf*.

If the Caller ID and/or agent ID are not found, this application will go to priority n+1, if it exists (where n is the current priority).

Returns 0 unless overridden by one of the options.

The *options* argument may include one or more of the following:

d

> Make this application return -1 if there is an error condition and there is no extension n+101.

c

> Change the Call Detail Record so that the source of the call is recorded as Agent/ agent_id.

n

> Don't generate warnings when there is no Caller ID or if the agent ID is not known. This option is useful if you want to have a shared context for agent and non-agent calls.

```
; record outbound calls for this agent, and change the CDR to reflect
; that the call is being made by an agent
exten => 123,1,AgentMonitorOutgoing(c)
```

See Also

AgentCallbackLogin(), *agents.conf*

AGI()   **Executes an AGI-compliant application**

[E]AGI(*program*[,*arguments*])

Executes an Asterisk Gateway Interface-compliant *program* on the current channel. AGI programs allow external programs (written in almost any language) to control the telephony channel by playing audio, reading DTMF digits, and so on. Asterisk communicates with the AGI program on STDIN and STDOUT. The specified *arguments* are passed to the AGI program.

The *program* must be set as executable in the underlying filesystem. The program path is relative to the Asterisk AGI directory, which by default is */var/lib/asterisk/agi-bin/*.

If you want to run an AGI when no channel exists (such as in an h extension), use the `DeadAGI()` application instead. You may want to use the `FastAGI()` application if you want to do AGI processing across the network.

If you want access to the inbound audio stream from within your AGI program, use `EAGI()` instead of `AGI()`. Inbound audio can then be read in on file descriptor 3.

If the channel hangs up prematurely, the process initiated by the `AGI` command will be sent a `HUP` signal to tell it that the channel has hung up. If your program does not catch this signal, it will be terminated. You can override this behavior by setting the channel variable `AGISIGHUP` to 0:

```
; call the demo AGI program
exten => 123,1,AGI(agi-test)
exten => 123,2,EAGI(eagi-test)
```

See Also

`DeadAGI()`, `FastAGI()`, Chapter 9

AlarmReceiver()   Provides support for receiving alarm reports from a burglar or fire alarm panel

`AlarmReceiver()`

Emulates an alarm receiver, and allows Asterisk to receive and decode special data from fire and/or burglar alarm panels. At this time, only the Ademco Contact ID format is supported.

When called, `AlarmReceiver()` will handshake with the alarm panel, receive events, validate them, handshake them, and store them until the panel hangs up. Once the panel hangs up, the application will run the command line specified by the `eventcmd` setting in *alarmreceiver.conf* and pipe the events to the standard input of the application. *alarmreceiver.conf* also contains settings for DTMF timing and for the loudness of the acknowledgment tones.

```
; set up Asterisk to answer a call from a supported fire alarm panel
exten => s,1,AlarmReceiver()
```

 This application is not guaranteed to be reliable, so don't depend on it unless you have extensively tested it. If you use this application without extensive testing, you may be putting your life and property at great risk.

See Also

alarmreceiver.conf

```
AMD([initialSilence[,greeting[,afterGreetingSilence[,totalAnalysisTime
[,minimumWordLength[,betweenWordsSilence[,maximumNumberOfWords
[,silenceThreshold]]]]]]]])
```

This application attempts to detect an answering machine, based on the timing patterns. This application is usually used by outbound calls originated from either call files or from the Asterisk manager Interface. This application sets AMDSTATUS variable is set to one of the following, to show what type of call was detected:

MACHINE
>   The called party is believed to be an answering machine.

HUMAN
>   The called party is believed to be a human being, and not an answering machine.

NOTSURE
>   The application was unable to tell whether the called party was a human or an answering machine.

HANGUP
>   A hangup occurred during the detection.

The AMD() application also sets a channel variable named AMDCAUSE with the cause that lead to the conclusion stated in the AMDSTATUS variable. The AMDCAUSE variable will be set to one of the following values:

>   TOOLONG-*total_time*
>   INITIALSILENCE-*silence_duration-initial_silence*
>   HUMAN-*silence_duration-after_greeting_silence*
>   MAXWORDS-*word_count-maximum_number_of_words*
>   LONGGREETING-*voice_duration-greeting*

The parameters to this application all help tune it so that it can more effectively tell the difference between a human and an answering machine. If the parameters are not passed to this application, Asterisk will read the default values as configured in *amd.conf*. The parameters are:

initialSilence
>   The maximum silence duration before the greeting. If exceeded, then the AMDSTATUS variable will be set to MACHINE.

greeting
>   The maximum length of the greeting. If exceeded, then the AMDSTATUS variable will set to MACHINE.

afterGreetingSilence
>   The maximum silence after detecting a greeting. If exceeded, then the AMDSTATUS variable will be set to MACHINE.

totalAnalysisTime

The maximum time allowed for the algorithm to decide whether the called party is a human or an answering machine.

minimumWordLength

If the duration of the voice activity is shorter than `minimumWordLength`, it will not be considered to be human speech.

betweenWordsSilence

The minimum duration of silence after a word to consider the audio that follows as a new word.

maximumNumberOfWords

The maximum number of words detected in the greeting. If exceeded, then the `AMDSTATUS` variable will set to `MACHINE`.

silenceThreshold

How sensitive the algorithm should be when detecting silence

```
; Use answering machine detection.  If the called party
; is human, connect them to Bob.  Otherwise, play a
; message and hang up
exten => 123,1,Answer()
exten => 123,n,AMD()
exten => 123,n,GotoIf($["${AMDSTATUS}" = "HUMAN"]?human:machine)
exten => 123,n(machine),WaitForSilence(2000)
exten => 123,n,Playback(asterisk-friend)
exten => 123,n,Hangup()
exten => 123,n(human),Verbose(3, We've got a human on the line!)
exten => 123,n,Playback(transfer)
exten => 123,n,Dial(SIP/bob)
exten => 123,n,Playback(im-sorry)
exten => 123,n,Hangup()
```

See Also

`WaitForSilence()`

Answer()

Answers a channel, if it is ringing

`Answer([delay])`

Causes Asterisk to answer the channel if it is currently ringing. If the current channel is not ringing, this application does nothing.

If a *delay* is specified, Asterisk will answer the call and then wait *delay* milliseconds before going on to the next priority in the dialplan.

It is often a good idea to use `Answer()` on the channel before calling any other applications, unless you have a very good reason not to. There are several key applications that require that the channel be answered before they are called, and may not work correctly otherwise:

```
exten => 123,1,Answer(750)
exten => 123,n,Playback(tt-weasels)
```

See Also

Hangup()

AppendCDRUserField() Appends a value to the user field of the Call Detail Record

AppendCDRUserField(*value*)

Appends *value* to the user field of the Call Detail Record (CDR). The user field is often used to store arbitrary data about the call, which may not be appropriate for any of the other fields:

```
; set the user field to 'abcde'
exten => 123,1,SetCDRUserField(abcde)
; now append 'xyz'
exten => 123,1,AppendCDRUserField(xyz)
```

 This application has been deprecated in favor of the CDR function.

```
exten => 123,1,Set(CDR(userfield)=${CDR(userfield)}12345)
```

See Also

SetCDRUserField(), ForkCDR(), NoCDR(), ResetCDR(), the CDR

Authenticate() Requires that the caller enter a correct password before continuing

Authenticate(*password*[,*options*[,*maxdigits*]])

Requires a caller to enter a given *password* in order to continue execution of the next priority in the dialplan. Authenticate() gives the caller three chances to enter the password correctly. If the password is not correctly entered after three tries, the channel is hung up.

If *password* begins with the / character, it is interpreted as a file that contains a list of valid passwords (one per line). Passwords may also be stored in the Asterisk database (AstDB); see the d option below.

The *maxdigits* parameter sets the maximum number of digits that may be entered by the caller. It not set, the application will accept an unlimited number of digits and will wait for the caller to press the # key after entering his authentication code.

A set of *options* may be provided, consisting of one or more of the letters in the following list:

a

Sets the CDR field named `accountcode` and the channel variable `ACCOUNTCODE` to the password that is entered

d

Interprets the path as the database key from the Asterisk database in which to find the password, not a literal file. When using a database key, the value associated with the key can be anything.

j

Supports jumping to priority n+101 if authentication fails

m

Interprets the given path as a file that contains a list of account codes and password hashes delimited with : (colon character), listed one per line in the file. When one of the passwords is matched, the channel will have its account code set to the corresponding account code in the file.

r

Removes the database key upon successful entry (valid with **d** only).

```
; force the caller to enter the password before continuing,
; and set the CDR field named 'accountcode' to the entered password
exten => 123,1,Answer()
exten => 123,n,Authenticate(1234,a)
exten => 123,n,Playback(pin-number-accepted)
exten => 123,n,SayDigits(${ACCOUNTCODE})
```

See Also

VMAuthenticate(), DISA(), Chapter 6

Background()   Plays a file while accepting touch-tone (DTMF) digits

Background(*filename1*[&*filename2*...][,*options*[,*language*]])

Plays the specified audio file(s) while waiting for the user to begin entering DTMF digits. Once the user begins to enter DTMF digits, the playback is terminated. Asterisk tries to find a matching extension in the destination *context* (or the current context if none is specified), and execution of the dialplan will continue at the matching extension as soon as an unambiguous match is found.

The *filename* should be specified without a file extension, as Asterisk will automatically find the file format with the lowest translation cost.

Valid *options* include one of the following:

s

Causes the playback of the message to be skipped if the channel is not in the "up" state (i.e., hasn't yet been answered). If **s** is specified, the application will return immediately should the channel not be off-hook.

---

n

Does not answer the channel before playing the specified file. Without this option, the channel will automatically be answered before the sound is played. Not all channels support playing messages before being answered.

m

Only break if a digit hit matches a one-digit extension in the destination context.

The *language* argument may be used to specify a language to use for playing the prompt, if it differs from the current language of the channel.

```
exten => 123,1,Answer()
exten => 123,2,Background('exter-ext-of-person');
```

See Also

ControlPlayback(), WaitExten(), BackgroundDetect(), TIMEOUT

BackgroundDetect()   Plays a file in the background and detects talking

BackgroundDetect(*filename*[,*sil*[,*min*[,*max*]]])

Similar to Background(), but attempts to detect talking.

During the playback of the file, audio is monitored in the receive direction. If a period of non-silence that is greater than *min* milliseconds yet less than *max* milliseconds and is followed by silence for at least *sil* milliseconds occurs, the audio playback is aborted and processing jumps to the **talk** extension, if available.

If unspecified, *sil*, *min*, and *max* default to 1,000 ms, 100 ms, and infinity, respectively.

```
exten => 123,1,BackgroundDetect(tt-monkeys)
exten => 123,2,Playback(im-sorry)
exten => talk,1,Playback(yes-dear)
```

See Also

Playback(), Background()

Busy()   Indicates a busy condition to the channel

Busy([*timeout*])

Requests that the channel indicate the busy condition and then waits for the user to hang up or for the optional *timeout* (in seconds) to expire.

This application signals a busy condition only to the bridged channel. Each particular channel type has its own way of communicating the busy condition to the caller. You can use Playtones(busy) to play a busy tone to the caller.

```
exten => 123,1,Playback(im-sorry)
exten => 123,2,Playtones(busy)
exten => 123,3,Busy()
```

See Also

Congestion(), Progress(), Playtones(), Hangup()

ChangeMonitor()   Changes the monitoring filename of a channel

ChangeMonitor(*filename_base*)

Changes the name of the recorded file created by monitoring a channel with the Monitor() application. This application has no effect if the channel is not monitored. The argument *filename_base* is the new filename base to use for monitoring the channel.

```
; start recording this channel with a basename of 'sample'
exten => 123,1,Monitor(sample)
; change the filename base to 'example'
exten => 123,2,ChangeMonitor(example)
```

See Also

Monitor(), StopMonitor(), MixMonitor()

ChanIsAvail()   Finds out if a specified channel is currently available

ChanIsAvail(*technology1/resource1*[*&technology2/resource2*...][*,option*])

Checks to see if any of the requested channels are available. This application also sets the following channel variables:

AVAILCHAN
: The name of the available channel, including the call session number used to perform the test

AVAILORIGCHAN
: The canonical channel name that was used to create the channel—that is, the channel name without any session number

AVAILSTATUS
: The status code for the channel

If the option s (which stands for "state") is specified, Asterisk will consider the channel unavailable whenever it is in use, even if it can take another call.

If the j option is specified, and none of the requested channels are available, the new priority will be n+101 (where n is the current priority), if that priority exists.

```
; check both Zap/1 and Zap/2 to see if they're available
exten => 123,1,ChanIsAvail(Zap/1&Zap/2)
```

```
; print the available channel name to the Asterisk CLI
exten => 123,2,Verbose(0,${AVAILORIGCHAN})
```

 This application does not work correctly on MGCP channels.

ChannelRedirect()
Redirects a channel to a new location in the dialplan

ChannelRedirect(*channel*,[[*context*,]*extension*,]*priority*)

This application redirects the specified *channel* to a new *priority* in the dialplan. If *extension* is not specified, the current extension is assumed. If *context* is not specified, the current context will be assumed:

```
; Transfer SIP/Bob to hold music when extension 123 is dialed
exten => 123,1,ChannelRedirect(SIP/Bob,124,1)

exten => 124,1,Answer()
exten => 124,2,MusicOnHold()
```

See Also

Transfer()

ChanSpy()
Listens to the audio on a channel, and optionally whisper to the calling channel

ChanSpy([*chanprefix*[,*options*]])

This application is used to listen to the audio going to and from an Asterisk channel. If the *chanprefix* parameter is specified, only channels beginning with this value will be spied upon.

While a channel is being spied upon, the following actions may be performed:

- Dialing # cycles the volume level.
- Dialing * will cause the application to spy on the next available channel.
- Dialing a series of digits followed by # builds a channel name (which will be appended to *chanprefix*). For example, placing ChanSpy(Zap) and then dialing the digits 42# while spying will begin spying on the channel Zap/42.

The *options* parameter may contain zero or more of the following options:

b
    Only spy on channels that are involved in a bridged call.

g(*group*)
    Only spy on channels that contain a channel variable named SPYGROUP, which should contain *group* in an optional colon-delimited list.

---

q

Quiet mode. Tells the application not to beep or read the selected channel's name when spying begins.

r[(*basename*)]

Records the channel audio to the monitor spool directory (usually */var/spool/aster isk/monitor*). An optional *basename* set the base filename of the recordings, which defaults to chanspy.

v([*value*])

Adjusts the volume of the audio being listened to. The *value* must be in the range of **4** to **-4**. A negative *value* will make the volume quieter, while a positive value will make it louder.

w

Whisper mode. This allows the spying channel to talk to the spied-upon channel, without any other bridged channel being able to hear the audio.

W

Private whisper mode. This enables the spying channel to speak to the spied-upon channel without being able to hear the audio from the spied-upon channel.

```
; Spy on the Zap channels in whisper mode
exten => 123,1,ChannelSpy(Zap,w)
```

See Also

ExtenSpy()

Congestion()   Indicates congestion on the channel

Congestion([*timeout*])

Requests that the channel indicate congestion and then waits for the user to hang up or for the optional *timeout* (in seconds) to expire.

This application signals congestion only to the far end; it doesn't actually play a congestion tone to the user. Use Playtones(congestion) to play a congestion tone to the caller.

 If you use this command without a timeout, you run the risk of having a channel get stuck in this state. This is not really needed when you want to indicate congestion to a user. Just use Playtones(congestion) so they hear the fast-busy, and then Hangup().

Always exits abnormally:

```
; if the Caller ID number is 555-1234, always play congestion
exten => 123,1,GotoIf($[${CALLERID(num)} = 5551234]?5:2)
exten => 123,2,Playtones(congestion)
exten => 123,3,Congestion(3)
```

```
exten => 123,4,Hangup()
exten => 123,5,Dial(Zap/1)
```

See Also

Busy(), Progress(), Playtones(), Hangup()

ContinueWhile()   Restart a While() loop

ContinueWhile()

Return to the top of a While loop and re-evaluate the conditional.

See Also

While(), ExitWhile()

ControlPlayback()   Plays a file, with the ability to fast forward and rewind the file

ControlPlayback(*file*[,*skipms*[,*ff*[,*rew*[,*stop*[,*pause*[,*restart*[,*options*]]]]]]])

Plays back a given *file* (without the file extension), while allowing the caller to move forward and backward through the file by pressing *ff* and *rew* keys. By default, you can use * and # to rewind and fast-forward the playback of the file, respectively.

The *skipms* option specifies how far forward or backward to jump in the file with each press of *ff* or *rew*.

If *stop* is specified, the application will stop playback when *stop* is pressed.

A *pause* argument may also be specified, which when pressed will pause playback of the file. Pressing *pause* again will continue the playback of the file.

If the *restart* parameter is specified, the specified key may be used to restart the playback of the *file*.

If the *options* parameter is set to j, and the *file* does not exist, the application jumps to priority n+101, if present (where n is the current priority number).

The ControlPlayback() application sets a channel variable named CPLAYBACKSTATUS upon completion. The CPLAYBACKSTATUS variable will be set to one of the following values:

```
SUCCESS
USERSTOPPED
ERROR

; allow the caller to control the playback of this file
exten => 123,1,ControlPlayback(tt-monkeys|3000|#|*|5|0)
```

See Also

Playback(), Background(), Dictate(),

DateTime()
Says the date and/or time in the user-specified format

DateTime([*unixtime*[,*timezone*[,*format*]]])

If the *unixtime* parameter is specified, this application says that date and time. Otherwise, it says the current date and time. If a *timezone* is specified, the date and time is calculated according to that time zone. Otherwise, the time zone setting of the Asterisk server is used. If the *format* parameter is specified, the date and time will be said according to that format. (See the sample *voicemail.conf* file for more information on the date and time format.)

```
; say the current date and time in several time zones
exten => 123,1,DateTime(,America/New_York)
exten => 123,2,DateTime(,America/Chicago)
exten => 123,3,DateTime(,America/Denver)
exten => 123,4,DateTime(,America/Los_Angeles)
```

DBdel()
Deletes a key from the AstDB

DBdel(*family*/*key*)

Deletes the key specified by *key* from the key family named *family* in the AstDB.

```
exten => 123,1,DBput(test/name=John) ; add name to AstDB
exten => 123,2,DBget(NAME=test/name) ; retrieve name from AstDB
exten => 123,3,DBdel(test/name)      ; delete from AstDB
```

> This application is deprecated and the functionality has been replaced with the DB_DELETE() function.

See Also

DB_DELETE(), DBdeltree(), DB

DBdeltree()
Deletes a family or key tree from the AstDB

DBdeltree(*family*[/*keytree*])

Deletes the specified *family* or *keytree* from the AstDB.

```
; create a couple of entries in the AstDB
exten => 123,1,DBput(test/blue)
exten => 123,2,DBput(test/green)
```

```
; now delete the key family named test
exten => 123,3,DBdeltree(test)
```

See Also

DB_DELETE(), DBdel(), DB

DeadAGI()
Executes an AGI-compliant script on a dead (hung-up) channel

DeadAGI(*program*,*args*)

Executes an AGI-compliant *program* on a dead (hung-up) channel. AGI allows Asterisk to launch external programs written in almost any language to control a telephony channel, play audio, read DTMF digits, and so on by communicating with the AGI protocol on STDIN and STDOUT. The arguments specified by *args* will be passed to the program.

This application has been written specifically for dead channels, as the normal AGI interface doesn't work correctly if the channel has been hung up.

Use the show agi command on the command-line interface to list all of the available AGI commands.

```
exten => h,1,DeadAGI(agi-test)
```

See Also

AGI(), FastAGI()

Dial()
Attempts to connect channels

Dial(*tech/username:password@hostname/extension*[*&tech2/peer2...*]
    [*,ring-timeout*[*,flags*[*,URL*]]])

Allows you to connect together all of the various channel types.[*] Dial() is the most important application in Asterisk; you'll want to read through this section a few times.

Any valid channel type (such as SIP, IAX2, H.323, MGCP, Local, or Zap) is acceptable to Dial(), but the parameters that need to be passed to each channel will depend on the information the channel type needs to do its job. For example, a SIP channel will need a network address and user to connect to, whereas a Zap channel is going to want some sort of phone number.

When you specify a channel type that is network-based, you can pass the destination host (name or IP address), username, password, and remote extension as part of the

---

[*] The fact that Asterisk will happily connect IAX, SIP, H.323, Skinny, PRI, FX(O/S), and anything else is amazing, but possibly the most amazing of all is the Local channel. By allowing a single Dial() command to connect to multiple Local channels, one Dial() event can trigger a multitude of completely independent and unique actions in other parts of the dialplan. The power of this concept is truly revolutionary and has to be experienced to be believed.

options to `Dial()`, or you can refer to the name of a channel entry in the appropriate *.conf* file; all the required information will then need to be obtained from that file. The username and password can be replaced with the name contained within square brackets (`[]`) of the channel configuration file. The hostname is optional.

This is a valid `Dial` statement:

```
exten => s,1,Dial(SIP/sake:arigato@thathostoverthere.tld)
```

This is effectively identical:

```
exten => s,1,Dial(SIP/some_SIP_friend)
```

but will work only if there is a channel defined in *sip.conf* as [`some_SIP_friend`], whose channel definition contains `fromuser=sake`, `password=arigato`, and `host=thathostoverthere.tld`.

An extension number is often attached after the address information, like this:

```
exten => s,1,Dial(IAX2/user:pass@otherend.com/500)
```

This asks the far end to connect the call to extension 500 in the context in which the channel arrived. The extension is not required by `Dial()`, as the information in the remote end's channel configuration file may be used, or the remote server will pass the call to the **s** extension in the context in which the call came in. Ultimately, the far end controls what happens to the call; you can only request a specific treatment.

If no *ring-timeout* is specified, the channel will ring indefinitely. This is not always a bad thing, so don't feel you need to set it; just be aware that "indefinitely" could mean a very long time. *ring-timeout* is specified in seconds. The ring timeout always follows the addressing information, like this:

```
exten => s,1,Dial(IAX2/user:pass@otherend.com/500,ring-timeout)
```

Much of the power of the `Dial()` application is in the flags. These are assigned following the addressing and timeout information, like this:

```
exten => s,1,Dial(IAX2/user:pass@otherend.com/500,60,flags)
```

 If you don't have a timeout specified, and you want to assign flags, you must still assign a spot for the timeout. You do this by adding an extra comma in the spot where the timeout would normally go, like this:

```
exten => s,1,Dial(IAX2/user:pass@otherend.com/500,,flags)
```

The valid flags that may be used with the `Dial()` application are:

A(*x*)

Plays an announcement to the called party; *x* is the filename of the sound file to play as the announcement.

C

Resets the Call Detail Record for the call. Since the CDR time is set to when you Answer() the call, you may wish to reset the CDR so the end user is not billed for the time prior to the Dial() application being invoked.

d

Allows the user to dial a one-digit extension while waiting for a call to be answered. The call will then exit to that extension (either in the current context, if it exists, or in the context specified by the EXITCONTEXT variable).

D([called][: calling])

Sends DTMF digits after the call has been answered, but before the call is bridged. The *called* parameter is passed to the called party, and the *calling* parameter is passed to the calling party. Either parameter may be used individually.

f

Forces the Caller ID of the *calling* party to be set as the extension associated with the channel using a dialplan hint. This is often used when a provider doesn't allow the Caller ID to be set to anything other than a number that is assigned to you. For example, if you had a PRI, you would use the f flag to override any Caller ID set locally on a SIP phone.

g

Execution of the dialplan goes on in the current context if the destination channel hangs up.

G(context ^ extension ^ priority)

When the call is answered, the calling party is transferred to the specified priority and the called party to the specified priority+1. You cannot use any additional action post-answer options in conjunction with this option.

h

Allows the called user to hang up the channel by pressing the * key.

H

Allows the calling user to hang up the channel by pressing the * key.

i

Causes Asterisk to ignore any forwarding requests it may receive on this dial attempt.

j

Causes Asterisk to jump to priority n+101 if all the requested channels were busy (where n is the current priority).

L(x [: y][: z])

Limits the call to *x* milliseconds, warning when *y* milliseconds are left and repeating every *z* milliseconds until the limit is reached. The *x* parameter is required; the *y* and *z* parameters are optional. The following special variables may also be set to provide additional control:

LIMIT_PLAYAUDIO_CALLER=yes|no

> Specifies whether to play sounds to the caller. Defaults to yes.

LIMIT_PLAYAUDIO_CALLEE=yes|no

> Specifies whether to play sounds to the callee.

LIMIT_TIMEOUT_FILE= *filename*

> Specifies which file to play when time is up.

LIMIT_CONNECT_FILE= *filename*

> Specifies which file to play when call begins.

LIMIT_WARNING_FILE= *filename*

> Specifies the file to play if the argument *y* is defined. Defaults to saying the time remaining.

m[*class*]

> Provides music to the calling party until the call is answered. You may also optionally indicate the music-on-hold *class*, as defined in *musiconhold.conf*.

M(*x* [^*arg*])

> Executes the macro *x* upon the connection of a call, optionally passing arguments delimited by ^. The macro can also set the MACRO_RESULT channel variable to one of the following values, to determine what should happen after the macro has finished:

ABORT

> Hangs up both legs of the call.

CONGESTION

> Acts as if the line encountered congestion.

BUSY

> Acts as if the line was busy. If the *j* option is specified, it sends the call to priority n+101, where n is the current priority.

CONTINUE

> Hangs up the called party and continues on in the dialplan.

GOTO:<context>^<extension>^<priority>

> Transfers the call to the specified destination.

 You cannot use any additional action post-answer options in conjunction with this option. Also, PBX services are not run on the called channel, so you will not be able to set timeouts via the TIMEOUT function in this macro.

n

> This option is a modifier for the screen/privacy mode. It specifies that no introductions are to be saved in the *priv-callerintros* directory.

N

This option is a modifier for the screen/privacy mode. It tells Asterisk not to screen the call if Caller ID is present.

o

Uses the Caller ID received on the inbound leg of the call for the Caller ID on the outbound leg of the call. This is useful if you are accepting a call and then forwarding it to another destination, but you wish to pass the Caller ID from the inbound leg of the call instead of overwriting it with the local Caller ID settings. This was the default behavior on Asterisk versions prior to 1.0.

O[x]

This option turns on Operator Services mode on a Zaptel channel. If this option is used on a non-Zaptel interface, it will be ignored. When the destination answers (presumably an operator services station), the originator no longer has control of her line. She may hang up, but the switch will not release her line until the destination party (the operator) hangs up. Specified without an arg, or with 1 as an arg, the originator hanging up will cause the phone to ring back immediately. With a 2 specified as the argument, when the "operator" flashes the trunk, it will ring the caller's phone.

p

This option enables screening mode. This is basically Privacy mode without memory.

P[(x)]

Sets the privacy mode, optionally specifying x as the family/key value in the local AstDB database. This option is useful for accepting calls based on a blacklist (explicitly denying calls from listed numbers) or whitelist (explicitly accepting calls from listed numbers). See also LookupBlacklist().

r

Indicates ringing to the calling party, without passing any audio until the call is answered. This flag is not normally required to indicate ringing, as Asterisk will signal ringing if a channel is actually being called.

S(x)

Hangs up the call x seconds after the *called* party has answered the call.

t

Permits the called party to transfer a call by pressing the # key. Please note that if this option is used, reinvites are disabled, as Asterisk needs to monitor the call to detect when the called party presses the # key.

T

Permits the caller to transfer a connected call by pressing the # key. Again, note that if this option is used, reinvites are disabled, as Asterisk needs to monitor the call to detect when the caller presses the # key.

w

Permits the called user to start and stop recording the call audio to disk by pressing the `automon` sequence (as configured in *features.conf*). If the variable `TOUCH_MONI TOR` is set, its value will be passed as the arguments to the `Monitor()` application when recording is started. If it is not set, the default values of `WAV||m` are passed to `Monitor()`.

W

Permits the calling user to record the call audio to disk by pressing the `automon` sequence (as configured in *features.conf*).

k

Permits the called party to park the call by sending the DTMF sequence defined for call parking in *features.conf*.

K

Permits the calling party to park the call by sending the DTMF sequence defined for call parking in *features.conf*.

If the *URL* argument is included, that URL will be sent to the channel (if supported).

 If the channel variable named `OUTBOUND_GROUP` is set before `Dial()` is called, all peer channels created by this application will be put in to that call group. In the following example, all peer channels created by the `Dial()` application will be part of the `test` call group:

```
; using OUTBOUND_GROUP
exten => 123,1,Set(OUTBOUND_GROUP=test)
exten => 123,n,Dial(IAX2/anotherbox/12345)
```

If the `OUTBOUND_GROUP_ONCE` variable is set, all peer channels created by this application will be put in to that group. Unlike `OUTBOUND_GROUP`, however, the variable will be unset after use.

The `Dial()` application sets the following variables upon exiting:

DIALEDTIME

The total time elapsed from execution of `Dial()` until completion.

ANSWEREDTIME

The total time elapsed during the call.

DIALSTATUS

The status of the call, set as one of the following values:

CHANUNAVAIL

The channel is unavailable.

CONGESTION

The channel returned a congestion signal, usually indicating that it was unable to complete the connection.

NOANSWER

The channel did not answer in the time indicated by the ring-timeout option.

BUSY

The dialed channel is currently busy.

ANSWER

The channel answered the call.

CANCEL

The call was cancelled.

DONTCALL

The call was set to DONTCALL by the screening or privacy options.

TORTURE

The call was set to TORTURE by the screening or privacy options.

INVALIDARGS

Invalid arguments were passed to the Dial() application.

```
; dial a seven-digit number on Zap channel 4
exten => 123,1,Dial(Zap/4/2317154)

; dial the same number, but this time only have it ring for 10 seconds
; before continuing on with the dialplan
exten => 124,1,Dial(Zap/4/2317154,10)
exten => 124,2,Playback(im-sorry)
exten => 124,3,Hangup()

; dial the same number, but this time with no timeout, and using the
; t, T, and m flags
exten => 125,1,Dial(Zap/4/2317154,,tTm)

; dial extension 500 at a remote host (over the IAX protocol), using
; the specified username and password
exten => 126,1,Dial(IAX2/username:password@remotehost/500)

; dial a number, but limit the call to 5 minutes (300,000 milliseconds)
; start warning the caller 4 minutes (240,000 milliseconds) in to the call,
; and repeat the warning every 30 seconds (30,000 milliseconds)
exten => 127,1,Dial(Zap/4/2317154,,L[300000:240000:30000])
```

See Also

RetryDial()

Dictate()   Virtual dictation machine

Dictate([*base_dir*[,*filename*]])

This application allows the recording and playback of files, similar to a traditional dictation machine. The *base_dir* parameter specifies the directory in which Asterisk will write the recorded files. If not specified, it defaults to the *dictate* subdirectory of the Asterisk spool directory (as defined in *asterisk.conf*).

If the *filename* parameter is specified, it will be used when the file is written. If not specified, Asterisk will prompt the caller for a numeric filename for the file.

 Asterisk writes the files in raw, headerless, signed-linear format. If you'd like to convert the file to another format, you can use an outside utility such as sox, or use the `file convert` command from the Asterisk command-line interface.

The `Dictate()` application has two main modes: recording mode and playback mode. The caller can press the 1 key to switch between these modes. In both modes, the 0 key can be used to get help. The * key is used to pause or unpause the recording or playback. The # key allows the caller to choose a new filename.

In recording mode, the 8 key can be used to erase the entire recording and start over.

In playback mode, the 7 key rewinds the recording a few frames, and the 8 key forwards the recording a few frames. The 2 key is used to toggle the playback speed (either 1x, 2x, 3x, or 4x).

```
; begin dictating, and save the files in the /tmp/dictate directory
exten => 123,1,Dictate(/tmp/dictate)
```

See Also

`Playback()`, `Background()`, `ControlPlayback()`,

Directory()   Provides a dialable directory of extensions

`Directory(vm-context[,dial-context[,options]])`

Presents users with a directory of extensions from which they may select by name. The list of names and extensions is discovered from *voicemail.conf*. The **vm-context** argument is required; it specifies the context of *voicemail.conf* to use.

The *dial-context* argument is the context to use for dialing the users, and it defaults to **vm-context** if unspecified. If the *options* argument is set to f, Asterisk will find a directory match based on the first name in *voicemail.conf* instead of the last name. If the e option is specified, Asterisk will read the extension of the directory match as well as the person's name.

If the user enters 0 (zero) and there exists an extension o (the lowercase letter o) in the current context, the call control will go to that extension. Entering * will exit similarly, but to the a extension, much like `Voicemail()`'s behavior.

```
exten => *,1,Directory(default,incoming)
exten => #,1,Directory(default,incoming,f)
exten => 9,1,Directory(default,incoming,fe)
```

See Also

voicemail.conf

DISA() Direct Inward System Access: allows inbound callers to make outbound calls

DISA(*password*[,*context*[,*callerid*[,*mailbox*[@*vmcontext*]]]])
DISA(*password-file*[,*callerid*[,*mailbox*[@*vmcontext*]]])

Allows outside callers to obtain an "internal" system dial tone and to place calls from it as if they were placing calls from within the switch. The user is given a dial tone, after which she should enter her passcode, followed by #. If the passcode is correct, the user is then given a system dial tone on which a call may be placed.

 Obviously, this type of access has serious security implications, and *extreme* care must be taken not to compromise the security of your phone system.

The *password* argument is a numeric passcode that the user must enter to be able to make outbound calls. Using this syntax, all callers to this extension will use the same password. To allow users to use DISA() without a password, use the string no-pass word instead of the password.

The *context* argument specifies the context in which the user will be dialing. If no context is specified, the DISA() application defaults the context to disa.

The *callerid* argument specifies a new Caller ID string that will be used on the outbound call.

The *mailbox* argument is the mailbox number (and optional voicemail context, *vmcontext*) of a voicemail box. The caller will hear a stuttered dial tone if there are any new messages in the specified voicemail box.

Additionally, you may use an alternate syntax and pass the name of a global password file instead of the *password* and *context* arguments. On each line, the file may contain either a passcode, or a passcode and context, separated by a pipe character (|). If a context is not specified, the application defaults to the context named disa.

If the user login is successful, the application parses the dialed number in the specified *context*:

```
; allow outside callers to call 1-800 numbers, as long
; as they know the passcode. Set their Caller IDs to make
; it appear that they are dialing from within the company
[incoming]
exten => 123,1,DISA(4569,disa,"Company ABC" <(234) 123-4567>)
```

```
[disa]
exten => _1800NXXXXXX,1,Dial(Zap/4/${EXTEN})
```

See Also

Authenticate(), VMAuthenticate()

DumpChan()   Dumps information about the calling channel to the console

DumpChan([*min_verbose_level*])

Displays information about the calling channel, as well as a listing of all channel variables. If *min_verbose_level* is specified, output is displayed only when the verbosity
level is currently set to that number or greater.

 If you have many channel variables set, DumpChan() will show only the
first 1,024 characters of your channel variable listing.

```
exten => s,1,Answer()
exten => s,2,DumpChan()
exten => s,3,Background(enter-ext-of-person)
```

See Also

NoOp(), Verbose()

EAGI()

See AGI().

Echo()   Echoes inbound audio back to the caller

Echo()

Echoes audio read from the channel back to the channel. This application is often used
to test the latency and voice quality of a VoIP link. The caller may press the # key to exit.

```
exten => 123,1,Echo()
exten => 123,2,Playback(vm-goodbye)
```

See Also

Milliwatt()

EndWhile()

EndWhile()

Returns to the previously called While() application. See While() for a complete de-scription of how to use a while loop.

```
exten => 123,1,Set(COUNT=1)
exten => 123,2,While($[ ${COUNT} < 5 ])
exten => 123,3,SayNumber(${COUNT})
exten => 123,4,Set(COUNT=$[${COUNT} + 1]
exten => 123,5,EndWhile()
```

See Also

While(), ExitWhile(), GotoIf()

Exec()

Exec(*appname(arguments)*)

Allows an arbitrary application to be invoked even when not hard-coded in to the dialplan. Exits exactly the same as the underlying *application*, or abnormally, if the underlying *application* cannot be found. The *arguments* are passed to the called *application*.

This application allows you to dynamically call applications by pulling them from a database or other external source.

```
exten => 123,1,Set(MYAPP=SayDigits(12345))
exten => 123,2,Exec(${MYAPP})
```

See Also

The EVAL, TryExec(), ExecIf()

ExecIf()

ExecIf(*expression,application,arguments*)

If *expression* is true, executes the given *application* with *arguments* as its arguments, and returns the result. For more information on Asterisk expressions, see Chapter 6 or the *channelvariables.txt* file in the *doc/* subdirectory of the Asterisk source.

If *expression* is false, execution continues at the next priority.

```
exten => 123,1,ExecIf($[ ${CALLERIDNUM} = 101 ],SayDigits,12345)
exten => 123,2,SayDigits(6789)
```

See Also

The EVAL, Exec(), TryExec()

ExitWhile()   Exit from a While() loop, whether or not the conditional has been satisfied

ExitWhile()

Will cause a While() loop to exit whether or not the conditional expression has been satisfied.

```
exten => 123,1,Set(COUNT=1)
exten => 123,n,While($[${COUNT} < 5])
exten => 123,n,GotoIf($[${COUNT} != 3]?continue)
exten => 123,n,ExitWhile()
exten => 123,n(continue),NoOp()
exten => 123,n,SayNumber(${COUNT})
exten => 123,n,Set(COUNT=$[${COUNT} + 1])
exten => 123,n,EndWhile()
```

See Also

While(), ContinueWhile(), EndWhile()

ExtenSpy()   Listen to the audio on an extension, and optionally whisper to the calling channel

ExtenSpy([*exten@context*[,*options*]])

This application is used to listen to the audio going to and from an Asterisk channel. Only channels created by outgoing calls for the specified extension will be selected for spying.

While a channel is being spied upon, the following actions may be performed:

- Dialing # cycles the volume level
- Dialing * will cause the application to spy on the next available channel

The *options* parameter may contain zero or more of the following options:

b

Only spy on channels that are involved in a bridged call.

g(*group*)

Only spy on channels that contain a channel variable named SPYGROUP, which should contain *group* in an optional colon-delimited list.

q

Quiet mode. Tells the application not to beep or read the selected channel's name when spying begins.

r[(*basename*)]

Records the channel audio to the monitor spool directory (usually */var/spool/aster isk/monitor*). An optional *basename* set the base filename of the recordings, which defaults to chanspy.

v([*value*])

Adjusts the volume of the audio being listened to. The *value* must be in the range of 4 to -4. A negative *value* will make the volume quieter, while a positive value will make it louder.

w

Whisper mode. This allows the spying channel to talk to the spied-upon channel, without any other bridged channel being able to hear the audio.

W

Private whisper mode. This enables the spying channel to speak to the spied-upon channel without being able to hear the audio from the spied-upon channel.

```
; Spy on channels created by extension 125 in the lab context
    exten => 123,1,ExtenSpy(125@lab,w)
```

See Also

ChanSpy()

ExternalIVR()   Interfaces with an external IVR application

ExternalIVR(*command*[,*arg1*[,*arg2*...]])

Forks a process to run the specified ExternalIVR-compliant *command*, and starts a generator on the channel. The generator's play list is controlled by the external application, which can add and clear entries via simple commands issued over STDOUT. The external application will receive notifications of all DTMF events received on the channel, and notification if the channel is hung up. The application will not be forcibly terminated when the channel is hung up.

See *doc/externalivr.txt* in the Asterisk source code for the specification of the External-IVR interface.

```
; Run a test external IVR program, passing an argument
    exten => 123,1,ExternalIVR(test_program,${MYARGUMENT})
```

See Also

AGI()

FastAGI()   Executes an AGI-compliant script across a network connection

FastAGI(agi://*hostname*[:*port*][/*script*],*args*)

Executes an AGI-compliant program across the network. This application is very similar to AGI(), except that it calls a specially written FastAGI script across a network connection. The main purposes for using FastAGI are to offload CPU-intensive AGI scripts to remote servers and to help reduce AGI script startup times (the FastAGI program is already running before Asterisk connects to it).

FastAGI() tries to connect directly to the running FastAGI program, which must already be listening for connections on the specified *port* on the server specified by *hostname*. If *port* is not specified, it defaults to port 4573. If *script* is specified, it is passed to the FastAGI program as the agi_network_script variable. The arguments specified by *args* will be passed to the program.

 See *agi/fastagi-test* in the Asterisk source directory for a sample FastAGI script. This should serve as a good roadmap for writing your own FastAGI programs.

Returns -1 if the application requested a hangup, or 0 on a non-hangup exit.

```
; connect to the sample fastagi-test program, which must already be running
; on the local machine
exten => 123,1,Answer()
exten => 123,2,FastAGI(agi://localhost)

; connect to a FastAGI script on a host named "calvin" on port 8000, and
pass along
; a script name of "testing", with the argument "12345"
exten => 124,1,Answer()
exten => 124,2,FastAGI(agi://calvin:8000/testing,12345)
```

See Also
AGI(), DeadAGI()

Festival() Uses the Festival text-to-speech engine to read text to the caller

Festival(*text*[,*intkeys*])

Connects to the locally running Festival server, sends it the text specified by *text*, and plays the resulting sound file back to the user. This application allows the caller to press a key (specified by *intkeys*) to immediately stop the playback and return the value of *intkeys*. If *intkeys* is set to any, Festival() will send control of the channel to the extension entered by the user.

See Chapter 14 for more in-depth information on using Festival with Asterisk and the *README.festival* file located in the *contrib/* subdirectory of the Asterisk source.

You must start the Festival server before starting Asterisk, and you must use the Answer() application to answer the channel before calling Festival().

```
exten => 123,1,Answer()
exten => 123,2,Festival('This is sample speech from Festival',#)
```

Flash()

```
Flash()
```

Sends a flash on a Zap channel. This is only a hack for people who want to perform transfers and other actions that require a flash via an AGI script. It is generally quite useless otherwise.

Returns 0 on success, or -1 if this is not a Zap trunk.

```
exten => 123,1,Flash()
```

FollowMe()

```
FollowMe(followmeid[,options])
```

This application attempts to locate the callee by dialing many different destinations either serially or in parallel, as defined in *followme.conf*.

The *followmeid* identifies the section of *followme.conf* that specifies how this callee should be found. The *options* parameter can be zero or more of the following:

s

Playback the incoming status message prior to starting the follow-me step(s)

a

Record the caller's name so it can be announced to the callee on each step

n

Playback the unreachable status message if we've run out of steps to reach the callee or the callee has elected not to be reachable

```
exten => 123,1,Answer()
exten => 123,2,FollowMe(123,san)
exten => 123,3,VoiceMail(123,u)
```

ForkCDR()

```
ForkCDR([options])
```

Creates an additional Call Detail Record for the remainder of the current call.

This application is often used in calling-card applications to distinguish the inbound call (the original CDR) from the billable call time (the second CDR).

If the v option is specified, all the CDR variables from the current record will be inherited by the new CDR record.

```
exten => 123,1,Answer()
exten => 123,2,ForkCDR(v)
exten => 123,3,Playback(tt-monkeys)
exten => 123,4,Hangup()
```

See Also

CDR function, NoCDR(), ResetCDR()

GetCPEID()

Gets the CPE ID from an ADSI-capable telephone

GetCPEID()

Obtains the CPE ID and other information and displays it on the Asterisk console. This information is often needed in order to properly set up *zapata.conf* for on-hook operations with ADSI-capable telephones.

Returns -1 on hangup only.

```
; use this extension to get the necessary information to set up ADSI
; telephones
exten => 123,1,GetCPEID()
```

See Also

ADSIProg(), *adsi.conf*, *zapata.conf*

Gosub()

Branches to a new location, saving the return address

```
Gosub(context,extension,priority)
Gosub(extension,priority)
Gosub(priority)
```

Branches to the location specified, similar to Goto(), except that Gosub() saves the return location, to be returned to later by invoking Return().

See Also

GosubIf(), Macro(), Goto(), Return(), StackPop()

GosubIf()

Conditionally branches to a new location, saving the return address

GosubIf(condition?labeliftrue:labeliffalse)

Based upon the evaluation of *condition*, Gosub will branch execution either to *labeliftrue()* or *labeliffalse*. You may return to this same place in the dialplan by later calling Return.

The word label is often used to denote that you may specify a *priority*; an *extension* and a *priority*; or a *context*, an *extension* and a *priority*. We use the word label to avoid having to spell out all of the possible options each time.

```
; Specify a default outgoing Caller*ID if one is not set by a specific channel.
exten => _NXXXXXX,1,GosubIf($["${CALLERID(num)}" = ""]?setcallerid,1)
exten => _NXXXXXX,n,Dial(Zap/g1/${EXTEN})
exten => _1NXXNXXXXXX,1,GosubIf($["${CALLERID(num)}" = ""]?setcallerid,1)
exten => _1NXXNXXXXXX,n,Dial(Zap/g1/${EXTEN})
exten => setcallerid,1,Set(CALLERID(num)=6152345678)
exten => setcallerid,n,Return
```

See Also

Gosub(), Return(), MacroIf(), IF, GotoIf(),

Goto()   Sends the call to the specified priority, extension, and context

Goto([[*context*,]*extension*,]*priority*)
Goto(*named_priority*)

Sends control of the current channel to the specified *priority*, optionally setting the destination *extension* and *context*.

Optionally, you can use the application to go to the named priority specified by the *named_priority* argument. Named priorities work only within the current extension.

```
exten => 123,1,Answer()
exten => 123,2,Set(COUNT=1)
exten => 123,3,SayNumber(${COUNT})
exten => 123,4,Set(COUNT=$[ ${COUNT} + 1 ])
exten => 123,5,Goto(3)

; same as above, but using a named priority
exten => 124,1,Answer()
exten => 124,2,Set(COUNT=1)
exten => 124,3(repeat),SayNumber(${COUNT})
exten => 124,4,Set(COUNT=$[ ${COUNT} + 1 ])
exten => 124,5,Goto(repeat)
```

See Also

GotoIf(), GotoIfTime(), Gosub(), Macro()

GotoIf()   Conditionally goes to the specified priority

GotoIf(*condition*?*label1*:*label2*)

Sends the call to *label1* if *condition* is true or to *label2* if *condition* is false. Either *label1* or *label2* may be omitted (in that case, we just don't take the particular branch), but not both.

A label can be any one of the following:

- A priority, such as `10`
- An extension and a priority, such as `123,10`
- A context, extension, and priority, such as `incoming,123,10`
- A named priority within the same extension, such as `passed`

Each type of label is explained in this example:

```
[globals]
; set TEST to something else besides 101 to see what GotoIf()
; does when the condition is false
TEST=101
;
[incoming]
; set a variable
; go to priority 10 if ${TEST} is 101, otherwise go to priority 20
exten => 123,1,GotoIf($[ ${TEST} = 101 ]?10:20)
exten => 123,10,Playback(the-monkeys-twice)
exten => 123,20,Playback(tt-somethingwrong)
;
; same thing as above, but this time we'll specify an extension
; and a priority for each label
exten => 124,1,GotoIf($[ ${TEST} = 101 ]?123,10:123,20)
;
; same thing as above, but these labels have a context, extension, and
; priority
exten => 125,1,GotoIf($[ ${TEST} = 101 ]?incoming,123,10:incoming,123,20)
;
; same thing as above, but this time we'll go to named priorities
exten => 126,1,GotoIf($[ ${TEST} = 101 ]?passed:failed)
exten => 126,15(passed),Playback(the-monkeys-twice)
exten => 126,25(failed),Playback(the-monkeys-twice)
```

See Also

`Goto()`, `GotoIfTime()`, `GosubIf()`, `MacroIf()`

GotoIfTime()   Conditionally branches, depending on the time and day

`GotoIfTime(times,days_of_week,days_of_month,months?label)`

Branches to the specified extension, if the current time matches the specified time. Each of the elements may be specified either as * (for always) or as a range.

The arguments to this application are:

times

> Time ranges, in 24-hour format

days_of_week

> Days of the week (mon, tue, wed, thu, fri, sat, sun)

days_of_month

> Days of the month (1-31)

months

> Months (jan, feb, mar, apr, etc.)

```
; If we're open, then go to the open context
; We're open from 9am to 6pm Monday through Friday
exten => s,1,GotoIfTime(09:00-17:59,mon-fri,*,*?open,s,1)
;
; We're also late on Tuesday and Thursday
exten => s,n,GotoIfTime(09:00-19:59,tue&thru,*,*?open,s,1)
;
; We're also open from 9am to noon on Saturday
exten => s,n,GotoIfTime(09:00-11:59,sat,*,*?open,s,1)
;
; Otherwise, we're closed
exten => s,n,Goto(closed,s,1)
```

See Also

GotoIf(), IFTIME

Hangup()
Unconditionally hangs up the current channel

Hangup(*cause-code*)

Unconditionally hangs up the current channel. If supported on the channel, *cause-code* will be specified to the remote end as the reason for ending the call. *cause-code* defaults to 16 (normal call clearing). Acceptable values for **cause-code** are the following:

16

> Normal call clearing

17

> Busy

19

> No answer

21

> Rejected

34

> Congestion

```
exten => 123,1,Answer()
exten => 123,2,Playback(im-sorry)
exten => 123,3,Hangup()
```

See Also

Answer(), Busy(), Congestion()

HasNewVoicemail()

Checks to see if there is new voicemail in the indicated voicemail box

HasNewVoicemail(*vmbox*[@*context*][:*folder*][,*varname*[,*options*]])

 The application has been deprecated in favor of the VMCOUNT() function.

Similar to HasVoicemail(). This application sets the VMSTATUS to 1 or 0, to indicate whether there is new (unheard) voicemail in the voicemail box indicated by *vmbox*. The *context* argument corresponds to the voicemail context, and *folder* corresponds to a voicemail folder. If the voicemail folder is not specified, it defaults to the *INBOX* folder. If the *varname* argument is present, HasNewVoicemail() assigns the number of messages in the specified folder to that variable.

If the *options* argument is set to the letter j, then Asterisk will send the call to priority n +101 if there is new voicemail.

```
; check to see if there's unheard voicemail in INBOX of mailbox 123
; in the default voicemail context
exten => 123,1,Answer()
exten => 123,n,HasNewVoicemail(123@default)
exten => 123,n,GotoIf($[${HASVMSTATUS} > 0]?newvm)
exten => 123,n,Playback(vm-youhave)
exten => 123,n,Playback(vm-no)
exten => 123,n,Playback(vm-messages)
exten => 123,n,Goto(done)
exten => 123,n(newvm),Playback(vm-youhave)
exten => 123,n,SayNumber(${HASVMSTATUS})
exten => 123,n,Playback(vm-INBOX)
exten => 123,n,Playback(vm-messages)
exten => 123,n(done),NoOp()
```

See Also

HasVoicemail(), MailboxExists(), VMCOUNT

HasVoicemail()

Indicates whether there is voicemail in the indicated voicemail box

HasVoicemail(*vmbox*[@*context*][:*folder*][|*varname*[,*options*]])

Sets the HASVMSTATUS channel variable to indicate whether there is voicemail in the voicemail box indicated by *vmbox*. The *context* argument corresponds to the voicemail context, and *folder* corresponds to a voicemail folder. If the folder is not specified, it defaults to the *INBOX* folder. If the *varname* argument is passed, this application assigns the number of messages in the specified folder to that variable.

If the *options* argument is set to the letter j, then Asterisk will send the call to priority n+101 if there is voicemail in the specified *folder*.

```
; check to see if there's any voicemail at all in INBOX of mailbox 123
; in the default voicemail context
exten => 123,1,Answer()
exten => 123,2,HasVoicemail(123@default,COUNT)
exten => 123,3,GotoIf(${VMSTATUS}?1000)
exten => 123,4,Playback(vm-youhave)
exten => 123,5,Playback(vm-no)
exten => 123,6,Playback(vm-messages)
exten => 123,1000,Playback(vm-youhave)
exten => 123,1001,SayNumber($COUNT)
exten => 123,1002,Playback(vm-messages)
```

See Also

The HasVoicemail(), MailboxExists()

IAX2Provision()   Provisions a calling IAXy device

IAX2Provision([*template*])

Provisions a calling IAXy device (assuming that the calling entity is an IAXy) with the given *template*. If no template is specified, the default template is used. IAXy provisioning templates are defined in the *iaxprov.conf* configuration file.

```
; provision IAXy devices with the default template when they dial this extension
exten => 123,1,IAX2Provision(default)
```

ICES()   Streams audio to an Icecast server

ICES([*config*])

Streams audio to an Icecast (or compatible) server using the ices application.

See *contrib/asterisk-ices.xml* for a sample *config* file.

```
exten => 123,1,Answer()
exten => 123,n,ICES(/tmp/my-ices-config.xml)
```

ImportVar()   Sets a variable based on a channel variable from a different channel

ImportVar(*newvar*=*channel*,*variable*)

Sets variable *newvar* to *variable* as evaluated on the specified *channel* (instead of the current channel). If *newvar* is prefixed with _, single inheritance is assumed. If prefixed with _ _, infinite inheritance is assumed.

```
; read the Caller ID information from channel Zap/1
exten => 123,1,Answer()
exten => 123,n,ImportVar(cidinfo=Zap/1,CALLERID(all))
```

See Also

Set()

Log()

Log(*level*|*message*)

Sends a custom message to the logfiles from the dialplan. This application can be useful to log an exceptional condition to the logfiles, for later examination. *Level* may be one of the following:

DEBUG
: Debugging message. This is generally not logged on a production system.

NOTICE
: An informational message.

WARNING
: A condition that may be serious, but is not a definite error.

ERROR
: Something went terribly wrong.

See Also

NoOp(), Verbose()

LookupBlacklist()

LookupBlacklist([*options*])

> This application has been deprecated in favor of GotoIf(${BLACKLIST ()}?*context*|*extension*|*priority*)

Looks up the Caller ID number on the active channel in the Asterisk database (family blacklist). If the Caller ID number is found in the blacklist, Asterisk sets the LOOKUPBL STATUS channel variable to FOUND. Otherwise, the variable is set to NOTFOUND.

If the j option is used in the *options* parameter, and the number is found, and if there exists a priority n+101 (where n is the priority of the current instance), the channel will be set up to continue at that priority level.

To add to the blacklist from the Asterisk CLI, type **database put blacklist** *name* **/** *number*.

```
; send blacklisted numbers to an endless loop
; otherwise, dial the number defined by the variable ${JOHN}
exten => 123,1,Answer()
exten => s,2,LookupBlacklist()
    ; if the Caller ID number is found in the blacklist, jump to the "goaway" label
    exten => 123,n,GotoIf($["${LOOKUPBLSTATUS}" = "FOUND"]?goaway)
    ; otherwise, go ahead and call John
exten => 123,n,Dial(${JOHN})
exten => 123,n(goaway),Busy(5)
exten => 123,n,Hangup()
```

See Also

BLACKLIST

LookupCIDName()

Performs a lookup of a Caller ID name from the AstDB

```
LookupCIDName()
```

 This application has been deprecated in favor of Set(CALLERID(*name*)= ${DB(cidname/${CALLERID(*num*)})})

Uses the Caller ID number on the active channel to retrieve the Caller ID name from the AstDB (family cidname). This application does nothing if no Caller ID was received on the channel. This is useful if you do not subscribe to Caller ID name delivery, or if you want to change the Caller ID names on some incoming calls.

```
; look up the Caller ID information from the AstDB, and pass it along
; to Jane's phone
exten => 123,1,Answer()
exten => 123,2,LookupCIDName()
exten => 123,3,Dial(SIP/Jane)
```

See Also

DB

Macro()

Calls a previously defined dialplan macro

```
Macro(macroname,arg1,arg2...)
```

Executes a macro defined in the context named macro- *macroname*, jumping to the s extension of that context and executing each step, then returning when the steps end.

The calling extension, context, and priority are stored in ${MACRO_EXTEN}, ${MACRO_CON TEXT}, and ${MACRO_PRIORITY}, respectively. Arguments *arg1*, *arg2*, etc. become ${ARG1}, ${ARG2}, etc. in the macro context.

Macro() exits abnormally if any step in the macro exited abnormally or indicated a hangup. If ${MACRO_OFFSET} is set at termination, this application will attempt to continue at priority MACRO_OFFSET+n+1 if such a step exists, and at n+1 otherwise. (In both cases, n stands for the current priority.)

If you call the Goto() application inside of the macro to specify a context outside of the currently executing macro, the macro will terminate and control will go to the destination of the Goto().

```
; define a macro to count down from the specified value
[macro-countdown]
exten => s,1,Set(COUNT=${ARG1})
exten => s,2,While($[ ${COUNT} > 0])
exten => s,3,SayNumber(${COUNT})
exten => s,4,Set(COUNT=$[ ${COUNT} - 1 ])
exten => s,5,EndWhile()

; call our macro with two different values
[example]
exten => 123,1,Macro(countdown,10)
exten => 124,1,Macro(countdown,5)
```

 While a macro is being executed, it becomes the current context. This means that if a hangup occurs, for instance, the macro will be searched for an h extension, *not* the context from which the macro was called. So, make sure to define all appropriate extensions in your macro (you can use catch in AEL).

 Because of the way Macro() is implemented (it executes the priorities contained within it via sub-engine), and a fixed per-thread memory stack allowance, macros are limited to seven levels of nesting (macro calling macro calling macro, etc.); It may be possible that stack-intensive applications in deeply nested macros could cause Asterisk to crash earlier than this limit.

See Also

MacroExit(), Goto(), Gosub(), Chapter 6

MacroExclusive()   Runs a macro, exclusive of any other channel

MacroExclusive(*macroname*[,*arguments*])

Runs the specified macro, but ensures that only one channel is running inside that macro at one time. If another channel is already executing that macro, then `MacroExclusive()` will pause the channel until that channel has exited the macro.

See Also

`Macro()`

MacroExit()

Explicitly returns from a macro

`MacroExit()`

Explicitly return from a macro. Normally, `Macro()` automatically exits when it runs out of priorities. `MacroExit()` provides a method by which a macro may be terminated early.

See Also

`Macro()`

MacroIf()

Conditionally calls a previously defined macro

`MacroIf(condition?macroiftrue,args:macroiffalse,args)`

Evaluates *condition*, then executes one of either *macroiftrue* or *macroiffalse*. Once beyond the *condition*, however, note that `MacroIf()` behaves identically to `Macro()`.

```
; define a macro to count down from the specified value
[macro-countdown]
exten => s,1,Set(COUNT=${ARG1})
exten => s,2,While($[ ${COUNT} > 0])
exten => s,3,SayNumber(${COUNT})
exten => s,4,Set(COUNT=$[ ${COUNT} - 1])
exten => s,5,EndWhile()

; define a macro to count up to the specified value
[macro-countup]
exten => s,1,Set(COUNT=1)
exten => s,2,While($[ ${COUNT} < ${ARG1}])
exten => s,3,SayNumber(${COUNT})
exten => s,4,Set(COUNT=$[ ${COUNT} + 1])
exten => s,5,EndWhile()

; call our macro with two different values
[example]
exten => 123,1,MacroIf($[ ${foo} < 5 ]?countup,${foo}:countdown,${foo})
```

See Also

`GotoIf()`, `GosubIf()`, IF, Chapter 6

MailboxExists()

```
MailboxExists(mailbox[@context[,options]])
```

Checks to see if the mailbox specified by the *mailbox* argument exists in the Asterisk voicemail system. You may pass a voicemail *context* if the mailbox is not in the default voicemail context.

This application sets a channel variable named VMBOXEXISTSSTATUS. If the mailbox exists, it will be set to SUCCESS. Otherwise, it will be set to FAILED.

If the j option is passed as the *options* parameter, the application will jump to priority n +101 (where n is the current priority) if the voicemail box specified by the *mailbox* argument exists.

```
exten => 123,1,Answer()
exten => 123,n,Set(MYMAILBOX=123@default)
exten => 123,n,MailboxExists(${MYMAILBOX})
exten => 123,n,GotoIf($["${VMBOXEXISTSSTATUS}" = "SUCCESS"]?exists)
exten => 123,n,Playback(im-sorry)
exten => 123,n,Hangup()
exten => 123,n(exists),Voicemail(u123)
```

See Also

```
HasVoicemail(), HasNewVoicemail()
```

MeetMe()

```
MeetMe([confno[,options[,PIN]]])
```

Places the caller in to the audio conference bridge specified by the *confno* argument. If the conference number is omitted, the user will be prompted to enter one.

If the *PIN* argument is passed, the caller must enter that PIN number to successfully enter the conference.

The *options* string may contain zero or more of the characters in the following list:

a

Sets admin mode.

A

Sets marked mode.

b

Runs the AGI script specified in ${MEETME_AGI_BACKGROUND}; default: *conf-background.agi*. (Note: this does not work with non-Zap channels in the same conference.)

c

Announces user(s) count upon joining a conference.

d

Dynamically adds conference.

D

Dynamically adds conference, prompting for a PIN.

e

Selects an empty conference.

E

Selects an empty Pinless conference.

F

Passes DTMF digits through the conference to other participants. DTMF digits used to enable conference features will not be passed through.

i

Announces user join/leave with review.

I

Announces user join/leave without review.

l

Sets listen only mode (listen only, no talking).

m

Sets the participant as initially muted.

M

Enables music on hold when the conference has a single caller.

o

Turns on talker optimization. With talker optimization, Asterisk treats talkers who aren't speaking as being muted, meaning that no encoding is done on transmission and that received audio that is not registered as talking is omitted, causing no buildup in background noise.

p

Allows user to exit the conference by pressing **#**.

P

Always prompts for the PIN even if it is specified.

q

Sets quiet mode. In quiet mode, Asterisk won't play sounds as conference participants enter or leave.

r

Records conference (as ${MEETME_RECORDINGFILE} using format ${MEETME_RECORDINGFORMAT}). The default filename is *meetme-conf-rec-*${CONFNO}-{UNIQUEID} and the default format is *.wav*.

s

Presents the menu (user menu or admin menu, depending on whether the caller is marked as an administrator) when * is received.

t

Sets talk-only mode (talk only, no listening).

T

Sets talker detection. Asterisk will sends events on the Manager Interface identifying the channel that is talking. The talker will also be identified on the output of the `meetme list` CLI command.

w[(seconds)]

Waits for a marked admin to join the conference. If *seconds* is not specified, the conference will wait indefinitely for the admin to join. If *seconds* is specified, the conference will wait the specified number of seconds. If the admin still hasn't joined, the call will continue on with the next priority in the dialplan.

x

Closes the conference when the last marked user exits.

X

Allows user to exit the conference by entering a valid single-digit extension (set via the variable `${MEETME_EXIT_CONTEXT}`), or the number of an extension in the current context if that variable is not defined.

1

Doesn't play initial message when the first person joins the conference.

```
exten => 123,1,Answer()
; add the caller to conference number 501 with pin 1234
exten => 123,2,MeetMe(501,DpM,1234)
```

 A suitable Zaptel timing interface must be installed for MeetMe conferencing to work.

See Also

MeetMeAdmin(), MeetMeCount()

MeetMeAdmin() Performs MeetMe conference administration

MeetMeAdmin(*confno*,*command*[,*user*])

Runs the specified MeetMe administration *command* on the specified conference. On some commands, you may specify the *user* on which to run the specified command. The *command* may be one of the following:

e

Ejects the last user that joined.

k

Kicks the specified *user* out of the conference.

K

Kicks all users out of the conference.

l

Unlocks the conference.

L

Locks the conference.

m

Unmutes the specified *user*.

M

Mutes the specified *user*.

n

Unmutes the entire conference.

N

Mutes all non-admin participants in the conference.

r

Resets the volume settings for the specified *user*.

R

Resets the volume settings for all participants.

s

Lowers the speaking volume for the entire conference.

S

Raises the speaking volume for the entire conference.

t

Lowers the speaking volume for the specified *user*.

T

Raises the speaking volume for the specified *user*.

u

Lowers the listening volume for the specified *user*.

U

Raises the listening volume for the specified *user*.

v

Lowers the listening volume for the entire conference.

V

Raises the listening volume for the entire conference.

```
; mute conference 501
exten => 123,1,MeetMeAdmin(501,N)

; kick user 1234 from conference 501
exten => 124,1,MeetMeAdmin(501,k,1234)
```

You can find a list of users in the conference by using the `meetme list` command from the Asterisk CLI, or by using the Asterisk Manager Interface.

See Also

MeetMe(), MeetMeCount()

MeetMeCount()   Counts the number of participants in a MeetMe conference

MeetMeCount(*confno*[,*variable*])

Plays back the number of users in the MeetMe conference identified by *confno*. If a variable is specified by the *variable* argument, playback will be skipped and the count will be assigned to *variable*.

```
; count the number of users in conference 501, and assign that number
to ${COUNT}
exten => 123,1,MeetMeCount(501,COUNT)
```

See Also

MeetMe(), MeetMeAdmin()

Milliwatt()   Generates a 1,000 Hz tone

Milliwatt()

This application generates a constant 1,000 Hz tone at 0 dbm (µlaw). This application is often used for testing the audio properties of a particular channel.

```
; generate a 1000HZ tone
exten => 123,1,Milliwatt()
```

 Please note that there is a service that the carriers use to test circuits for loss, that folks in the industry have nicknamed "1,000-cycles." The thing is, the tone the carrier equipment sends is actually 1,004 Hz, so if you want to test circuit loss on an analog channel from Asterisk, Milliwatt() may not give you exactly what you want.

See Also

Echo(), Playtones()

MixMonitor()   Records a channel in the background, mixing both directions synchronously

MixMonitor(*filename.ext*,*options*,*command*)

Records the audio on the current channel to the specified file. If the filename is an absolute path, `MixMonitor()` uses that path; otherwise it creates the file in the configured monitoring directory from *asterisk.conf*.

If *command* is specified, it will be run when recording ends, either by hangup or by calling `StopMixMonitor()`.

The *options* parameter can contain zero or more of the following options:

a

Append to the file, instead of overwriting it.

b

Only save audio when the channel is bridged.

 This does not include conferences or sounds played to each bridged party.

v(*x*)

Adjust the heard volume by a factor of *x* (range **-4** to **4**).

V(*x*)

Adjust the spoken volume by a factor of *x* (range **-4** to **4**).

W(*x*)

Adjust both the heard and the spoken volumes by a factor of *x* (range **-4** to **4**).

```
; Record channel
exten => 123,1,MixMonitor(/var/lib/asterisk/sounds/123.wav)
```

See Also

`Monitor()`, `StopMixMonitor()`, `PauseMonitor()`, `UnpauseMonitor()`

Monitor()   Monitors (records) the audio on the current channel

`Monitor([file_format[:urlbase][,fname_base][,options]])`

Starts monitoring a channel. The channel's input and output voice packets are logged to files until the channel hangs up or monitoring is stopped by the `StopMonitor()` application.

`Monitor()` takes the following arguments:

file_format

Specifies the file format. If not set, defaults to `wav`.

fname_base

If set, changes the filename used to the one specified.

---

options

One of two options can be specified:

m

> When the recording ends, mix the two leg files in to one and delete the original leg files. If the variable ${MONITOR_EXEC} is set, the application referenced in it will be executed instead of *soxmix*, and the raw leg files will *not* be deleted automatically. *soxmix* (or ${MONITOR_EXEC}) is handed three arguments: the two leg files and the filename for the target mixed file, which is the same as the leg filenames but without the in/out designator. If ${MONITOR_EXEC_ARGS} is set, the contents will be passed on as additional arguments to ${MONITOR_EXEC}. Both ${MONITOR_EXEC} and the m flag can be set from the administrator interface.

b

> Don't begin recording unless a call is bridged to another channel.

```
exten => 123,1,Answer()
; record the current channel, and mix the audio channels at the end of
; recording
exten => 123,2,Monitor(wav,monitor_test,mb)
exten => 123,3,SayDigits(12345678901234567890)
exten => 123,4,StopMonitor()
```

See Also

ChangeMonitor(), StopMonitor(), MixMonitor(), PauseMonitor(), UnpauseMonitor()

MorseCode() Plays Morse code

MorseCode(*string*)

Plays the *string*, encoded in International Morse Code. The following channel variables will affect the playback:

MORSEDITLEN

> The length, in milliseconds, of a DIT. Defaults to 80 ms.

 All of the other tone and silence lengths are defined in the International Morse Code standard with respect to the length of a DIT, and therefore, each of the other lengths will be adjusted suitably.

MORSETONE

> The tone, in Hertz (Hz), which will be used. Defaults to 800 Hz.

```
; dah-dit-dah dit-dit dit-dit-dit-dit-dah dah-dit-dah dit-dit-dah dit-dah
exten => 123,1,Answer()
exten => 123,2,MorseCode(KI4KUA)
```

SayAlpha(), SayPhonetic()

MP3Player()

Plays an MP3 file or stream

MP3Player(*location*)

Uses the *mpg123* program to play the given *location* to the caller. The specified *location* can be either a filename or a valid URL. The caller can exit by pressing any key.

The correct version of *mpg123* must be installed for this application to work properly. Asterisk currently works best with *mpg123-0.59r*. Other versions may give less than desirable results.

```
exten => 123,1,Answer()
exten => 123,2,MP3Player(test.mp3)

exten => 123,1,Answer()
exten => 123,2,MP3Player(http://example.com/test.mp3)
```

MusicOnHold()

Plays music on hold indefinitely

MusicOnHold(*class*)

Plays hold music specified by *class*, as configured in *musiconhold.conf*. If omitted, the default music class for the channel will be used. You can use the MUSICCLASS dialplan function to set the default music class for the channel.

```
; transfer telemarketers to this extension to keep them busy
exten => 123,1,Answer()
exten => 123,n,Playback(tt-allbusy)
exten => 123,n,MusicOnHold(default)
```

See Also

SetMusicOnHold(), WaitMusicOnHold(), MUSICCLASS

NBScat()

Plays an NBS local stream

NBScat()

Uses the nbscat8k program to listen to the local Network Broadcast Sound (NBS) stream. (For more information, see the *nbs* module in Digium's Subversion server.) The caller can exit by pressing any key.

Returns -1 on hangup; otherwise, does not return.

```
exten => 123,1,Answer()
exten => 123,2,NBScat()
```

NoCDR()

```
NoCDR()
```

Disables CDRs for the current call.

```
; don't log calls to 555-1212
exten => 5551212,1,Answer()
exten => 5551212,2,NoCDR()
exten => 5551212,3,Dial(Zap/4/5551212)
```

See Also

AppendCDRUserField(), ForkCDR(), SetCDRUserField()

NoOp()

```
NoOp(text)
```

Does nothing—this application is simply a placeholder. This application is often used as a debugging tool. Whenever Asterisk's core verbosity level is set to 3 or above, Asterisk evaluates and prints each line of the dialplan before executing it. This means that any arguments passed to the NoOp() application (in the *text* parameter) are printed to the console. By watching the console output, a skilled Asterisk administrator can use this output to debug problems in the dialplan.

```
exten => 123,1,NoOp(CallerID is ${CALLERID})
```

 You don't have to place quotes around the text. If quotes are placed within the brackets, they will show up on the console.

When to use NoOp() and Verbose()

The difference between Verbose() and NoOp() is subtle. Here are some suggestions as to how you can discern when to use each. The Verbose() application is handy when you want to output something to the console. Using the set verbose command (followed by the level of verbosity you desire—0 to 4), you can set the output to a level that will not show you all of the activity on the system, but rather only those things that are equal to or less than that the current level. (Actually, you can set the verbosity to whatever you want. The set verbose 999 will work just fine, but we have not been able to find anything that outputs at a level higher than 4, so anything from 4 to infinity will be effectively the same for the time being.) This means that you can display all

---

kinds of information pertaining to a section of code you are testing, without having to see all of the other activity in the system. If you set the following in your dial plan:

```
exten => _X.,n,Verbose(2, ${SOME_VAR})
```

You can then use the CLI to set the verbosity to 2 or less (`core set verbose 2`), and you will see output from the various calls to `Verbose()`, but very little else.

Read the section on `Verbose()` later in this appendix for more on how to use it. The `NoOp()` application is best used as a place holder. For example, if you are setting a `Goto()` point in your dialplan that is using a priority label, you can use `NoOp()` as the destination point for that goto. For example,

```
exten => _X.,n(call_forward),NoOp()
```

is an excellent marker for pointing a jump in your dialplan to a spot. From that point you can carry on with whatever logic you wanted to apply to that part of the extension (judging by the label, it'd have something to do with call forwarding). The value of the `NoOp()` is that when you don't really know what sort of things you might want to move around in relation to what follows that label, you can be sure you'll never have to recode the label itself. It will never do anything other than provide a destination for the `Goto()`, so you can put it wherever you like and be sure it won't introduce any unexpected behavior.

If this seems confusing, it is due to our inability to describe it right. Experiment with `Verbose()` and `NoOp()` all in your dialplan (you can use them anywhere), and you will quickly gain an understanding of how they can help you (especially if you are like us and cause a lot of syntax errors).

See Also

`Verbose()`, `Log()`

Page()   Opens one-way audio to multiple phones

`Page(Tech/chan1[&Tech/chan2]&[...][&Tech/chanN][|options])`

Places outbound calls to the given technology/resource and dumps them in to a conference bridge as muted participants. The original caller is dumped in to the conference as a speaker, and the room is destroyed when the original caller leaves. The following options may be specified:

d

Full duplex audio. Allow the paged persons to respond to the caller.

q

Quiet. Do not play a beep to the caller.

r

Record the page. See the r option to `MeetMe` for more information.

```
exten => 123,1,Page(SIP/101&SIP/102&IAX2/iaxy123)
```

See Also

MeetMe()

Park() Parks the current call

Park()

Parks the current call (typically in combination with a supervised transfer to determine the parking space number). This application is always registered internally and does not need to be explicitly added in to the dialplan, although you should include the parkedcalls context. Parking configuration is set in *features.conf*.

```
; explicitly park the caller
include => parkedcalls
exten => 123,1,Answer()
exten => 123,n,Park()
```

See Also

ParkAndAnnounce(), ParkedCall()

ParkAndAnnounce() Parks the current call and announces the call over the specified channel

ParkAndAnnounce(*template,timeout,channel*[,*return_context*])

Parks the current call in the parking lot and announces the call over the specified *channel*. The *template* is a colon-separated list of files to announce; the word PARKED is replaced with the parking space number of the call. The *timeout* argument is the time in seconds before the call returns to the *return_context*. The *channel* argument is the channel to call to make the announcement. Console/dsp calls the console. The *return_context* argument is a Goto()-style label to jump the call back in to after timeout, which defaults to n+1 (where n is the current priority) in the *return_context* context.

```
include => parkedcalls
exten => 123,1,Answer()
exten => 123,2,ParkAndAnnounce(vm-youhave:a:pbx-transfer:at:vm-extension:PARKED,120,
Console/dsp)
exten => 123,3,Playback(vm-nobodyavail)
exten => 123,4,Playback(vm-goodbye)
exten => 123,5,Hangup()
```

See Also

Park(), ParkedCall()

ParkedCall() Answers a parked call

ParkedCall(*parkingslot*)

Connects the caller to the parked call in the parking space identified by *parkingslot*. This application is always registered internally and does not need to be explicitly added in to the dialplan, although you should include the `parkedcalls` context.

```
; pick up the call parked in parking space 701
exten => 123,1,Answer()
exten => 123,2,ParkedCall(701)
```

See Also

Park(), ParkAndAnnounce()

PauseMonitor()   Suspends monitoring of a channel

PauseMonitor()

Temporarily suspends the monitoring (recording) of the current channel

```
exten => 123,1,Answer()
exten => 123,n,Monitor(wav,monitor_test)
exten => 123,n,Playback(demo-congrats)
      ; temporarily pause the monitoring while we gather some secret info
exten => 123,n,PauseMonitor()
exten => 123,n,Read(NEWPASS,vm-newpassword)
exten => 123,n,SayDigits(${NEWPASS})
exten => 123,n,UnpauseMonitor()
exten => 123,n,Dial(${JOHN})
```

See Also

Monitor(), StopMonitor(), UnpauseMonitor()

PauseQueueMember()   Temporarily blocks a queue member from receiving calls

PauseQueueMember([*queuename*],*interface*[,*options*])

Pauses the specified queue *interface*. This prevents any calls from being distributed from the queue to the *interface* until it is unpaused by the UnpauseQueueMember() application or the Manager Interface. If no *queuename* is given, the interface is paused in every queue it is a member of.

This application sets a channel variable named PQMSTATUS to either PAUSED or NOTFOUND upon completion.

If the *options* parameter is set to j and the *interface* is not in the named queue, or if no queue is given and the *interface* is not in any queue, it will jump to priority n+101 (where n is the current priority), if it exists.

```
exten => 123,1,PauseQueueMember(,SIP/300)
exten => 124,1,UnpauseQueueMember(,SIP/300)
```

See Also

UnpauseQueueMember()

Pickup() Answers a ringing call from another phone

Pickup(*extension*[@*context*][&*extension2*[@*context2*][...])

Picks up any ringing channel that is ringing the specified *extension*. If multiple extensions are specified, Pickup() will retrieve the first matching call found. If no *context* is specified, the current context will be used.

There is also a special *context* name of PICKUPMARK. If specified, Pickup will find the first ringing channel with the channel variable PICKUPMARK set, with a value corresponding to the value of *extension*.

Playback() Plays the specified audio file to the caller

Playback(*filename*[&*filename2*...][,*options*])

Plays back a given filename to the caller. The filename should not contain the file extension, as Asterisk will automatically choose the audio file with the lowest conversion cost. Zero or more *options* may also be included. The skip option causes the playback of the message to be skipped if the channel is not in the "up" state (i.e., it hasn't yet been answered). If skip is specified, the application will return immediately should the channel not be off-hook. Otherwise, unless noanswer is specified, the channel will be answered before the sound file is played. (Not all the channels support playing messages while still on-hook.) If j is passed as one of *options* and the file does not exist, this application jumps to priority n+101 (where n is the current priority), if it exists.

```
exten => 123,1,Answer()
exten => 123,n,Playback(tt-weasels)
```

See Also

Background(), ControlPlayback()

Playtones() Plays a tone list

Playtones(*tonelist*)

Plays a tone list. Execution immediately continues with the next step, while the tones continue to play. The *tonelist* is either the tone name defined in the *indications.conf* configuration file, or a specified list of frequencies and durations. See *indications.conf* for a description of the specification of a tone list.

Use the StopPlaytones() application to stop the tones from playing.

```
; play a busy signal for two seconds, and then a congestion tone
for two seconds
exten => 123,1,Playtones(busy)
exten => 123,2,Wait(2)
exten => 123,3,StopPlaytones()
exten => 123,4,Playtones(congestion)
exten => 123,5,Wait(2)
exten => 123,6,StopPlaytones()
exten => 123,7,Goto(1)
```

See Also

StopPlaytones(), *indications.conf*, Busy(), Congestion(), Progress(), Ringing()

PrivacyManager()   Requires a caller to enter his phone number, if no Caller ID information is received

PrivacyManager([*maxretries*[,*minlength*[,*options*]]])

If no Caller ID is received, this application answers the channel and asks the caller to enter his phone number. By default, the caller is given three attempts. PrivacyManager() sets a channel variable named PRIVACYMGRSTATUS to either SUCCESS or FAILURE. If Caller ID is received on the channel, PrivacyManager() does nothing.

If the *options* parameter is set to j and the caller fails to enter his Caller ID number, the call will continue at priority n+101 (where n is the current priority).

The *privacy.conf* configuration file changes the functionality of the PrivacyManger() application. It contains the following two lines:

maxretries
: Specifies the maximum number of attempts the caller is allowed to input a Caller ID number (default: 3).

minlength
: Specifies the minimum allowable digits in the input Caller ID number (default: 10).

The *maxretries* and *minlength* settings may also be passed to the application as parameters. Parameters passed to the application override any settings in *privacy.conf*.

```
exten => 123,1,Answer()
exten => 123,n,PrivacyManager()
exten => 123,n,GotoIf($["${PRIVACYMGRSTATUS}" = "FAILURE"]?bad)
exten => 123,n,Dial(Zap/1)
exten => 123,n,Hangup()
exten => 123,n(bad),Playback(im-sorry)
exten => 123,n,Playback(vm-goodbye)
exten => 123,n,Hangup()
```

See Also

Zapateller()

Progress()

```
Progress()
```

Requests that the channel indicate that in-band progress is available to the user. Each channel type in Asterisk has its own way of signaling progress on the call.

```
; indicate progress to the calling channel, wait 5 seconds,
; and then answer the call
exten => 123,1,Progress()
exten => 123,n,Wait(5)
exten => 123,n,Answer()
```

See Also

Busy(), Congestion(), Ringing(), Playtones()

Queue()

```
Queue(queuename[,options[,URL[,announceoverride[,timeout[,AGI]]]]])
```

Places an incoming call in to the call queue specified by *queuename*, as defined in *queues.conf*.

The *options* argument may contain zero or more of the following characters:

d

Specifies a data-quality (modem) call (minimum delay).

h

Allows callee to hang up by hitting *.

H

Allows caller to hang up by hitting *.

i

Ignores call forward requests from queue members and does nothing when they are requested.

n

Disallows retries on the timeout; exits this application and goes to the next step.

r

Rings instead of playing music on hold.

t

Allows the called user to transfer the call.

T

Allows the calling user to transfer the call.

w

Allows the called user to write the conversation to disk.

W
> Allows the calling user to write the conversation to disk.

In addition to being transferred, a call may be parked and then picked up by another user.

The *announceoverride* argument overrides the standard announcement played to queue agents before they answer the specified call.

The optional *URL* will be sent to the called party if the channel supports it.

The *timeout* will cause the queue to fail out after a specified number of seconds, checked between each *queues.conf* *timeout* and *retry* cycle. The call will continue on with the next priority in the dialplan.

This application sets a channel variable named `QUEUESTATUS` upon completion. It will take one of the following values:

TIMEOUT
> The call was in the queue too long, and timed out. See the *timeout* parameter.

FULL
> The queue was already full. See the `maxlen` setting for the queue in *queues.conf*.

JOINEMPTY
> The caller could not join the queue, as there were no queue members to answer the call. See the `joinempty` setting for the queue in *queues.conf*.

LEAVEEMPTY
> The caller joined the queue, but then all queue members left. See the `leavewhenempty` setting for the queue in *queues.conf*.

JOINUNAVAIL
> The caller could not join the queue, as there were no queue members available to answer the call. See the `joinempty` setting for the queue in *queues.conf*.
>
> The caller joined the queue, but then all of the queue members became unavailable. See the `leavewhenempty` setting for the queue in *queues.conf*.

```
; place the caller in the techsupport queue
exten => 123,1,Answer()
exten => 123,2,Queue(techsupport,t)
```

See Also

AddQueueMember(), RemoveQueueMember(), PauseQueueMember(), UnpauseQueueMember(), AgentLogin(), *queues.conf*, QUEUE_MEMBER_COUNT, QUEUE_MEMBER_LIST, QUEUE_WAITING_COUNT

QueueLog()   Writes arbitrary queue events to the queue log

QueueLog(*queuename,uniqueid,member,event*[*,additionalinfo*])

Writes an arbitrary queue event to the queue log. The *queuename* parameter specifies the name of the queue. The *uniqueid* parameter specifies the unique identifier for the channel. The *member* parameter specifies which queue member the event pertains to. The *event* and *additionalinfo* parameters may be set to arbitrary data, as needed.

```
; Write an arbitrary event to the queue log
exten => 123,1,QueueLog(myqueue,${UNIQUEID},Agent/123,MyTestEvent)
```

See Also

Queue()

Random()   Conditionally branches, based upon a probability

Random([*probability*]:[[*context*,]*extension*,]*priority*)

 This application has been deprecated in favor of:

```
GotoIf($[${RAND(1,100)} > num]?label)
```

Conditionally jumps to the specified *priority* (and optional *extension* and *context*), based on the specified *probability*. *probability* should be an integer between 1 and 100. The application will jump to the specified destination *priority* percent of the time.

```
; choose a random number over and over again
exten => 123,1,SayNumber(${RAND(1|10)})
exten => 123,n,Goto(1)
```

See Also

RAND

Read()   Reads DTMF digits from the caller and assigns the result to a variable

Read(*variable*[,*filename*[,*maxdigits*[,*option*[,*attempts*[,*timeout*]]]]])

Reads a #-terminated string of digits from the user in to the given *variable*.

Other arguments include:

filename
    Specifies the file to play before reading digits.

maxdigits
    Sets the maximum acceptable number of digits. If this argument is specified, the application stops reading after *maxdigits* have been entered (without requiring the

user to press the # key). Defaults to 0 (no limit, wait for the user to press the # key). Any value below 0 means the same. The maximum accepted value is 255.

option

Zero or more of the following options:

s

Return immediately if the line is not answered.

i

Interpret the filename as an indication tone setting from *indications.conf*.

n

Read digits even if the line has not been answered.

attempts

If greater than 1, that many attempts will be made in the event that no data is entered.

timeout

If greater than 0, that value will override the default timeout.

```
; read a two-digit number and repeat it back to the caller
exten => 123,1,Read(NUMBER,,2)
exten => 123,2,SayNumber(${NUMBER})
exten => 123,3,Goto(1)
```

See Also

SendDTMF()

ReadFile()

Reads the contents of a file in to a variable

ReadFile(*variable=filename,length*)

ReadFile captures the contents of *filename*, with a maximum size of *length*.

```
; read the first 80 characters of a file in to a variable
exten => 123,1,Answer()
exten => 123,n,ReadFile(TEST=/tmp/test.txt,80)
exten => 123,n,SayAlpha(${TEST})
```

See Also

System(), Read()

RealTime

Looks up information from the RealTime configuration handler

RealTime(*family,colmatch,value[,prefix]*)

Uses the RealTime configuration handler system to read data in to channel variables. All unique column names (from the specified *family*) will be set as channel variables, with an optional *prefix* to the name (e.g., a prefix of `var_` would make the column name become the variable ${var_name}).

```
; retrieve all columns from the sipfriends table where the name column
; matches "John", and prefix all the variables with "John_"
exten => 123,1,RealTime(sipfriends,name,John,John_)
; now, let's read the value of the column named "port"
exten => 123,n,SayNumber(${John_port})
```

See Also

RealTimeUpdate()

RealTimeUpdate()   Updates a value via the RealTime configuration handler

RealTimeUpdate(*family*,*colmatch*,*value*,*newcol*,*newval*)

Uses the RealTime configuration handler system to update a value. The column *newcol* in *family* matching column *colmatch* = *value* will be updated to *newval*.

A channel variable named `REALTIMECOUNT` will be set with the number of rows updated or `-1` if an error occurs.

```
; this will update the port column in the sipfriends table to a new
; value of 5061, where the name column matches "John"
exten => 123,1,RealTimeUpdate(sipfriends,name,John,port,5061)
```

See Also

RealTime

Record()   Records channel audio to a file

Record(*filename.format*[,*silence*[,*maxduration*[,*options*]]])

Records audio from the channel in to the given *filename*. If the file already exists, it will be overwritten.

Optional arguments include:

format
    Specifies the format of the file type to be recorded.

silence
    Specifies the number of seconds of silence to allow before ending the recording and continuing on with the next priority in the dialplan.

maxduration
    Specifies the maximum recording duration, in seconds. If not specified or `0`, there is no maximum.

options
    May contain any of the following letters:

a

Append to the recording, instead of overwriting it.

n

Do not answer, but record anyway if the line is not yet answered.

q

Quiet mode; do not play a beep tone at the beginning of the recording.

s

Skip recording if the line is not yet answered.

t

Use the alternate * terminator key instead of the default #.

x

Ignore all termination keys and keep recording until hangup.

If the *filename* contains %d, these characters will be replaced with a number incremented by one each time the file is recorded.

The user can press # to terminate the recording and continue to the next priority in the dialplan.

```
; record the caller's name
exten => 123,1,Playback(pls-rcrd-name-at-tone)
exten => 123,n,Record(/tmp/name.gsm,3,30)
exten => 123,n,Playback(/tmp/name)
```

RemoveQueueMember()   Dynamically removes queue members

RemoveQueueMember(*queuename*[,*interface*[,*options*]])

Dynamically removes the specified *interface* from the *queuename* call queue. If *interface* is not specified, this application removes the current channel from the queue.

If the *options* parameter is set to j, and the interface is not in the queue and there exists a priority n+101 (where n is the current priority), the application will jump to that priority.

```
; remove SIP/3000 from the techsupport queue
exten => 123,1,RemoveQueueMember(techsupport,SIP/3000)
```

See Also

Queue(), AddQueueMember(), PauseQueueMember(), UnpauseQueueMember()

ResetCDR()   Resets the Call Detail Record

ResetCDR([*options*])

Causes the Call Detail Record to be reset for the current channel. The *options* parameter can be zero or more of the following options:

a

Store any stacked records.

w

Store the current CDR record before resetting it.

v

Save CDR variables.

```
; write a copy of the current CDR record, and then reset the CDR
exten => 123,1,Answer()
exten => 123,2,Playback(tt-monkeys)
exten => 123,3,ResetCDR(wv)
exten => 123,4,Playback(tt-monkeys)
```

See Also

ForkCDR(), NoCDR()

RetryDial() Attempts to place a call, and retries on failure

RetryDial(*announce,sleep,loops,technology/resource*[*&Technology2/resource2*...]
[,*timeout*][,*options*][,*URL*])

Attempts to place a call. If no channel can be reached, plays the file defined by *announce*, waiting *sleep* seconds to retry the call. If the specified number of attempts matches *loops*, the call will continue with the next priority in the dialplan. If *loops* is set to 0, the call will retry endlessly.

While waiting, a one-digit extension may be dialed. If that extension exists in either the context defined in ${EXITCONTEXT} (if defined) or the current one, the call will transfer to that extension immediately.

All arguments after *loops* are passed directly to the Dial() application.

```
; attempt to dial the number three times via IAX, retrying every five
seconds
exten => 123,1,RetryDial(priv-trying,5,3,IAX2/VOIP/8885551212,30)
; if the caller presses 9 while waiting, dial the number on the Zap/4
channel
exten => 9,1,RetryDial(priv-trying,5,3,Zap/4/8885551212,30)
```

See Also

Dial()

Return() Returns from a Gosub() or GosubIf()

Return()

Returns from a previously invoked Gosub() or GosubIf(). If there was no previous invocation of Gosub() or GosubIf(), Return() exits abnormally.

See Also

Gosub(), StackPop()

Ringing()

Ringing()

Requests that the channel indicate ringing tone to the user. It is up to the channel driver to specify exactly how ringing is indicated.

Note that this application does not actually provide audio ringing to the caller. Use the Playtones() application to do this.

```
; indicate that the phone is ringing, even though it isn't
exten => 123,1,Ringing()
exten => 123,2,Wait(5)
exten => 123,3,Playback(tt-somethingwrong)
```

See Also

Busy(), Congestion(), Progress(), Playtones()

SayAlpha()

SayAlpha(*string*)

Spells out the specified *string*, using the current language setting for the channel. See the CHANNEL function for more information on changing the language for the current channel.

```
exten => 123,1,SayAlpha(ABC123XYZ)
```

See Also

SayDigits(), SayNumber(), SayPhonetic(), CHANNEL

SayDigits()

SayDigits(*digits*)

Says the specified *digits*, using the current language setting for the channel. See the CHANNEL function for more information on changing the language for the current channel.

```
exten => 123,1,SayDigits(1234)
```

See Also

SayAlpha(), SayNumber(), SayPhonetic(), CHANNEL

SayNumber()

SayNumber(*digits*[,*gender*])

Says the specified number, using the current language setting for the channel. See the CHANNEL function for more information on changing the language for the current channel.

If the current language supports different genders, you can pass the *gender* argument to change the gender of the spoken number. You can use the following *gender* arguments:

- Use the *gender* arguments f for female, m for male, and n for neuter in European languages such as Portuguese, French, Spanish, and German.
- Use the *gender* argument c for commune and n for neuter in Nordic languages such as Danish, Swedish, and Norwegian.
- Use the *gender* argument p for plural enumerations in German.

```
; say the number in English
exten => 123,1,Set(CHANNEL(language)=en)
exten => 123,2,SayNumber(1234)
```

For this application to work in languages other than English, you must have the appropriate sounds for the language you wish to use.

See Also

SayAlpha(), SayDigits(), SayPhonetic(), CHANNEL

SayPhonetic()

SayPhonetic(*string*)

Spells the specified *string* using the NATO phonetic alphabet.

```
exten => 123,1,SayPhonetic(asterisk)
```

See Also

SayAlpha(), SayDigits(), SayNumber()

SayUnixTime()

SayUnixTime([*unixtime*][,[*timezone*][,*format*]])

Speaks the specified time according to the specified time zone and format. The arguments are:

unixtime

>The time, in seconds, since January 1, 1970. May be negative. Defaults to now.

timezone

>The time zone. See */usr/share/zoneinfo/* for a list. Defaults to the machine default.

format

>The format in which the time is to be spoken. See *voicemail.conf* for a list of formats. Defaults to "ABdY 'digits/at' IMp".

```
exten => 123,1,SayUnixTime(,,IMp)
```

See Also

STRFTIME, STRPTIME, IFTIME

SendDTMF()   Sends arbitrary DTMF digits to the channel

SendDTMF(*digits*[,*timeout_ms*])

Sends the specified DTMF digits on a channel. Valid DTMF digits include 0-9, *, #, and A-D. You may also use the letter w as a digit, which indicates a 500 millisecond wait. The *timeout_ms* argument is the amount of time between digits, in milliseconds. If not specified, *timeout_ms* defaults to 250 milliseconds.

```
exten => 123,1,SendDTMF(3212333w222w366w3212333322321,250)
```

See Also

Read()

SendImage()   Sends an image file

SendImage(*filename*,*options*)

Sends an image on a channel, if image transport is supported. This application sets a channel variable named SENDIMAGESTATUS to either OK or NOSUPPORT upon completion.

If the *options* parameter is set to j and the channel does not support image transport, and there exists a priority n+101 (where n is the current priority), execution will continue at that step.

```
exten => 123,1,SendImage(logo.jpg)
```

See Also

SendText(), SendURL()

SendText()   Sends text to the channel

SendText(*text*,*options*)

Sends *text* on a channel, if text transport is supported. Upon completion, a channel variable named SENDTEXTSTATUS will be set to one of the following values:

SUCCESS
> The transmission of the text was successful.

FAILURE
> The transmission of the text failed.

NOSUPPORT
> The underlying channel does not support the transmission of text.

If the *options* parameter is set to j and the channel does not support text transport, and there exists a priority n+101 (where n is the current priority), execution will continue at that step.

```
exten => 123,1,SendText(Welcome to Asterisk)
```

See Also

SendImage(), SendURL()

SendURL()   Sends the specified URL to the channel (if supported)

SendURL(*URL*[,*options*])

Requests that the client go to the specified URL. The application also sets a channel variable named SENDURLSTATUS to one of the following values upon completion:

SUCCESS
> The transmission of the URL was successful.

FAILURE
> The transmission of the URL failed.

NOLOAD
> The is web-enabled but failed to load the URL.

NOSUPPORT
> The underlying channel does not support the transmission of the URL.

If the *options* parameter contains the wait option, execution will wait for an acknowledgment that the URL has been loaded before continuing.

If the *options* parameter contains the j option and the client does not support HTML transport, and there exists a priority n+101 (where n is the number of the current priority), execution will continue at that step.

```
exten => 123,1,SendURL(www.asterisk.org,wait)
```

See Also

SendImage(), SendText()

Set()

```
Set(n=value,[n2=value2...[,options]])
```

Sets the variable *n* to the specified **value**. Also sets the variable *n2* to the value of **value2**. If the variable name is prefixed with _, single inheritance is assumed. If the variable name is prefixed with _ _, infinite inheritance is assumed. Inheritance is used when you want channels created from the current channel to inherit the variable from the current channel.

If the *options* parameter is set to **g**, the variables will be set as global variables instead of channel variables.

```
; set a variable called DIALTIME, then use it
exten => 123,1,Set(DIALTIME=20)
exten => 123,1,Dial(Zap/4/5551212,,${DIALTIME})
```

 The setting of multiple variables and the use of the **g** option have been deprecated. Please use multiple calls to Set() and the GLOBAL() dialplan function instead.

See Also

GLOBAL, SET, ENV, *channelvariables.txt*

SetAMAFlags()

```
SetAMAFlags(flag)
```

Sets the AMA flags in the Call Detail Record for billing purposes, overriding any AMA settings in the channel configuration files. Valid choices are default, omit, billing, and documentation.

```
exten => 123,1,SetAMAFlags(billing)
```

See Also

SetCDRUserField(), AppendCDRUserField()

SetCallerID()

```
SetCallerID(clid[,a])
```

This application has been deprecated in favor of:

```
Set(CALLERID(all)=Some Name <1234>)
```

Sets the Caller ID on the channel to a specified value. If the **a** argument is passed, ANI is also set to the specified value.

```
; override the Caller ID for this call
exten => 123,1,Set(CALLERID(all)="John Q. Public <8885551212>")
```

See Also

The CALLERID

SetCallerPres()

Sets Caller ID presentation flags

```
SetCallerPres(presentation)
```

Sets the Caller ID presentation flags on a Q931 PRI connection.

Valid presentations are:

allowed_not_screened
: Presentation allowed, not screened

allowed_passed_screen
: Presentation allowed, passed screen

allowed_failed_screen
: Presentation allowed, failed screen

allowed
: Presentation allowed, network number

prohib_not_screened
: Presentation prohibited, not screened

prohib_passed_screen
: Presentation prohibited, passed screen

prohib_failed_screen
: Presentation prohibited, failed screen

prohib
: Presentation prohibited, network number

unavailable
: Number unavailable

```
exten => 123,1,SetCallerPres(allowed_not_screened)
exten => 123,2,Dial(Zap/g1/8885551212)
```

See Also

CALLERID()

SetCDRUserField()

SetCDRUserField(*value*)

Sets the CDR user field to the specified *value*. The CDR user field is an extra field that you can use for data not stored anywhere else in the record. CDR records can be used for billing purposes or for storing other arbitrary data about a particular call.

```
exten => 123,1,SetCDRUserField(testing)
exten => 123,2,Playback(tt-monkeys)
```

 This application has been deprecated in favor of the CDR() function.

```
exten => 123,1,Set(CDR(userfield)=54321)
```

See Also

AppendCDRUserField(), SetAMAFlags()

SetGlobalVar()

SetGlobalVar(*n=value*)

 This application has been deprecated in favor of:

```
Set(GLOBAL(var)=...)
```

Sets a global variable called *n* to the specified *value*. Global variables are available across channels.

```
; set the NUMRINGS global variable to 3
exten => 123,1,SetGlobalVar(NUMRINGS=3)
```

See Also

Set()

SetMusicOnHold()

SetMusicOnHold(*class*)

This application has been deprecated in favor of:

```
Set(CHANNEL(musicclass)=...)
```

Sets the default *class* for music on hold for the current channel. When music on hold is activated, this class will be used to select which music is played. Classes are defined in the configuration file *musiconhold.conf*.

```
exten=s,1,Answer()
exten=s,2,SetMusicOnHold(default)
exten=s,3,WaitMusicOnHold()
```

See Also

WaitMusicOnHold(), *musiconhold.conf*, MusicOnHold()

SetTransferCapability()
Sets the ISDN transfer capability of a channel

SetTransferCapability(*transfercapability*)

This application sets the ISDN transfer capability of the current channel to a new value. Valid values for *transfercapability* are:

SPEECH
: 0x00, speech (default, voice calls)

DIGITAL
: 0x08, unrestricted digital information (data calls)

RESTRICTED_DIGITAL
: 0x09, restricted digital information

3K1AUDIO
: 0x10, 3.1kHz Audio (fax calls)

DIGITAL_W_TONES
: 0x11, unrestricted digital information with tones/announcements

VIDEO
: 0x18, video

This application is deprecated and the functionality has been replaced with Set(CHANNEL(transfercapability)=*transfercapability*) syntax.

```
exten => 123,1,Set(CHANNEL(transfercapability)=SPEECH)
```

SIPAddHeader()
Adds a SIP header to the outbound call

SIPAddHeader(*Header: Content*)

Adds a header to a SIP call placed with the `Dial()` application. A nonstandard SIP header should begin with X-, such as `X-Asterisk-Accountcode:`. Use this application with care —adding the wrong headers may cause any number of problems.

For more flexibility, see the `SIP_HEADER()` dialplan function.

```
exten => 123,1,SIPAddHeader(X-Asterisk-Testing: Just testing!)
exten => 123,2,Dial(SIP/123)
```

See Also

SIP_HEADER

SIPDtmfMode()   Changes the DTMF method for a SIP call

`SIPDtmfMode(method)`

Changes the DTMF method for a SIP call. The *method* can be either `inband`, `info`, or `rfc2833`.

```
exten => 123,1,SIPDtmfMode(rfc2833)
exten => 123,2,Dial(SIP/123)
```

See Also

Appendix A

SLAStation()   Shared line appearance station

`SLAStation(station)`

This application should be executed by an SLA station. The format of the *station* parameter depends on how the call was initiated. If the phone was just taken off hook, then the *station* parameter should contain only the station name. If the call was initiated by pressing a line key, then the station name should be preceded by an underscore and the trunk name associated with that line button (`station1_line2`, for example).

For more information on shared line appearances, see the *doc/sla.pdf* file in the Asterisk source.

```
exten => 123,1,SLAStation(station1)

    exten => 124,1,SLAStation(station1_line2)
```

See Also

SLATrunk(), *sla.conf*

SLATrunk()   Shared line appearance trunk

`SLATrunk(trunk)`

This application should be executed by an SLA trunk on an inbound call. The channel calling this application should correspond to the SLA trunk specified by the *trunk* parameter.

For more information on shared line appearances, see the *doc/sla.pdf* file in the Asterisk source.

```
exten => 123,1,SLATrunk(line2)
```

See Also

SLAStation(), *sla.conf*

SoftHangup()

Performs a soft hangup of the requested channel

SoftHangup(*technology/resource*,*options*)

Hangs up the requested channel. The *options* argument may contain the letter a, which causes all channels on the specified device to be hung up (currently, the *options* argument may contain only a).

```
; hang up all calls using Zap/4 so we can use it
exten => 123,1,SoftHangup(Zap/4,a)
exten => 123,2,Wait(2)
exten => 123,3,Dial(Zap/4/5551212)
```

See Also

Hangup()

StackPop()

Removes last address from Gosub() stack

StackPop()

Removes the last address from the Gosub() stack. This is frequently used when handling error conditions within Gosub() routines when it is no longer appropriate to return control of the dialplan back to where the Gosub() routine was called from.

```
exten => s,1,Read(input,get-input)
exten => s,n,Gosub(validate,1)
exten => s,n,Dial(SIP/${input})
; Ensure that input is between 400 and 499
exten => validate,1,GotoIf($[ ${input} > 499 ]?error,1)
exten => validate,n,GotoIf($[ ${input} < 400 ]?error,1)
exten => validate,n,Return
exten => error,1,StackPop()
exten => error,2,Goto(s,1)
```

See Also

Return(), Gosub()

StartMusicOnHold()

StartMusicOnHold([*class*])

Plays hold music specified by *class*, as configured in *musiconhold.conf*. If omitted, the default music class for the channel will be used. You can use the CHANNEL(musicclass) function to set the default music class for the channel.

Returns immediately.

```
; transfer telemarketers to this extension to keep them busy
exten => 123,1,Answer()
exten => 123,2,Playback(tt-allbusy)
exten => 123,3,StartMusicOnHold(default)
exten => 123,4,Wait(600)
exten => 123,5,StopMusicOnHold()
```

See Also

WaitMusicOnHold(), StopMusicOnHold()

StopMixMonitor()

StopMixMonitor()

Stops monitoring (recording) a channel. This application has no effect if the channel is not currently being monitored.

```
exten => 123,1,Answer()
exten => 123,2,MixMonitor(monitor_test.wav)
exten => 123,3,SayDigits(12345678901234567890)
exten => 123,4,StopMixMonitor()
```

See Also

MixMonitor()

StopMonitor()

StopMonitor()

Stops monitoring (recording) a channel. This application has no effect if the channel is not currently being monitored.

```
exten => 123,1,Answer()
exten => 123,2,Monitor(wav,monitor_test,mb)
exten => 123,3,SayDigits(12345678901234567890)
exten => 123,4,StopMonitor()
```

See Also

ChangeMonitor()

StopPlaytones()

```
StopPlaytones()
```

Stops playing the currently playing tone list.

```
exten => 123,1,Playtones(busy)
exten => 123,2,Wait(2)
exten => 123,3,StopPlaytones()
exten => 123,4,Playtones(congestion)
exten => 123,5,Wait(2)
exten => 123,6,StopPlaytones()
exten => 123,7,Goto(1)
```

See Also

Playtones(), *indications.conf*

StopMusicOnHold()

```
StopMusicOnHold()
```

Ends playing music on hold on a channel. If music on hold was not already playing, has no effect.

```
; transfer telemarketers to this extension to keep them busy
exten => 123,1,Answer()
exten => 123,2,Playback(tt-allbusy)
exten => 123,3,StartMusicOnHold(default)
exten => 123,4,Wait(600)
exten => 123,5,StopMusicOnHold()
```

See Also

WaitMusicOnHold(), StartMusicOnHold()

System()

```
System(command)
```

Executes a *command* in the underlying operating system. This application sets a channel variable named SYSTEMSTATUS to either FAILURE or SUCCESS, depending on whether or not Asterisk was successfully able to run the command.

This application is very similar to the TrySystem() application, except that it will return -1 if it is unable to execute the system command, whereas the TrySystem() application will always return 0.

```
exten => 123,1,System(echo hello > /tmp/hello.txt)
```

See Also

TrySystem()

Transfer()   Transfers the caller to a remote extension

Transfer([*Technology/*]*destination*[,*options*)

Requests that the remote caller be transferred to the given (optional *Technology* and) *destination*. If the *Technology* is set to IAX2, SIP, Zap, etc., then transfer will happen only if the incoming call is of the same channel type.

Upon completion, this application sets a channel variable named TRANSFERSTATUS to one of the following values:

SUCCESS
> The transfer was successful.

FAILURE
> The transfer was not successful.

UNSUPPORTED
> The transfer was not supported by the underlying channel driver.

If the *options* parameter is set to the letter j and the transfer is *not* supported or successful, and there exists a priority n+101 (where n is the current priority), that priority will be taken next.

```
; transfer calls from extension 123 to extension SIP/123@otherserver
exten => 123,1,Transfer(SIP/123@otherserver)
```

TryExec()   Tries to execute an Asterisk application

TryExec(*app(args)*)

Attempts to run the specified Asterisk application.

This application is very similar to the Exec() application, except that it always returns normally, whereas the Exec() application will act as if the underlying application was natively called, including exit status. This application can be used to catch a condition that would normally cause the underlying application to exit abnormally.

```
exten => 123,1,TryExec(VMAuthenticate(@default))
```

See Also

Exec()

TrySystem()   Tries to execute an operating system command

TrySystem(*command*)

Attempts to execute a *command* in the underlying operating system. The result of the *command* will be placed in the SYSTEMSTATUS channel variable and will be one of the following:

FAILURE

> The specified *command* could not be executed.

SUCCESS

> The specified *command* executed successfully.

APPERROR

> The specified *command* executed, but returned an error code.

This application is very similar to the System() application, except that it always returns normally, whereas the System() application will return abnormally if it is unable to execute the system command.

```
exten => 123,1,TrySystem(echo hello > /tmp/hello.txt)
```

See Also

System()

UnpauseMonitor()   Resumes monitoring of a channel

UnpauseMonitor()

Resumes monitoring a channel, after being suspended by PauseMonitor().

```
exten => 123,1,Answer()
exten => 123,n,Monitor(wav,monitor_test)
exten => 123,n,Playback(demo-congrats)
      ; temporarily pause the monitoring while we gather some secret info
exten => 123,n,PauseMonitor()
exten => 123,n,Read(NEWPASS,vm-newpassword)
exten => 123,n,SayDigits(${NEWPASS})
      ; unpause the recording and continue recording the call
exten => 123,n,UnpauseMonitor()
exten => 123,n,Dial(${JOHN})
```

See Also

Monitor(), StopMonitor(), Page()

UnpauseQueueMember()   Unpauses a queue member

UnpauseQueueMember([*queuename,*]*interface*[,*options*])

Unpauses (resumes calls to) a queue member. This is the counterpart to PauseQueue Member(), and it operates exactly the same way, except it unpauses instead of pausing the given interface.

Upon completion, this application sets a channel variable named UPQMSTATUS to either UNPAUSED or NOTFOUND.

If the *options* parameter is set to the letter j and the queue member can not be found, and there exists a priority n+101 (where n is the current priority), control of the call will continue at that priority

```
exten => 123,1,PauseQueueMember(myqueue,SIP/300)
exten => 124,1,UnpauseQueueMember(myqueue,SIP/300)
```

See Also

PauseQueueMember()

UserEvent()   Sends an arbitrary event to the Manager Interface

UserEvent(*eventname*[,*body*])

Sends an arbitrary event to the Manager Interface, with an optional body representing additional arguments. The format of the event is:

```
Event: UserEvent
UserEvent: eventname
            body
```

If the body is not specified, only the Event and UserEvent fields will be present in the Manager event.

```
exten => 123,1,UserEvent(BossCalled,${CALLERID(name)} has called the boss!)
exten => 123,2,Dial(${BOSS})
```

See Also

manager.conf, Asterisk Manager Interface

Verbose()   Sends arbitrary text to verbose output

Verbose([*level*,]*message*)

Sends the specified *message* to verbose output. The *level* must be an integer value. If not specified, *level* defaults to 0.

```
exten => 123,1,Verbose(Somebody called extension 123)
exten => 123,2,Playback(extension)
exten => 123,3,SayDigits(${EXTEN})
```

 The optional argument *level* is not so optional, if you include the delimiter | in your invocation of Verbose(). If the delimiter is found, Verbose() assumes that you meant to specify *level* (and chops off everything preceding the initial |). It is therefore probably best to get in the habit of always specifying the *level*.

See also

NoOp(), Log()

VMAuthenticate()   Authenticates the caller from voicemail passwords

VMAuthenticate([*mailbox*][@*context*[,*options*]])

Behaves identically to the Authenticate() application, with the exception that the passwords are taken from *voicemail.conf*.

If *mailbox* is specified, only that mailbox's password will be considered valid. If *mailbox* is not specified, the channel variable AUTH_MAILBOX will be set with the authenticated mailbox.

If the *options* parameter is set to the letter s, Asterisk will skip the initial prompts.

```
; authenticate off of any mailbox password in the default voicemail context
; and tell us the matching mailbox number
exten => 123,1,VMAuthenticate(@default)
exten => 123,2,SayDigits(${AUTH_MAILBOX})
```

See Also

Authenticate(), *voicemail.conf*

VoiceMail()   Leaves a voicemail message in the specified mailbox

VoiceMail(*mailbox*[@*context*][&*mailbox*[@*context*]][...]|*options*)

Leaves voicemail for a given *mailbox* (must be configured in *voicemail.conf*). If more than one mailbox is specified, the greetings will be taken from the first mailbox specified.

Options:

s

Skip the playback of instructions.

u

Play the user's unavailable greeting.

b

Play the user's busy greeting.

g(num)

Amplify the recording by *num* decibels (dB).

j

If the requested mailbox does not exist, and there exists a priority n+101 (where n is the current priority), that priority will be taken next.

If the caller presses 0 (zero) during the prompt, the call jumps to the o (lowercase letter o) extension in the current context, if operator=yes was specified in *voicemail.conf*.

If the caller presses * during the prompt, the call jumps to extension a in the current context. This is often used to send the caller to a personal assistant.

Upon completion, this application sets a channel variable named VMSTATUS. It will contain one of the following values:

SUCCESS
> The call was successfully sent to voicemail.

USEREXIT
> The caller exited from the voicemail system.

FAILED
> The system was not able to send the call to voicemail.

```
; send caller to unavailable voicemail for mailbox 123
exten => 123,1,VoiceMail(123@default,u)
```

See Also

VoiceMailMain(), *voicemail.conf*

VoiceMailMain()   Enters the voicemail system

VoiceMailMain([*mailbox*][@*context*][,*options*])

Enters the main voicemail system for the checking of voicemail. Passing the *mailbox* argument will stop the voicemail system from prompting the user for the mailbox number. You should also specify a voicemail *context*.

The *options* string can contain zero or more of the following options:

s
> Skip the password check.

p
> This option tells Asterisk to interpret the *mailbox* as a number that should be prepended to the mailbox entered by the caller. This is most often used when many voicemail boxes for different companies are hosted on the same Asterisk server.

g(*gain*)
> When recording voicemail messages, increase the volume by *gain*. This option should be specified in an integer number of decibels.

a(*folder*)
> Skip the folder prompt and go directly to the specified *folder*. This option defaults to the INBOX.

```
; go to voicemail menu for mailbox 123 in the default voicemail context
exten => 123,1,VoiceMailMain(123@default)
```

VoiceMail(), *voicemail.conf*

Wait()

Wait(*seconds*)

Waits for the specified number of *seconds*. You can pass fractions of a second. For example, setting *seconds* to **1.5** would make the dialplan wait 1.5 seconds before going on to the next priority in the dialplan.

```
; wait 1.5 seconds before playing the prompt
exten => s,1,Answer()
exten => s,2,Wait(1.5)
exten => s,3,Background(enter-ext-of-person)
```

WaitExten()

WaitExten([*seconds*][,*options*])

Waits for the user to enter a new extension for the specified number of seconds. You can pass fractions of a second (e.g., 1.5 = 1.5 seconds). If *seconds* is unspecified, the default extension timeout will be used. Most often, this application is used without specifying the *seconds* options.

The *options* parameter may be set to the following option:

m[(*class*)[
Provide music on hold to the caller while waiting for an extension. Optionally, you can specify the music on hold *class* within parentheses.

```
; wait 15 seconds for the user to dial an extension
exten => s,1,Answer()
exten => s,2,Playback(enter-ext-of-person)
exten => s,3,WaitExten(15)
```

See Also

Background(), TIMEOUT

WaitForRing()

WaitForRing(*timeout*)

Waits at least *timeout* seconds after the next ring has completed.

```
; wait five seconds for a ring, and then send some DTMF digits
exten => 123,1,Answer()
exten => 123,2,WaitForRing(5)
exten => 123,3,SendDTMF(1234)
```

See Also

WaitForSilence()

WaitForSilence() Waits for a specified amount of silence

WaitForSilence(*silencerequired*[,*repeat*[,*timeout*]])

Waits for *repeat* instances of *silencerequired* milliseconds of silence. If *repeat* is omitted, the application waits for a single instance of *silencerequired* milliseconds of silence.

If the *timeout* option is specified, this application will return to the next priority in the dialplan after the specified number of seconds, even if silence has not been detected.

 Please use the *timeout* with caution, as it may defeat the purpose of this application, which is to wait indefinitely until silence is detected on the line. You may want to set the timeout to a high value only to avoid an infinite loop in cases where silence is never detected.

This applications sets a channel variable named WAITSTATUS to either SILENCE or TIMEOUT.

```
; wait for three instances of 300 ms of silence
exten => 123,WaitForSilence(300,3)
```

See Also

WaitForRing()

WaitMusicOnHold() Waits the specified number of seconds, playing music on hold

WaitMusicOnHold(*delay*)

Plays hold music for the specified number of seconds. If no hold music is available, the delay will still occur but with no sound.

Returns 0 when done, or -1 on hangup.

```
; allow caller to hear Music on Hold for five minutes
exten => 123,1,Answer()
exten => 123,2,WaitMusicOnHold(300)
exten => 123,3,Hangup()
```

See Also

SetMusicOnHold(), *musiconhold.conf*

While() Starts a while loop

While(*expr*)

Starts a while loop. Execution will return to this point when EndWhile() is called, until *expr* is no longer true. If a condition is met causing the loop to exit, Asterisk continues execution of the dialplan on the next priority after the corresponding EndWhile().

```
exten => 123,1,Set(COUNT=1)
exten => 123,2,While($[ ${COUNT} < 5 ])
exten => 123,3,SayNumber(${COUNT})
exten => 123,4,Set(COUNT=$[${COUNT} + 1])
exten => 123,5,EndWhile()
```

See Also

EndWhile(), ExitWhile(), GotoIf()

Zapateller()

Uses a special information tone to block telemarketers

Zapateller(*options*)

Generates a special information tone to block telemarketers and other computer-dialed calls from bothering you.

The *options* argument is a pipe-delimited list of options. The following options are available:

answer

 Causes the line to be answered before playing the tone.

nocallerid

 Causes Zapateller to play the tone only if no Caller ID information is available.

```
; answer the line, and play the SIT tone if there is no Caller
ID information
exten => 123,1,Zapateller(answer|nocallerid)
```

See Also

PrivacyManager()

ZapBarge()

Barges in on (monitors) a Zap channel

ZapBarge([*channel*])

Barges in on a specified Zap *channel*, or prompts if one is not specified. The people on the channel won't be able to hear you and will have no indication that their call is being monitored.

If *channel* is not specified, you will be prompted for the channel number. Enter **4#** for Zap/4, for example.

```
exten => 123,1,ZapBarge(Zap/2)
exten => 123,2,Hangup()
```

See Also

ZapScan()

ZapRAS() Executes the Zaptel ISDN Remote Access Server

ZapRAS(*args*)

Executes an ISDN RAS server using *pppd* on the current channel. The channel must be a clear channel (i.e., PRI source) and a Zaptel channel to be able to use this function (no modem emulation is included).

Your *pppd* must be patched to be Zaptel-aware. *args* is a pipe-delimited list of arguments.

This application is for use only on ISDN lines, and your kernel must be patched to support ZapRAS(). You must also have *ppp* support in your kernel.

```
exten => 123,1,Answer()
exten => 123,1,ZapRas(debug|64000|noauth|netmask|255.255.255.0|
10.0.0.1:10.0.0.2)
```

ZapScan() Scans Zap channels to monitor calls

ZapScan([*group*])

Allows a call center manager to monitor Zap channels in a convenient way. Use # to select the next channel, and use * to exit. You may limit scanning to a particular channel group (as set with the GROUP() function) by setting the *group* argument.

```
exten => 123,1,ZapScan()
```

See Also

ZapBarge()

AGI Reference

ANSWER

Answers the channel (if it is not already in an answered state).

Return values:

-1

Failure

0

Success

CHANNEL STATUS

CHANNEL STATUS [*channelname*]

Queries the status of the channel indicated by *channelname* or, if no channel is specified, the current channel.

Return values:

0

Channel is down and available

1

Channel is down, but reserved

2

Channel is off-hook

3

Digits have been dialed

4

Line is ringing

5

Line is up

6

    Line is busy

DATABASE DEL

`DATABASE DEL` *`family key`*

Deletes an entry from the Asterisk database for the specified family and key.

Return values:

0

    Failure

1

    Success

DATABASE DELTREE

`DATABASE DELTREE` *`family`* [*`keytree`*]

Deletes a family and/or keytree from the Asterisk database.

Return values:

0

    Failure

1

    Success

DATABASE GET

`DATABASE GET` *`family key`*

Retrieves a value from the Asterisk database for the specified family and key.

Return values:

0

    Not set

1 (*`value`* **)**
    Value is set (and is included in parentheses)

DATABASE PUT

DATABASE PUT *family key value*

Adds or updates an entry in the Asterisk database for the specified family and key, with the specified value.

Return values:

0
　　Failure

1
　　Success

EXEC

EXEC *application options*

Executes the specified dialplan application, including options.

Return values:

-2
　　Failure to find the application

value
　　Return value of the application

GET DATA

GET DATA *filename* [*timeout*] [*max_digits*]

Plays the audio file specified by *filename* and accepts DTMF digits, up to the limit set by *max_digits*. Similar to the Background() dialplan application.

Return value:

value
　　Digits received from the caller

GET FULL VARIABLE

GET FULL VARIABLE *variablename* [*channelname*]

If the variable indicated by *variablename* is set, returns its value in parentheses. This command understands complex variable names and built-in variable names, unlike GET VARIABLE.

Return values:

0

No channel, or variable not set

1 (*value*)

Value is retrieved (and is included in parentheses)

GET OPTION

`GET OPTION` *filename escape_digits* [*timeout*]

Behaves the same as `STREAM FILE`, but has a *timeout* option (in seconds).

Return value:

value

ASCII value of digits received, in decimal

GET VARIABLE

`GET VARIABLE` *variablename*

If the variable is set, returns its value in parentheses. This command does not understand complex variables or built-in variables; use the `GET FULL VARIABLE` command if your application requires these types of variables.

Return values:

0

No channel, or variable not set

1 (*value*)

Value is retrieved (and is included in parentheses)

HANGUP

`HANGUP` [*channelname*]

Hangs up the specified channel or, if no channel is given, the current channel.

Return values:

-1

Specified channel does not exist

1

Hangup was successful

NoOp

NoOp [*text*]

Performs no operation. As a side effect, this command prints **text** to the Asterisk console. Usually used for debugging purposes.

Return value:

0
> No channel, or variable not set

RECEIVE CHAR

RECEIVE CHAR *timeout*

Receives a character of text on a channel. Specify a **timeout** in milliseconds as the maximum amount of time to wait for input, or set to 0 to wait infinitely. Note that most channels do not support the reception of text.

Return values:

-1 (hangup)
> Failure or hangup

char (timeout)
> Timeout

value
> ASCII value of character, in decimal

RECORD FILE

RECORD FILE *filename format escape_digits timeout* [*offset_samples*] [BEEP] [s=*silence*]

Records the channel audio to the specified file until the reception of a defined escape (DTMF) digit. The *format* argument defines the type of file to be recorded (wav, gsm, etc.). The *timeout* argument is the maximum number of milliseconds the recording can last, and can be set to -1 for no timeout. The *offset_samples* argument is optional; if provided, it will seek to the offset without exceeding the end of the file. The BEEP argument will play a beep to the user to signify the start of the record operation. The *silence* argument is the number of seconds of silence allowed before the function returns despite the lack of DTMF digits or reaching the timeout. The silence value must be preceded by s= and is also optional.

Return values:

-1
> Failure

0

> Successful recording

SAY ALPHA

`SAY ALPHA` *`number escape_digits`*

Says a given character string, returning early if any of the given DTMF digits are received on the channel.

Return values:

-1

> Error or hangup

0

> Playback completed without being interrupted by an escape digit

value

> ASCII value of digit (if pressed), in decimal

SAY DATE

`SAY DATE` *`date escape_digits`*

Says a given *date*, returning early if any of the given DTMF digits are received on the channel. The *date* is the number of seconds elapsed since 00:00:00 on January 1, 1970, Coordinated Universal Time (UTC).

Return values:

-1

> Error or hangup

0

> Playback completed without being interrupted by an escape digit

value

> ASCII value of digit (if pressed), in decimal

SAY DATETIME

`SAY DATETIME` *`datetime escape_digits`* [*`format`*] [*`timezone`*]

Says the given *datetime*, returning early if any of the given DTMF digits are received on the channel. The *datetime* is the number of seconds elapsed since 00:00:00 on January 1, 1970, Coordinated Universal Time (UTC). The optional *format* argument is the format in which the time should be spoken. (See *voicemail.conf* for a complete description of the format options.) *format* defaults to `"ABdY 'digits/at' IMp"`. Acceptable values for *timezone* can be found in */usr/share/zoneinfo/*. *timezone* defaults to the default time zone of the Asterisk server.

Return values:

-1

> Error or hangup

0

> Playback completed without being interrupted by an escape digit

value

> ASCII value of digit (if pressed), in decimal

SAY DIGITS

SAY DIGITS *number escape_digits*

Says a given digit string, returning early if any of the given DTMF digits are received on the channel.

Return values:

-1

> Error or hangup

0

> Playback completed without being interrupted by an escape digit

value

> ASCII value of digit (if pressed), in decimal

SAY NUMBER

SAY NUMBER *number escape_digits*

Says a given number, returning early if any of the given DTMF digits are received on the channel.

Return values:

-1

> Error or hangup

0

> Playback completed without being interrupted by an escape digit

value

> ASCII value of digit (if pressed), in decimal

SAY PHONETIC

SAY PHONETIC *string escape_digits*

Says a given character string with phonetics, returning early if any of the given DTMF digits are received on the channel.

Return values:

-1

    Error or hangup

0

    Playback completed without being interrupted by an escape digit

value

    ASCII value of digit (if pressed), in decimal

SAY TIME

SAY TIME *time escape_digits*

Says the indicated *time*, returning early if any of the given DTMF digits are received on the channel. The *time* is the number of seconds elapsed since 00:00:00 on January 1, 1970, Coordinated Universal Time (UTC).

Return values:

-1

    Error or hangup

0

    Playback completed without being interrupted by an escape digit

value

    ASCII value of digit (if pressed), in decimal

SEND IMAGE

SEND IMAGE *image*

Sends the given image on the current channel. Most channels do not support the transmission of images. Image names should not include extensions.

Return values:

-1

    Error or hangup

0

    Image sent, or channel does not support sending an image

SEND TEXT

SEND TEXT *"text_to_send"*

Sends the specified text on the current channel. Most channels do not support the transmission of text. Text consisting of more than one word should be placed in quotes, since the command accepts only a single argument.

Return values:

-1

Error or hangup

0

Text sent, or channel does not support sending text

SET AUTOHANGUP

SET AUTOHANGUP *time*

Causes the channel to automatically be hung up once *time* seconds have elapsed. Of course, it can be hung up before then as well. Setting *time* to 0 will cause the auto hangup feature to be disabled on this channel.

Return value:

0

Autohangup has been set

SET CALLERID

SET CALLERID *number*

Changes the Caller ID of the current channel.

Return value:

1

Caller ID has been set

SET CONTEXT

SET CONTEXT *context*

Sets the *context* for continuation upon exiting the AGI application.

Return value:

0

Context has been set

SET EXTENSION

`SET EXTENSION` *extension*

Changes the *extension* for continuation upon exiting the AGI application.

Return value:

0

Extension has been set

SET MUSIC ON

`SET MUSIC ON [on|off] [class]`

Enables/disables the music-on-hold generator. If `class` is not specified, the default music-on-hold class will be used.

Return value:

0

Always returns 0

SET PRIORITY

`SET PRIORITY` *priority*

Changes the priority for continuation upon exiting the AGI application. *priority* must be a valid priority or label.

Return value:

0

Priority has been set

SET VARIABLE

`SET VARIABLE` *variablename value*

Sets or updates the *value* for the variable name specified by *variablename*. If the variable does not exist, it is created.

Return value:

1

> Variable has been set

STREAM FILE

STREAM FILE *filename escape_digits* [*sample_offset*]

Play the audio file indicated by *filename*, allowing playback to be interrupted by the digits specified by *escape_digits*, if any. Use double quotes for the digits if you wish none to be permitted. If *sample_offset* is provided, the audio will seek to *sample_off-set* before playback starts.

Remember, the file extension must not be included in the filename.

Return values:

0

> Playback completed with no digit pressed

-1

> Error or hangup

value

> ASCII value of digit (if pressed), in decimal

TDD MODE

Enables and disables Telecommunications Devices for the Deaf (TDD) transmission/reception on this channel. Return values:

0

> Channel not TDD-capable

1

> Success

VERBOSE

VERBOSE *message level*

Sends *message* to the console via the verbose message system. The *level* argument is the minimum verbosity level at which the message will appear on the Asterisk command-line interface.

Return value:

0

> Always returns 0

WAIT FOR DIGIT

`WAIT FOR DIGIT` *timeout*

Waits up to *timeout* milliseconds for the channel to receive a DTMF digit. Use `-1` for the *timeout* value if you want the call to block indefinitely.

Return values:

`-1`

Error or channel failure

`0`

Timeout

value

ASCII value of digit (if pressed), in decimal

Configuration Files

 This appendix contains a reference to the configuration files not covered in the previous appendices. If you are looking for VoIP channel configurations, refer to Appendix A. For a dialplan reference, you'll want to use Appendix B.

A configuration file is required for each Asterisk module you wish to use. These *.conf* files contain channel definitions, describe internal services, define the locations of other modules, or relate to the dialplan. You do not need to configure all of them to have a functioning system, only the ones required for your configuration. Although Asterisk ships with samples of all of the configuration files, it is possible to start Asterisk without any of them. This will not provide you with a working system, but it clearly demonstrates the modularity of the platform.

If no *.conf* files are found, Asterisk will make some decisions with respect to modules. For example, the following steps are always taken:

- The Asterisk Event Logger is loaded, and events are logged to */var/log/asterisk/event_log*.
- Manager actions are registered.
- The PBX core is initialized.
- The RTP port range is allocated from 5,000 through 31,000.
- Several built-in applications are loaded, such as Answer(), Background(), GotoIf(), NoOp(), and Set().
- The dynamic loader is started; this is the engine responsible for loading modules defined in *modules.conf*.

This appendix starts with an in-depth look at the *modules.conf* configuration file. We'll then briefly examine all the other files that you may need to configure for your Asterisk system.

modules.conf

The *modules.conf* file controls which modules are loaded or not loaded at Asterisk startup. This is done through the use of the **load =>** or **noload =>** constructs.

 This file is a key component to building a secure Asterisk installation: best practice suggests that only required modules be loaded.

The *modules.conf* file always starts with the [**modules**] header. The **autoload** statement tells Asterisk whether to automatically load all modules contained within the modules directory or to load only those modules specifically defined by **load =>** statements. We recommend you manually load only those modules you need, but many people find it easier to let Asterisk attempt to **autoload** whatever it finds in */usr/lib/asterisk/modules*. You can then exclude* certain modules with **noload =>** statements.

Here's a sample *modules.conf* file:

```
[modules]
    autoload=no                  ; set this to yes and Asterisk will load any
                                 ; modules it finds in /usr/lib/asterisk/modules

    load => res_adsi.so
    load => pbx_config.so        ; Requires: N/A
    load => chan_iax2.so         ; Requires: res_crypto.so, res_features.so
    load => chan_sip.so          ; Requires: res_features.so
    load => codec_alaw.so        ; Requires: N/A
    load => codec_gsm.so         ; Requires: N/A
    load => codec_ulaw.so        ; Requires: N/A
    load => format_gsm.so        ; Requires: N/A
    load => app_dial.so          ; Requires: res_features.so, res_musiconhold.so
```

Since we assume Asterisk is built on Linux, all the module names we use end in a *.so* extension. However, this may not be the case if you have built Asterisk on a different operating system.

As of this writing, there are eight module types: *resources, applications, Call Detail Record database connectors, channels, codecs, formats, PBX modules,* and *standalone functions.* Let's take a look at each of them.

adsi.conf

The Analog Display Services Interface (ADSI) was designed to allow telephone companies to deliver enhanced services across analog telephone circuits. In Asterisk, you

---

* With the advent of the new menuselect system, this best practice may no longer be necessary if you are building only the modules you need in the first place

can use this file to send ADSI commands to compatible telephones. Please note that the phone must be directly connected to a Zapata channel. ADSI messages cannot be sent across a VoIP connection to a remote analog phone.

The *res_adsi.so* module is required for the `Voicemail()` application; however, the *adsi.conf* file is not necessarily used. Detailed information about ADSI is not publicly available, and documentation needs to be purchased from Telcordia.

adtranvofr.conf

Prior to Voice over IP, Voice over Frame Relay (VoFR) enjoyed brief fame as a means of carrying packetized voice. Supporting VoFR through Adtran equipment is part of the history of Asterisk.

This feature is no longer popular in the community, though, so it may be difficult to find support for it.

agents.conf

This file allows you to create and manage agents for your call center. If you are using the `Queue()` application, you may want to configure agents for the queue. The *agents.conf* file is used to configure the AGENT channel driver.

The `[general]` section in *agents.conf* currently contains only two parameters. The `persistentagents` parameter tells Asterisk whether or not to save the status of agents who use the callback feature of queues in the local Asterisk database. If set to **yes**, a logged-in remote agent will then remain logged in across a reboot (unless removed from the database through some other means). The `multiplelogin` parameter tells Asterisk whether or not multiple agents can log in from the same extension.

The following parameters, which are specified in the `[agents]` section, are used to define agents and the way the system interacts with them. The settings apply to all agents, unless otherwise specified in the individual agent definitions:

maxlogintries
: The maximum number of times an agent may attempt to log in. Defaults to 3.

autologoff
: Accepts an argument (in seconds) defining how long an agent channel should ring for before the agent is deemed unavailable and logged off.

autologoffunavail
: Define `autologoffunavail` to have agents automatically logged out when the `Dial()` application returns a `CHANUNAVAIL` status while trying to dial that agent. Default is "no".

ackcall

Accepts the arguments **yes** and **no**. If set to **yes**, requires a callback agent to acknowledge log in by pressing the # key after logging in. This works in conjunction with the `AgentCallbackLogin()` application.

endcall

If set to **yes**, allows an agent to hang up a call by pressing the * key. Defaults to **yes**. Set this to **no** to have Asterisk do nothing when the agent presses the * key.

wrapuptime

You can configure this parameter to allow agents a few seconds of downtime after completing a call before the queue presents them with another call. This setting is measured in milliseconds.

musiconhold => *class*

Accepts a music-on-hold class as its argument. This setting applies to all agents.

agentgoodbye

Defines the default goodbye sound for agents.

updatecdr

Accepts the arguments **yes** and **no**. Used to define whether the source channel in the CDRs should be set to `agent/agent_id` to determine which agent generated the calls.

group

Defines the groups to which an agent belongs, specified with integers. Specify that an agent belongs to multiple groups by separating the integers with commas.

recordagentcalls

Accepts the arguments **yes** and **no**. Defines whether or not agent calls should be recorded.

recordformat

Defines the format to record files in. The argument specified should be **wav**, **gsm**, or **wav49**. The default recording format is **wav**.

urlprefix

Accepts a string as its argument. The string can be formed as a URL and is appended to the start of the text to be added to the name of the recording.

savecallsin

Accepts a filesystem path as its argument. Allows you to override the default path of */var/spool/asterisk/monitor/* with one of your choosing.

Since the storage of calls will require a large amount of hard drive space, you will want to define a strategy to handle storing and managing these recordings.

This location should probably reside on a separate volume, one with very high performance characteristics.

`custom_beep`
> Accepts a filename as its argument. Can be used to define a custom notification tone to signal to an always-connected agent that there is an incoming call.

The final parameter is used to define agents. Just like in the *zapata.conf* file, configuration parameters are inherited from above the `agent =>` definition. Agents are defined with the following format:

```
agent => agent_id,agent_password,name
```

For example, we can define agent Happy Tempura with the agent ID 1000 and password 1234, as follows.

```
agent => 1000,1234,Happy Tempura
```

Be aware that an *agents.conf* file is a complement to the queue configuration process. The most critical configuration file for your queues is *queues.conf*. You can configure a very basic queue without *agents.conf*.

alarmreceiver.conf

 The `AlarmReceiver()` application is not approved by Underwriter's Laboratory (UL) and should not be used as the primary or sole means of receiving alarm messages or events. This application is not guaranteed to be reliable, so don't depend on it unless you have extensively tested it. Use of this application without extensive testing may place your life and/or property at risk.

The *alarmreceiver.conf* file is used by the `AlarmReceiver()` application, which allows Asterisk to accept alarms using the SIA (Ademco) Contact ID protocol. When a call is received from an alarm panel, it should be directed to a context that calls the `AlarmReceiver()` application. In turn, `AlarmReceiver()` will read the *alarmreceiver.conf* configuration file and perform the configured actions as required. All parameters are specified under the `[general]` heading.

The sample configuration file will contain the current settings for this application and is very well documented.

alsa.conf

The *alsa.conf* file is used to configure Asterisk to use the Advanced Linux Sound Architecture (ALSA) to provide access to a sound card, if desired. You can use this file to configure the CONSOLE channel, which is most commonly used to create an overhead paging system (although, as with any other channel, there are all kinds of creative ways this can be used). Keep in mind that the usefulness of the ALSA channel by itself is limited due to its lack of a user interface.[†]

amd.conf

This is the configuration for the answering matching detection application in Asterisk, called `AMD()`. It is used to adjust the different parameters for detecting an answering machine based on items such as initial silence, greeting length, silence after the greeting, and so forth.

asterisk.conf

The *asterisk.conf* file defines the locations for the configuration files, the spool directory, and the modules, as well as a location to write logfiles to. The default settings are recommended unless you understand the implications of changing them. The *asterisk.conf* file is generated automatically when you run the `make samples` command, based on information it collects about your system. It will contain a `[directories]` section such as the following:

```
[directories]
astetcdir => /etc/asterisk
astmoddir => /usr/lib/asterisk/modules
astvarlibdir => /var/lib/asterisk
astdatadir => /var/lib/asterisk
astagidir => /var/lib/asterisk/agi-bin
astspooldir => /var/spool/asterisk
astrundir => /var/run
astlogdir => /var/log/asterisk
```

Additionally, you can specify an `[options]` section, which will allow you to define startup options (command-line switches) in the configuration file. The following example shows the available options and the command-line switches that they effectively enforce:

```
[options]
;Under "options" you can enter configuration options
;that you also can set with command line options

verbose = 0               ; Verbosity level for logging (-v)
debug = 3                 ; Debug: "No" or value (1-4)
nofork=yes | no           ; Background execution disabled (-f)
alwaysfork=yes | no       ; Always background, even with -v or -d (-F)
console= yes | no         ; Console mode (-c)
highpriority = yes | no   ; Execute with high priority (-p)
initcrypto = yes | no     ; Initialize crypto at startup (-i)
nocolor = yes | no        ; Disable ANSI colors (-n)
dumpcore = yes | no       ; Dump core on failure (-g)
quiet = yes | no          ; Run quietly (-q)
timestamp = yes | no      ; Force timestamping in CLI verbose output (-T)
```

---

† Yes, we are aware that the user interface to the channel interface is the Asterisk CLI; however, this is not usable as a telephone and therefore does not meet the criteria of an interface from the perspective of a telephone user.

```
runuser = asterisk      ; User to run asterisk as (-U) NOTE: will require changes to
                        ; directory and device permissions
rungroup = asterisk     ; Group to run asterisk as (-G)
internal_timing = yes | no   ; Enable internal timing support (-I)
                        ; These options have no command line equivalent
cache_record_files = yes | no   ; Cache record() files in another directory until
                        ; completion
record_cache_dir = <dir>
transcode_via_sln = yes | no        ; Build transcode paths via SLINEAR
transmit_silence_during_record = yes | no ; send SLINEAR silence while channel is
                        ; being recorded
maxload = 1.0           ; The maximum load average we accept calls for
maxcalls = 255          ; The maximum number of concurrent calls you want
                        ; to allow
execincludes = yes | no    ; Allow #exec entries in configuration files
dontwarn = yes | no        ; Don't over-inform the Asterisk sysadm, he's a guru
systemname = <a_string>    ; System name. Used to prefix CDR uniqueid and to
                        ; fill ${SYSTEMNAME}
languageprefix = yes | no ; Should language code be last component of sound file
                        ; name or first?
                        ; When off, sound files are searched as
                        ; <path>/<lang>/<file>
                        ; When on, sound files are search as <lang>/<path>/<file>
                        ; (only affects relative paths for sound files)
```

cdr.conf

The *cdr.conf* file is used to enable call detail record logging to a flat file or a database. Storing call records is useful for all sorts of purposes including billing, fraud prevention, QoS evaluations, and more. The *cdr.conf* file contains some general parameters that are not specific to any particular database, but rather indicate how Asterisk should handle the passing of information to the database. All options are under the [general] heading of the *cdr.conf* file:

enable

Accepts the arguments **yes** and **no**. Specifies whether or not to use CDR logging. If set to **no**, this will override any CDR module explicitly loaded. The default is **yes**.

batch

Accepts the arguments **yes** and **no**. Allows Asterisk to write data to a buffer instead of writing to the database at the end of every call, to reduce load on the system.

 Note that if the system dies unexpectedly when this option is set to **yes**, data loss may occur.

size

> Sets the maximum number of CDRs to accumulate in the buffer before posting to the backend CDR storage systems. This setting only takes effect if the `batch` setting is set to `yes`. This setting defaults to 100 records.

time

> Accepts an integer (in seconds) as its argument. Sets the number of seconds before Asterisk flushes the buffer and writes the CDRs to the database, regardless of the number of records in the buffer (as defined by `size`). The default is 300 seconds (5 minutes).

scheduleronly

> Accepts the arguments `yes` and `no`. If you are generating a massive volume of CDRs on a system that is pushing them to a remote database, setting `scheduleronly` to `yes` may be of benefit. Since the scheduler cannot start a new task until the current one is finished, slow CDR writes may adversely affect other processes needing the scheduler. This setting will instruct Asterisk to handle CDR writes in a new thread, essentially assigning a dedicated scheduler to this function. In normal operation, this would yield very little benefit.

safeshutdown

> Accepts the arguments `yes` and `no`. Setting `safeshutdown` to `yes` will prevent Asterisk from shutting down completely until the buffer is flushed and all information is written to the database. If this parameter is set to `no` and you shut down Asterisk with information still residing in the buffers, that information will likely be lost.

endbeforehexten

> Normally, CDR records are not closed out until after all extensions are finished executing. By enabling this option, the CDR will be ended before executing the `h` extension so that CDR values such as `end` and `billsec` may be retrieved inside of this extension. Defaults to `no`.

The rest of *cdr.conf* contains setup for several of the backend CDR engines. See the sample *cdr.conf* for more information.

cdr_manager.conf

The *cdr_manager.conf* file simply contains a `[general]` heading and a single option, `enabled`, which you can use to specify whether or not the Asterisk Manager API generates CDR events. If you want CDR events to be generated, you will need the following lines in your *cdr_manager.conf* file:

```
[general]
enabled=yes
```

The Manager API will then output CDR events containing the following fields:

```
Event: Cdr
AccountCode:
```

```
Source:
Destination:
DestinationContext:
CallerID:
Channel:
DestinationChannel:
LastApplication:
LastData:
StartTime:
AnswerTime:
EndTime:
Duration:
BillableSeconds:
Disposition:
AMAFlags:
UniqueID:
UserField:
```

cdr_odbc.conf

Asterisk can store CDR data in a local or remote database via the ODBC interface. The *cdr_odbc.conf* file contains the information Asterisk needs to connect to the database. The *cdr_odbc.so* module will attempt to load the *cdr_odbc.conf* file, and if information is found for connecting to a database, the CDR data will be recorded there.

> If you are going to use a database for storing CDR data, you will have to select *one* of the many that are available. Asterisk does not like having multiple CDR databases to connect to, so do not have extra *cdr_engine.conf* files hanging about your Asterisk configuration directory.

cdr_pgsql.conf

Asterisk can store CDR data in a PostgreSQL database via the *cdr_pgsql.so* module. When the module is loaded the necessary information will be read from the *cdr_pgsql.conf* file, and Asterisk will connect to the PostgreSQL database to write and store CDR data.

cdr_tds.conf

Asterisk can also store CDR data to a FreeTDS database (including MS SQL) with the use of the *cdr_tds.so* module. The configuration file *cdr_tds.conf* is read once the module is loaded. Upon a successful connection, CDR data will be written to the database.

codecs.conf

Most codecs do not have any configurable parameters; they are what they are, and that's all they are.

Some codecs, however, are capable of behaving in different ways. This primarily means that they can be optimized for a particular goal, such as cutting down on latency, making best use of a network, or perhaps delivering high quality audio.

The *codecs.conf* file is fairly new in Asterisk, and as of this writing it allows configuration of Speex parameters only. The settings are self-explanatory, as long as you are familiar with the Speex protocol (see *http://www.speex.org*).

codecs.conf also allows you to configure Packet Loss Concealment (PLC). You need to define a `[plc]` section and indicate `genericplc => true`. This will cause Asterisk to attempt to interpolate any packets that are missed. (Enabling this functionality will incur a small performance penalty.)

dnsmgr.conf

This file is used to configure whether Asterisk should perform DNS lookups on a regular basis, and how often those lookups should be performed.

dundi.conf

The DUNDi protocol is used to dynamically look up the VoIP address of a phone number on a network, and to connect to that address. Unlike the ENUM standard, DUNDi has no central authority. The *dundi.conf* file contains DUNDi extensions used to control what is advertised; it also contains the peers to whom you will submit lookup requests and from whom you will accept lookup requests. The DUNDi protocol was explored in Chapter 14.

enum.conf

The Electronic Numbering (ENUM) system is used in conjunction with the Internet's DNS system to map E.164 ITU standard (ordinary telephone) numbers to email addresses, web sites, VoIP addresses, and the like. An ENUM number is created in DNS by reversing the phone number, separating each digit with a period, and appending *e164.arpa* (the primary DNS zone). If you want Asterisk to perform ENUM lookups, configure the domain(s) in which to perform the lookups within the *enum.conf* file. In addition to the official *e164.arpa* domain, you can have Asterisk perform lookups in the publicly accessible *e164.org* domain.

extconfig.conf

Asterisk can write configuration data to and load configuration data from a database using the external configuration engine (also known as *realtime*). This enables you to map external configuration files (static mappings) to a database, allowing the information to be retrieved from the database. It also allows you to map special runtime entries that permit the dynamic creation and loading of objects, entities, peers, and so on without a reload. These mappings are assigned and configured in the *extconfig.conf* file, which is used by both *res_odbc* and *realtime*.

extensions.conf

At the center of every good universe is a dialplan. The *extensions.conf* file is the means by which you tell Asterisk how you want calls to be handled. The dialplan contains a list of instructions that, unlike traditional telephony systems, is entirely customizable. The dialplan is so important that rather than defining it in this appendix, we have dedicated all of Chapters 5 and 6, as well as Appendix B, to this topic. Go forth, read, and enjoy!

extensions.ael

This file is the equivalent of *extensions.conf*, only it's for dialplans written in the AEL language. When Asterisk loads the dialplan, it reads the AEL dialplan from *extensions.ael* and merges it with the dialplan from *extensions.conf*.

features.conf

features.conf, the file formally known as *parking.conf*, contains configuration information related to call parking and call transfers. Call parking configuration options include:

- The extension to dial to park calls (`parkext =>`)
- The extension range to park calls in (`parkpos =>`)
- Which context to park calls in (`context =>`)
- How long a call can remain parked for before ringing the extension that parked it (`parkingtime =>`)
- The sound file played to the parked caller when the call is removed from parking (`courtesytone =>`)
- ADSI parking announcements (`asdipark=yes|no`)

In addition to the call parking options, in this file you can configure the button mappings for blind transfers, attended transfers, one-touch recording, disconnections, and the pickup extension (which allows you to answer a remotely ringing extension).

festival.conf

The Festival text-to-speech engine allows Asterisk to read text files to the end user with a computer-generated voice. Festival is covered in Chapter 14.

followme.conf

The term findme/follow me is intended to give the impression that a PBX system has the intelligence to locate a user wherever they may be, such that their calls will find them and follow them. The *followme.conf* file is used to configure the FollowMe() dialplan application.

func_odbc.conf

The func_odbc dialplan function was one of the most anticipated capabilities to have been added to Asterisk 1.4. This function provides a simple mechanism with which to connect to ODBC databases through the dialplan. The SQL queries are defined in this configuration file, and a dialplan function is automatically created.

gtalk.conf

This configuration file is used to specify the parameters for connectivity with Google Talk.

http.conf

Asterisk has a very simple HTTP daemon built into it, which is used by the Asterisk GUI and AJAM. This functionality is discussed in Chapter 11.

iax.conf

Similar to *sip.conf*, the *iax.conf* file is where you configure options related to the IAX protocol. Your end devices and service providers are also configured here. The *iax.conf* file is covered in detail in Appendix A.

iaxprov.conf

This file is used by Asterisk to allow the system to provision and upgrade the firmware on an IAXy device.

indications.conf

The *indications.conf* file is used to tell Asterisk how to generate the various telephone sounds common in different parts of the world; a dial tone in England sounds very different from a dial tone in Canada, but your Asterisk system will be pleased to make the sounds you want to hear. This file consists of a list of sounds a telephone system might need to produce (dial tone, busy signals, and so forth), followed by the frequencies used to generate those sounds.

By default (and without an *indications.conf* file), Asterisk will use the tones common in North America. You can change the default country for your system by specifying the two-letter country code in the [general] section. Supported country codes are listed in the *indications.conf.sample* file located in */usr/src/asterisk/configs*. If you have the required information, your country can easily be added. Here's what the configuration for North America looks like:

```
[general]
country=us
;
[us]
description = United States / North America
ringcadance = 2000,4000
dial = 350+440
busy = 480+620/500,0/500
ring = 440+480/2000,0/4000
congestion = 480+620/250,0/250
callwaiting = 440/300,0/10000
dialrecall = !350+440/100,!0/100,!350+440/100,!0/100,!350+440/100,!0/100,350+440
record = 1400/500,0/15000
info = !950/330,!1400/330,!1800/330,0
```

jabber.conf

The *jabber.conf* file specifies the information needed to allow Asterisk to interact with an XMPP (Jabber) server.

logger.conf

The *logger.conf* file specifies the type and verbosity of messages logged to the various logfiles in the */var/log/asterisk/* directory. It has two sections, [general] and [logfile].

[general]

Settings under the [general] section are used to customize the output of the logs (and can safely be left blank, as the defaults serve most people very well). However, if you love to customize such things, read on.

You can define exactly how you want your timestamps to look through the use of the dateformat parameter:

```
dateformat=%F %T
```

The Linux manpage for strftime(3) lists all of the ways you can do this.

If you want to append your system's hostname to the names of the logfiles, set appendhostname=yes. This can be useful if you have a lot of systems delivering logfiles to you.

If for some reason you do not want to log events from your queues, you can set queue_log=no.

If generic events do not interest you, instruct Asterisk to omit them from the logfiles by setting event_log=no.

[logfiles]

The [logfiles] section defines the types of information you wish to log. There are multiple ranks for the various bits of information that will be logged, and it can be desirable to separate log entries into different files. The general format for lines in the [logfiles] section is *filename* => *levels*, where *filename* is the name of the file to save the logged information to and *levels* are the types of information you wish to save.

Using console for the *filename* is a special exception that allows you to control the type of information sent to the Asterisk console.

A sample [logfiles] section might look like this:

```
[logfiles]
console => notice,warning,error
messages => notice,warning,error
```

You can specify logging of the following types of information:

debug
> Enabling debugging gives far more detailed output about what is happening in the system. For example, with debugging enabled, you can see what DTMF tones the users entered while accessing their voicemail boxes. Debugging information should be logged only when you are actually debugging something, as it will create massive logfiles very rapidly.

verbose

> When you connect to the Asterisk console and set a verbosity of 3 or higher, you'll see output on the console showing what Asterisk is doing. You can save this output to a logfile by adding a line such as `verbose_log => verbose` to your *logger.conf* file. Note that a high amount of verbosity can quickly eat up hard drive space.

notice

> A *notice* is used to inform you of minor changes to the system, such as when a peer changes state. It is normal to see these types of messages, and the events they indicate generally have no adverse effects on the server.

warning

> A *warning* happens when Asterisk attempts to do something and is unsuccessful. These types of errors are usually not fatal, but they should be investigated, especially if a lot of them are seen.

error

> *Errors* are often related to Out of Memory errors. They generally indicate serious problems that may lead to Asterisk to crash or freeze.

manager.conf

The Asterisk Manager Interface is an API that external programs can use to communicate with and control Asterisk, much as you would do from the Asterisk console.

> The Manager gives programs the ability to run commands and request information from the Asterisk server. However, it is not very secure; its authentication mechanism defaults to using plain-text passwords, and all connected terminals default to receiving all events. The Asterisk Manager should be used only on a trusted local area network, or locally on the box. The `permit` and `deny` constructs allow you to restrict access to certain extensions or subnets.

Many of the available graphical interfaces to Asterisk—such as the Flash Operator Panel—use the Manager to pull data and determine the status of applications. The *manager.conf* file defines the way programs authenticate with the Manager.

The Manager commands (which you can list by typing `show manager commands` at the Asterisk console) have varying degrees of privilege. You can control the read and write permissions for these commands with the use of the `read` and `write` options in the *manager.conf* file.

Here's a sample *manager.conf* file:

```
[general]
enabled = no
port = 5038
bindaddr = 0.0.0.0
```

```
[oreilly]
secret = notvery
deny=0.0.0.0/0.0.0.0
permit= 192.168.1.0/255.255.255.0
read = system,call,log,verbose,command,agent,user,config
write = system,call,log,verbose,command,agent,user,config
```

For more information on the Asterisk Manager Interface, see Chapter 10.

meetme.conf

MeetMe is one of the more remarkable applications in Asterisk. It allows you to set up predefined audio conference rooms. This rather simple concept has proven to be extremely expensive to implement in every other PBX, but what seems like a big deal to them is simple to Asterisk. Whether by using a dedicated server, or through the use of a service, Asterisk now delivers this functionality as a standard application.

MeetMe conferences can be created either dynamically, with the d flag in the Dial() application, or statically in the *meetme.conf* file. The format for creating conference rooms is as follows:

```
conf => conference_number[,pin][,administrator_pin]
```

All conferences must be defined under the [rooms] section header.

```
[rooms]
conf => 4569
conf => 5060,54377017
conf => 3389,4242,1337
conf => 333,,2424
```

mgcp.conf

The Media Gateway Control Protocol (MGCP) has only primitive support in Asterisk. This is likely due to the fact that SIP has stolen the limelight from every other VoIP protocol (except IAX, of course). Because of this, you should attempt to use Asterisk's MCGP channel in a production environment only if you are prepared to perform extensive testing, are willing to pay to have features and patches implemented within your time frames, and have in-house expertise with the protocol.

Having said that, we are not prepared to pronounce MGCP dead. SIP is not yet the panacea it has been touted as, and MGCP has proven itself to be very useful in carrier backbone environments. Many believe MGCP will fill a niche or void that has not yet been discovered, and we remain interested in it.

modem.conf

The *modem.conf* file is used by Asterisk to communicate with ISDN-BRI interfaces through the ISDN4Linux driver. Since ISDN4Linux lacks many core ISDN features, it is not generally used. For BRI, the most popular add-on seems to be *chan_capi*, available from *http://www.junghanns.net*.

musiconhold.conf

The *musiconhold.conf* file is used to configure different classes of music and their locations for use in music-on-hold applications. Asterisk can play hold music in any native file formats. Asterisk can also makes use of a certain version of *mpg123* to play MP3 files, but this is discouraged. You can specify arguments for a class, allowing you to use an external application to stream music either locally or over a network.

osp.conf

The Open Settlement Protocol (OSP) is officially documented in ETSI TS 101 321, a European Telecommunication Standards Institute (ETSI) document that came out of the work of the TIPHON working group. As far as we can tell, OSP is another attempt to apply old-style telecom thinking to disruptive technologies.

oss.conf

The *oss.conf* file is used to configure Asterisk to use the Open Sound System (OSS) driver to allow communications with the sound card via the CONSOLE channel. Note that ALSA is now the preferred interface for the CONSOLE channel.

phone.conf

The *phone.conf* file is used to configure a Quicknet PhoneJACK card. The PhoneJACK card seems to provide something like an FXS interface, in that you can plug an analog telephone into it and pass calls through Asterisk.

privacy.conf

The *privacy.conf* file is used to control the maximum number of tries a user has to enter his 10-digit telephone number in the `PrivacyManager()` application. The `PrivacyManager()` application determines if a Caller ID is set for the incoming call. The user enters his 10-digit number within the number of tries configured in *privacy.conf*, the application sets the `PRIVACYMGRSTATUS` channel variable to either `SUCCESS` or `FAILED`. If the Caller ID is set, the application does nothing.

The `PrivacyManager()` application can also accept the arguments in the dialplan. This allows the value to remain in memory, instead of an I/O operation to the disk to read the configuration file, which you would only worry about with high usage of this application (many calls per second).

queues.conf

Asterisk provides basic call center functionality via its queueing system, but those who are using it in more mission-critical environments often report that their solutions required customization. You can do this customization in the *queues.conf* file.

The [`general`] section of *queues.conf* contains settings that will apply to all queues. If the `persistentmembers` parameter is set to `yes`, a member that is added to the system via the `AddQueueMember()` application or through the Asterisk Manager Interface will be stored in the AstDB, and therefore retained across a restart.

The `autofill` parameter allows Asterisk to be more efficient in the way it distributes calls to queue members, especially if there are multiple callers in the queue and multiple queue members available to receive the call. It is recommended you set `autofill` to `yes`.

Another general parameter of *queues.conf* is `MonitorType`. If set to `MixMonitor`, it will mix the inbound and outbound audio streams. If set to `Monitor`, it will revert to the older method of recording the inbound and outbound audio in separate files.

Next you can define one or more queues by placing its name inside of square brackets ([]). Within each queue, the following parameters are available:

musiconhold
> This parameter allows you to configure which music-on-hold class (configured in *musiconhold.conf*) to use for the queue.

announce
> When a call is presented to a member of the queue, the prompt specified by announce will be played to that agent before the caller is connected. This can be useful for agents who are logged in to more than one queue. You can specify either the full path to the file, or a path relative to */var/lib/asterisk/sounds/*.

strategy
> Asterisk can use six strategies to distribute calls to agents:
>
> ringall
>> The queue rings every available agent and connects the call to whichever agent answers first (this is the default).
>
> roundrobin (deprecated)
>> The queue cycles through the agents until it finds one who is available to take the call. roundrobin does not take into account the workload of the agents. Also, because roundrobin always starts with the first agent in the queue, this

strategy is suitable only in an environment where you want your higher-ranked agents to handle all calls unless they are busy, in which case the lower-ranked agents may get a call.

leastrecent
> The call is presented to the agent who has not been presented a call for the longest period of time.

fewestcalls
> The call is presented to the agent who has received the least amount of calls. This strategy does not take into account the actual agent workloads; it considers only the number of calls they have taken (for example, an agent who has had 3 calls that each lasted for 10 minutes will be preferred over an agent who has had 5 calls each lasting 2 minutes).

random
> As its name suggests, the random strategy chooses an agent at random. In a small call center, this strategy may prove to be the most fair.

rrmemory
> The queue cycles through the list of queue members, keeping track of which member last received a call. The next time a call needs to be distributed, Asterisk will continue from this point in the list of queue members. (This strategy is known as *round-robin memory*). This ensures that call presentation cycles through the agents as fairly as possible.

servicelevel
> In a call center, the service level represents the maximum amount of time a caller should ideally have to wait before being presented to an agent. For example, if servicelevel is set to 60 and the service level percentage is 80 percent, that means 80 percent of the calls that came into the queue were presented to an agent in less than 60 seconds.

context
> If a context is assigned to a queue, the caller will be able to press a single digit to exit to the corresponding extension within the configured context, if it exists. This action takes the caller out of the queue, which means that she will lose her place in line—be aware of this when you use this feature.

timeout
> The timeout value defines the maximum amount of time (in seconds) to let an agent's phone ring before deeming the agent unavailable and placing the call back into the queue.

retry
> When a timeout occurs, the retry value specifies how many seconds to wait before presenting the call again to an available agent.

weight

> The `weight` parameter assigns a rank to the queue. If calls are waiting in multiple queues, those queues with the highest `weight` values will be presented to agents first. When you are designing your queues, be aware that this strategy can prevent a call in a lower-weighted queue from ever being answered. Always ensure that calls in lower-weighted queues eventually get promoted to higher-weighted queues to ensure that they don't have to hold forever.

maxlen

> `maxlen` is the maximum number of calls that can be added to the queue before the call goes to the next priority of the current extension.

announce-frequency

> The `announce-frequency` value (defined in seconds) determines how often to announce to the caller his place in the queue and estimated hold time.

announce-holdtime

> There are three possible values for this parameter: `yes`, `no`, and `once`. The `announce-holdtime` parameter determines whether or not to include the estimated hold time within the position announcement. If set to `once`, it will be played to the caller only once.

monitor-format

> This parameter accepts three possible values: `wav`, `gsm`, and `wav49`. By enabling this option, you are telling Asterisk that you wish to record all completed calls in the queue in the format specified. If this option is not specified, no calls will be recorded.

monitor-join

> The `Monitor()` application in Asterisk normally records either end of the conversation in a separate file. Setting `monitor-join` to `yes` instructs Asterisk to merge the files at the end of the call. This should be set only if the `MonitorType` parameter is set to `Monitor`.

joinempty

> This parameter accepts three values: `yes`, `no`, and `strict`. It allows you to determine whether callers can be added to a queue based on the status of the members of the queue. The `strict` option will not allow callers to join the queue if all members are unavailable.

leavewhenempty

> This parameter determines whether you want your holding callers to be removed from the queue when the conditions preventing a caller from joining exist (i.e., when all of your agents log out and go home).

eventwhencalled

> Set `eventwhencalled` to `yes` if you wish to have queue events presented on the Manager Interface.

`eventmemberstatusoff`
> Setting this parameter to no will generate extra information pertaining to each queue member.

`reportholdtime`
> If you set this parameter to **yes**, the amount of time the caller held before being connected will be announced to the queue member answering the call.

`memberdelay`
> This parameter defines whether a delay will be inserted between the time when the queue identifies a free agent and the time when the call is connected to that agent.

`member => member_name`
> Members of a queue can be either channel types or agents. Any agents you list here must be defined in the *agents.conf* file.

res_odbc.conf

The purpose of the *res_odbc.so* module is to store configuration file information in a database and retrieve that information from the database. The *res_odbc.conf* file specifies how to access the table within the database. The *extconfig.conf* file is used to determine how to connect to the database.

res_snmp.conf

The *res_snmp.conf* file is used to configure SNMP support in Asterisk. There are two options in the [general]. The **subagent** option specifies whether *res_snmp* should run as a subagent or a full SNMP agent. Asterisk defaults to running as a subagent. The **enabled** option specifies whether SNMP support in Asterisk is enabled. This defaults to no, so you'll need to change this if you want SNMP support.

rpt.conf

The *rpt.conf* file is used to configure Jim Dixon's newest project, Jim's Radio Repeater Application (*app_rpt*), allows Asterisk to communicate using VoIP via radio repeater technology. This allows people to efficiently provide large-area coverage of wireless networking and routing information to the amateur radio public through their local high-speed Internet connections.

rtp.conf

The *rtp.conf* file controls the Real-time Transport Protocol (RTP) ports that Asterisk uses to generate and receive RTP traffic. The RTP protocol is used by SIP, H.323, MGCP, and possibly other protocols to carry media between endpoints.

The default *rtp.conf* file uses the RTP port range of 10,000 through 20,000. However, this is far more ports than you're likely to need, and many network administrators may not be comfortable opening up such a large range in their firewalls. You can limit the RTP port range by changing the upper and lower bound limits within the *rtp.conf* file.

For every bidirectional SIP call between two endpoints, five ports are generally used: port 5060 for SIP signaling, one port for the data stream and one port for the Real-Time Control Protocol (RTCP) in one direction, and an additional two ports for the data stream and RTCP in the opposite direction.

UDP datagrams contain a 16-bit field for a Cyclic Redundancy Check (CRC), which is used to verify the integrity of the datagram header and its data. It uses polynomial division to create the 16-bit checksum from the 64-bit header. This value is then placed into the 16-bit CRC field of the datagram, which the remote end can then use to verify the integrity of the received datagram.

Setting `rtpchecksums=no` requests that the OS not do UDP checksum creating/checking for the sockets used by RTP. If you add this option to the sample *rtp.conf* file, it will look like this:

```
[general]
rtpstart=10000
rtpend=20000
rtpchecksums=no
```

say.conf

The *say.conf* file is used to configure spoken language grammar rules for a number of applications, such as `SayNumber()`. If you're looking to use Asterisk in a language that isn't currently supported, you can script support through the configuration options in this file.

sip.conf

The *sip.conf* file defines all the SIP protocol options for Asterisk. The authentication for endpoints, such as SIP phones and service providers, is also configured in this file. Asterisk uses the *sip.conf* file to determine which calls you are willing to accept and where those calls should go in relation to your dialplan. Many SIP-related options are configured in *sip.conf*, which was covered in depth in Appendix A.

sip_notify.conf

Asterisk has the ability to remotely notify a SIP phone to recheck its configuration files or reboot by sending it a specially formatted, manufacturer-specific `NOTIFY` message

---

(defined in *sip_notify.conf*). Because each of these messages is manufacturer-specific, support varies from phone to phone.

skinny.conf

If you wish to connect to phones using Cisco's proprietary Skinny Client Control Protocol (SCCP), you can use the *skinny.conf* file to define the parameters and channels that will use it. However, since the SCCP protocol is proprietary, you may find that support in Asterisk for this protocol is less than perfect but steadily improving over time.

sla.conf

Even though Asterisk is a modern PBX, many people still want it to behave like an old key system with shared lines. Asterisk can emulate a key system by configuring shared lines in *sla.conf*.

smdi.conf

This file configures the Station Message Desk Interface. SMDI is a very useful addition to Asterisk, as it will allow it to act as a voicemail system for legacy PBXes that support the SMDI protocol.

udptl.conf

This file is used for the configuration of Asterisk's support for UDPTL packets. UDPTL packets are one of the transports used by T.38 faxing over IP connections.

users.conf

With the advent of the Asterisk GUI, the Asterisk developers found it would be helpful to create a configuration file where user accounts can be specified, instead of having different pieces spread across a myriad of files (such as *extensions.conf*, *sip.conf*, and *voicemail.conf*. This file is also updated by the Asterisk GUI when new users are added to the system or when user settings are modified.

voicemail.conf

The *voicemail.conf* file controls the Asterisk voicemail system (called Comedian Mail). It consists of three main sections. The first, called [general], sets the general system-wide settings for the voicemail system. The second, called [zonemessages], allows you to configure different voicemail zones, which are a collection of time and time zone settings. The third and final section is where you create one or more groups of voicemail boxes, each containing the mailbox definitions. For more information on adding voicemail capabilities to your dialplan, see Chapter 6.

General Voicemail Settings

The [general] section of *voicemail.conf* contains a plethora of options that affect the entire voicemail system:

format
> Lists the codecs that should be used to save voicemail messages. Codecs should be separated with the pipe character (|). The first format specified is the format used when attaching a voicemail message to an email. Defaults to wav49|gsm|wav. The reason that the voicemail might be saved in several different formats is to minimize the amount of transcoding that Asterisk does when the voicemail is played back.

serveremail
> Provides the email address from which voicemail notifications should be sent.

attach
> Specifies whether or not Asterisk should attach the voicemail sound file to the voicemail notification email.

maxmsg
> Sets the maximum number of messages that may be kept in any voicemail folder.

maxmessage
> Sets the maximum length of a voicemail message, in seconds.

minmessage
> Sets the minimum length of a voicemail message, in seconds.

maxgreet
> Sets the maximum length of voicemail greetings, in seconds.

skipms
> Specifies how many milliseconds to skip forward/back when the user skips forward or backward during message playback.

maxsilence
> Indicates how many seconds of silence to allow before ending the recording.

silencethreshold
> Sets the silence threshold (what we consider "silence"—the lower the threshold is, the more sensitive it is).

maxlogins
> Sets the maximum allowed number of failed login attempts.

userscontext
> Specifies which voicemail context the mailboxes defined in *users.conf* should be a part of. This defaults to the default voicemail context.

---

externnotify

Supplies the full path and filename of an external program to be executed when a voicemail is left or delivered, or when a mailbox is checked. It can also be set to smdi to use SMDI for external notification. If it is smdi, the smdiport should be set to a valid port as specified in *smdi.conf*.

smdiport

Specifies the communications port used by SMDI. This should correspond to a valid port as specified in *smdi.conf*. This is used when the externnotify setting is set to smdi.

externpass

Supplies the full path and filename of an external program to be executed whenever a voicemail password is changed.

directoryintro

If set, overrides the default introduction to the dial-by-name directory.

charset

Defines the character set for voicemail messages.

adsifdn

Specifies the ADSI feature descriptor number to download to.

adsisec

Sets the ADSI security lock code.

adsiver

Indicates the ADSI voicemail application version number.

pbxskip

Causes Asterisk not to add the string [PBX]: to the beginning of the subject line of a voicemail notification email.

fromstring:

Changes the From: string of voicemail notification email messages.

usedirectory

Permits a mailbox owner to select entries from the dial-by-name directory for forwarding and/or composing new voicemail messages.

odbcstorage

If support for ODBC voicemail storage has been compiled into Asterisk, this option allows you to specify which ODBC connection to use. ODBC connections are defined in *res_odbc.conf*.

odbctable

This option is used in conjunction with the odbcstorage option. This option specifies which database table to use for voicemail messages.

emailsubject

Specifies the email subject of voicemail notification email messages.

pagerfromstring

Changes the From: string of voicemail notification pager messages.

emailbody

Supplies the email body of voicemail notification email messages.

 Please note that all the emailsubject, emailbody, pagersubject, and pagerbody settings can use the following variables to provide more in-depth information about the voicemail:

- VM_NAME
- VM_DUR
- VM_MSGNUM
- VM_MAILBOX
- VM_CALLERID
- VM_CIDNUM
- VM_CIDNAME
- VM_DATE

emaildateformat

Specifies the format of the date and time for outbound email notifications. For more information on the format, see the strftime(3) manpage.

mailcmd

Supplies the full path and filename of the program Asterisk should use to send notification emails. This option is useful if you want to override the default email program.

nextaftercmd

Skips to the next message after the user hits 7 or 9 to delete or save the current message. This can be set only globally at this time, not on a per-mailbox basis.

Voicemail Zones

As voicemail users may be located in different geographical locations, Asterisk provides a way to configure the time zone and the way the time is announced for different callers. Each unique combination is known as a *voicemail zone*. You configure your voicemail zones in the [zonemessages] section of *voicemail.conf*. Later, you can assign your voicemail boxes to use the settings for one of these zones.

Each voicemail zone definition consists of a line with the following syntax:

    zonename=timezone | time_format

The *zonename* is an arbitrary name used to identify the zone. The *timezone* argument is the name of a system time zone, as found in */usr/share/zoneinfo*. The *time_format* ar-

---

gument specifies how times should be announced by the voicemail system. The *time_format* argument is made up of the following elements:

'*filename*'

The filename of a sound file to play (single quotes around the filename are required)

${*VAR*}

Variable substitution

A *or* a

The day of the week (Saturday, Sunday, etc.)

B *or* b *or* h

The name of the month (January, February, etc.)

d *or* e

The numeric day of the month (first, second... thirty-first)

Y

The year

I *or* l

The hour, in 12-hour format

H

The hour, in 24-hour format—single-digit hours are preceded by "oh"

k

The hour, in 24-hour format—single-digit hours are *not* preceded by "oh"

M

The minute

P *or* p

A.M. or .P.M.

Q

"today", "yesterday," or ABdY (note: not standard strftime value)

q

"" (for today), "yesterday", weekday, or ABdY (note: not standard strftime value)

R

24-hour time, including minutes

For example, the following example sets up two different voicemail zones, one for the Central time zone in 12-hour format, and a second in the Mountain time zone, in 24-hour format:

```
[zonemessages]
central=America/Chicago|'vm-received' Q 'digits/at' IMp
mountain24=America/Denver|'vm-received' q 'digits/at' H 'digits/hundred' M 'hours'
```

Defining Voicemail Contexts and Mailboxes

Now that the system-wide settings and voicemail zones have been set, you can define your voicemail contexts and individual mailboxes.

Voicemail contexts are used to separate out different groups of voicemail users. For example, if you are using Asterisk to host voicemail for more than one company, you should place each company's mailboxes in different voicemail contexts, to keep them separate. You might also use voicemail contexts to create per-department dial-by-name directories.

To define a new voicemail context, simply put the context name inside of square brackets, like this:

```
[default]
```

Inside a voicemail context, each mailbox definition takes the following syntax:

```
mailbox=password,name[,email[,pager_email[,options]]]
```

The *mailbox* argument is the mailbox number.

The *password* argument is the numeric code the mailbox owner must enter to access his voicemail. If the password is preceded by a minus sign (-), the password may not be changed by the mailbox owner.

 If the password is set to **d**, then this line is assumed to provide an alternate name that can be used for this mailbox in the dial-by-name directory. In the following example, people can reach extension **123** by searching for either **Robert** or **Bob**, along with a popular misspelling of his last name:

```
123 => 4444,Robert Schauerhamer
123 => d,Bob Schauerhamer
123 => d,Robert Showerhammer
```

The *email* and *pager_email* arguments are email addresses where voicemail notifications will be sent. These may be left blank if you don't want to send voicemail notifications via email. The message sent to the *pager_email* address is usually shorter and suitable for sending to a cell phone (via an email to SMS gateway) or to an alphanumeric pager.

The *options* argument is a pipe-separated list of voicemail options that may be specified for the mailbox. (These options may also be set globally by placing them in the [general] section.) Valid voicemail options include:

tz

Sets the voicemail zone from the [zonemessages] section earlier. This option is irrelevant if **envelope** is set to **no**.

attach

Attaches the voicemail to the notification email (but *not* to the pager email). May be set to either **yes** or **no**.

attachfmt

Specifies the format of the voicemail message that should be attached to the message sent to the notification email. Ordinarily, Asterisk uses the first format specified in the **format** parameter listed in **[general]** (covered earlier), but this may also be changed on a per-mailbox basis. This option may be set *only* on a per-mailbox basis.

This option is often used if Windows users want wav49 attachments but Linux users want their attachments in gsm format.

saycid

Says the Caller ID information before the message.

cidinternalcontexts

Sets the internal context for name playback instead of extension digits when saying the Caller ID information.

sayduration

Turns on/off the duration information before the message. Defaults to **on**.

saydurationm

Specifies the minimum duration to say when **sayduration** is **on**. Default is two minutes.

dialout

Specifies the context to dial out from (by choosing option 4 from the advanced menu). If not specified, dialing out from the voicemail system will not be permitted.

sendvoicemail

Specifies the context to send voicemail from (by choosing option 5 from the advanced menu). If not specified, sending messages from within the voicemail system will not be permitted.

searchcontexts

By default, Asterisk searches only the **default** context if no context is specified. To have Asterisk search all contexts, set this option to **yes**.

callback

Specifies the context to call back from. If not specified, calling the sender back from within the voicemail system will not be permitted.

review

Allows senders to review/re-record their messages before saving them. Defaults to **off**.

operator

Allows senders to hit **0** before, after, or while leaving a voicemail message to reach an operator. Defaults to **off**.

envelope

Turns on/off envelope playback before message playback. Defaults to **on**. This does not affect option 3,3 from the advanced options menu.

delete

Deletes voicemails from the server after notification is sent. This option may be set only on a per-mailbox basis; it is intended for use with users who wish to receive their voicemail messages *only* by email.

volgain

If your voicemail attachments to email are too quiet, you can set this option to increase the gain on the message before it is attached to the email notification.

 This option only works if the **sox** application has been installed on your Asterisk system.

forcename

Forces new users to record their names. A new user is determined by the password being the same as the mailbox number. Defaults to **no**.

forcegreetings

Forces new users to record greetings. A new user is determined by the password being the same as the mailbox number. Defaults to **no**.

hidefromdir

Hides the mailbox from the dial-by-name directory. Defaults to **no**.

tempgreetwarn

Warns users that their temporary greeting is still enabled.

You can specify multiple options by separating them with the pipe character, as shown in the definitions for mailboxes 102 and 103 below.

Here are some sample mailbox definitions:

```
[default]
; regular mailbox with email notification
101 => 4242,Example Mailbox,somebody@asteriskdocs.org

; more advanced mailbox with email and pager notification and a couple of
; special options
102 => 9855,Another User,another@asteriskdocs.org,pager@asteriskdocs.org,
attach=no|tz=central

; a mailbox with no email notification and lots of extra options
103 => 6522,John Q. Public,,,tz=central|attach=yes|saycid=yes|dialout=fromvm|
callback=fromvm|review=yes
```

vpb.conf

This file is used to configure Voicetronix cards with Asterisk. See the sample *vpb.conf* for more information.

zapata.conf

The *zapata.conf* file is used to define the relationship between Asterisk and the Zaptel driver. Because *zapata.conf* is specific to Asterisk, it is located with the other Asterisk configuration files in */etc/asterisk/*. As with *zaptel.conf*, the *zapata.conf* file contains a multitude of choices reflecting the multitude of hardware it supports, and we won't try to list all of the options here. In this book we've covered only the analog interfaces to the Zaptel driver, as described in Chapter 3.

zaptel.conf

The *zaptel.conf* file is not located with the other Asterisk *.conf* files; the Zaptel driver is available to any application that can make use of it, so it makes more sense to store it in a non-Asterisk-specific directory (*/etc/*). *zaptel.conf* is parsed by the *ztcfg* program to configure the TDM hardware elements in your system. You configure three main elements in the *zaptel.conf* file:

- A way of identifying the interfaces on the card within the dialplan
- The type of signaling the interface requires
- The tone language associated with a particular interface, as found in *zonedata.c*

 Be very careful not to plug your FXS module into a telephone line. The voltage associated with the phone line, especially during an incoming call, will be much too high for the module to handle and may permanently damage it, rendering it useless!

Within the *zaptel.conf* file, we define the type of signaling that the channel is going to use. We also define which channels to load. The options in the configuration file are the information that will be used to configure the channels with the `ztcfg` command.

The actual parameters available in the *zaptel.conf* file are quite extensive, as a wide variety of PSTN interfaces make use of the Zaptel telephony engine. Also, as this technology is rapidly evolving, anything we write now may not be accurate by the time you read it. Consequently, we won't try to list all of the options here.

In this book, we have focused on the Zaptel analog interfaces as provided by the Digium TDM400P card (see Chapter 3).

Asterisk Dialplan Functions

Dialplan functions are very powerful, and once you begin using them, you will wonder how you got along without them. Functions are used in the dialplan in a similar manner to variables. If it helps, you can think of them as intelligent variables (or for those of you from the database world, variables with triggers). When you invoke them, they perform a specific action, and their result becomes a part of the command in which you have included the function (in exactly the same way as a variable would).

AGENT   Returns information about an agent

AGENT(*agentid*[:*item*])

This function allows you to retrieve information pertaining to agents and may only be read, not set.

The valid items to retrieve are:

status *(default)*
  The status of the agent (LOGGEDIN | LOGGEDOUT)

password
  The password of the agent

name
  The name of the agent

mohclass
  Music-on-hold class

exten
  The callback extension for the Agent (AgentCallbackLogin)

channel
  The name of the active channel for the Agent (AgentLogin)

ARRAY

ARRAY(*var1*[|*var2*[...][|*varN*]])

The comma-separated list, which the function equals, will be interpreted as a set of values to which the comma-separated list of variable names in the argument should be set. This function may only be set, not read.

```
; Set var1 to 1 and var2 to 2.
exten => 123,1,Set(ARRAY(var1,var2)=1\,2)
```

 Remember to either backslash your commas in *extensions.conf* or quote the entire argument, since Set() can take multiple arguments itself.

See Also

Set()

BASE64_DECODE

BASE64_DECODE(*base64_string*)

Decodes a BASE64 string. This function may only be read, not set.

See Also

BASE64_ENCODE()

BASE64_ENCODE

BASE64_ENCODE(*string*)

Encodes a string in BASE64. This function may only be read, not set.

See Also

BASE64_DECODE()

BLACKLIST

When read, BLACKLIST() uses the AstDB to check if the Caller ID is in family blacklist. Returns 1 or 0.

This function may only be read, not set.

See Also
DB()

CALLERID

CALLERID(*datatype*[,*optional-CID*])

CALLERID() parses the Caller ID string within the current channel and returns all or part, as specified by *datatype*. The allowable datatypes are all, name, num, ani, dnid, or rdnis. Optionally, an alternative Caller ID may be specified if you wish to parse that string instead of the Caller ID set on the channel.

This function may be both read and set.

CDR

CDR(*fieldname*[,*options*])

Here is a list of all the available CDR field names:

clid
: Read-only. Use the CALLERID(all) function to set this value.

lastapp
: Read-only. Denotes the last application run.

lastdata
: Read-only. Denotes the arguments to the last application run.

src
: Read-only. Use the CALLERID(ani) function to set this value.

dst
: Read-only. Corresponds to the final extension in the dialplan.

dcontext
: Read-only. Corresponds to the final context in the dialplan.

channel
: Read-only. The name of the channel on which the call originated.

dstchannel
: Read-only. The name of the channel on which the call terminated.

disposition
: Read-only. Maximum reached state of the channel. If the u option is specified, this value will be returned as an integer, instead of a string: 1 = NO ANSWER, 2 = BUSY, 3 = FAILED, 4 = ANSWERED.

amaflags
: Read/write. Billing flags. If the u option is specified, this value will be returned as an integer, instead of a string: 1 = OMIT, 2 = BILLING, 3 = DOCUMENTATION.

accountcode

>Read/write. Billing account (19 char maximum).

userfield

>Read/write. User-defined field.

start

>Read-only. Time when the call started. If the u option is specified, this value will be returned as an integer (seconds since the epoch) instead of a formatted date/time string.

answer

>Read-only. Time when the call was answered (may be blank if the call is not yet answered). If the u option is specified, this value will be returned as an integer (seconds since the epoch) instead of a formatted date/time string.

end

>Read-only. Time when the call was completed (may be blank if the call is not yet complete). If the u option is specified, this value will be returned as an integer (seconds since the epoch) instead of a formatted date/time string.

duration

>Read-only. The difference between start and end, in seconds. May be 0, if the call is not yet complete.

billsec

>Read-only. The difference between answer and end, in seconds. May be 0, if the call is not yet complete.

uniqueid

>Read-only. A string that will be unique per-call within this Asterisk instance.

The following options may be specified:

l

>All results will be retrieved from the last Call Detail Record for the call, in the case of using multiple CDRs via ForkCDR().

r

>Custom CDR variables will be retrieved from the last Call Detail Record, but the standard fields will be retrieved from the first.

u

>The unparsed value will be returned. See the fieldname list above for entries that are affected by this flag.

You may also supply a *fieldname* not on the above list, and create your own variable, whose value can be changed with this function, and this variable will be stored in the CDR.

CHANNEL()

CHANNEL

Gets or sets various channel parameters

CHANNEL(item)

Standard items (provided by all channel technologies) are:

audioreadformat
> Read-only. Format currently being read.

audionativeformat
> Read-only. Format used natively for audio.

videonativeformat
> Read-only. Format used natively for video.

audiowriteformat
> Read-only. Format currently being written.

callgroup
> Read/write. Callgroups for call pickup.

channeltype
> Read-only. Technology used for channel.

language
> Read/write. Language used for sounds played and recorded.

musicclass
> Read/write. Class used (from *musiconhold.conf*) for hold music.

rxgain
> Read/write. Receive gain (in dB) on channel drivers that support it.

txgain
> Read/write. Transmit gain (in dB) on channel drivers that support it.

tonezone
> Read/write. Regional zone for indications played.

state
> Read-only. Current channel state.

transfercapability
> Read/write. What can be transferred on an ISDN circuit. Current valid values are:

SPEECH
> Speech (default, voice calls)

DIGITAL
> Unrestricted digital information (data calls)

```
RESTRICTED_DIGITAL
```
Restricted digital information

```
3K1AUDIO
```
3.1 kHz Audio (fax calls)

```
DIGITAL_W_TONES
```
Unrestricted digital information with tones/announcements

```
VIDEO
```
Video

Additional items may be available from the channel driver providing the channel; see its documentation for details. Any item requested that is not available on the current channel will return an empty string.

See Also

```
CDR()
```

CHECK_MD5 Validate an MD5 digest

```
CHECK_MD5(digest,data)
```

Returns 1 on success, 0 on failure.

This function is *deprecated* in favor of using the MD5() function with the built-in expression parser.

See Also

```
MD5()
```

CHECKSIPDOMAIN Checks if a domain is local

```
CHECKSIPDOMAIN(domain|IP)
```

This function checks if the domain in the argument is configured as a local SIP domain that this Asterisk server is configured to handle. Returns the domain name if it is locally handled, otherwise an empty string. Check the domain configuration option in *sip.conf*.

CURL Returns the data resulting from a GET or POST to a URI

```
CURL(url[|post-data])
```

By default, CURL() will perform an HTTP GET to retrieve the *url*. However, if *post-data* is specified, an HTTP POST will be performed instead.

CUT

Cuts a string based on a given delimiter

CUT(*varname,char-delim,range-s*)

CUT() works in a similar manner to the cut(1) Unix command-line tool, and is, in fact, designed based upon that tool.

In the dialplan, you may specify character offsets to select a substring of a variable based purely on the uniform length of characters (namely 1). CUT() is designed to help you work with data that may have multiple, variable-length sections, divided by a common delimiter.

The most common case is the name of a channel, which is composed of two parts, a base name and a unique identifier (e.g., *SIP/tom-abcd1234* or *SIP/bert-1a2b3c4d*). CUT() may be used to trim the unique identifier, no matter how long the base name may be:

```
; Trim the unique identifier from the current channel name
exten => 123,1,Set(chan=${CUT(CHANNEL,-,1)})
```

varname is the name of the variable that will be operated on. Note that CUT() operates on the *name* of a variable, rather than upon the *value* of a variable. CUT() is somewhat unique in this regard.

char-delim is the character that will serve as the delimiter ('-' is the default)

range-spec allows you to define which fields are returned. The *range-spec* can be specified as a range with - (e.g., 1-3) or a group of ranges and field numbers by separating each with & (e.g., 1&3-4). Note that if multiple field numbers are specified, the resulting value will have its fields separated by the same delimiter.

> *range-spec* uses a 1-based offset. That is, the first field is field 1 (as opposed to a 0-based offset, where the first field would be 0).

See Also

FIELDQTY()

DB

Read or write to AstDB

DB(*family/key*)

Will return the value of the entry in the database (or blank if it does not exist), or set the value in the database.

See Also

DBdel(), DB_DELETE(), DBdeltree(), DB_EXISTS()

DB_DELETE
Deletes a key or key family from the AstDB database

DB_DELETE(*family/key*)

Returns a value from the database and delete it.

See Also

DBdel(), DB(), DBdeltree()

DB_EXISTS
Checks AstDB for specified key

DB_EXISTS(*family/key*)

Check to see if a key exists in the Asterisk database.

See Also

DB()

DUNDILOOKUP
Queries DUNDi peers for a particular number

DUNDILOOKUP(*number*[|*context*[|*option*)

Does a DUNDi lookup of a phone number.

ENUMLOOKUP
Queries the ENUM database for a particular number

ENUMLOOKUP(*number*[|*Method-type*[|*options*[|*record#*[|*zone-suffix*]]]])

Allows for general or specific querying of NAPTR records or counts of NAPTR types for ENUM or ENUM-like DNS pointers.

ENV
References environment variables

ENV(*envname*)

Gets or sets the environment variable specified by *envname*.

EVAL

EVAL(*variable*)

EVAL() is one of the most powerful dialplan functions. It permits one to store variable expressions in a location other than *extensions.conf*, such as a database, yet evaluate them in the dialplan, as if they were included there all along. You can bet that EVAL() is a cornerstone in making a dialplan truly dynamic.

```
; We might store something like "SIP/${DB(ext2chan/123)}" in the
; database entry for extension/123, which tells us to look up yet
; another database entry.
exten => _XXX,1,Set(dialline=${DB(extension/${EXTEN})})
exten => _XXX,n,Dial(${EVAL(${dialline})})

; Real world example (taken from production code)
exten => _1NXXNXXXXXX,n(generic),Set(provider=${DB(rt2provider/${route})}-nanp)
exten => _1NXXNXXXXXX,n(provider),Dial(${EVAL(${DB(provider/${provider})})})
exten => _1NXXNXXXXXX,n,Goto(nextroute)
```

See Also

Exec()

EXISTS

EXISTS(*data*)

Existence test: returns 1 if non-blank, 0 otherwise

FIELDQTY

FIELDQTY(*varname*|*delim*)

Counts the fields, with an arbitrary delimiter

See Also

CUT()

FILTER

FILTER(*allowed-chars*|*string*)

Filters the *string* to include only the characters shown in *allowed-chars*:

```
; Ensure that the Caller*ID number contains only digits
exten => Set(CALLERID(num)=${FILTER(0123456789,${CALLERID(num)})})
```

This function may only be read, not set.

QUOTE()

GLOBAL

References global namespace

GLOBAL(*varname*)

Gets or sets the global variable specified.

GROUP

Associates the channel into a set group

GROUP([*category*])

Gets or sets the channel group.

```
; Permit only one user to access the paging system at once.
exten => 8000,1,Set(GROUP()=pager)
exten => 8000,n,GotoIf($[${GROUP_COUNT(pager)} > 1]?hangup)
exten => 8000,n,Page(SIP/101&SIP/102&SIP/103&SIP/104)
exten => 8000,n(hangup),Hangup
```

See Also

GROUP_COUNT(), GROUP_LIST(), GROUP_MATCH_COUNT()

GROUP_COUNT

Counts the number of channels in the specified group.

GROUP_COUNT([*groupname*][@*category*])

Counts the number of channels in the specified group. Will return the count of the current channel if the **groupname** is not specified.

See Also

GROUP(), GROUP_LIST(), GROUP_MATCH_COUNT()

GROUP_LIST

Lists channel groups

GROUP_LIST()([*groupname*][@*category*])

Gets a list of the groups set on a channel.

See Also

GROUP(), GROUP_COUNT(), GROUP_MATCH_COUNT()

GROUP_MATCH_COUNT

Counts channels in a matching group name

GROUP_MATCH_COUNT(*groupmatch*[@*category*])

Counts the number of channels in the groups matching the specified pattern.

See Also

GROUP(), GROUP_COUNT(), GROUP_LIST()

IAXPEER

Obtains IAX channel information

IAXPEER(*peername*[|*item*])
IAXPEER(CURRENTCHANNEL[|*item*])

Gets IAX peer information.

If peername is specified, valid items are:

ip
> The IP address of this peer. If *item* is not specified, the IP address will be given.

status
> The peer's status (if qualify=yes).

mailbox
> The peer's configured mailbox.

context
> The peer's configured context.

expire
> The epoch time of the next registration expiration for this peer.

dynamic
> Does this peer register with Asterisk? (yes/no)

callerid_name
> The Caller ID name configured on this peer.

callerid_num
> The Caller ID number configured on this peer.

codecs
> The peer's configured codecs.

codec[*x*]
> Preferred codec index number *x* (beginning with zero).

See Also

SIPPEER()

IF

IF(*expr*?[*true*][:*false*])

Conditional: returns the data following ? if true, otherwise the data following :.

```
; Returns foo
exten => 123,1,Set(something=${IF($[2 > 1]?foo:bar)})
; Returns bar
exten => 123,n,Set(something=${IF($[2 < 1]?foo:bar)})
```

See Also

GotoIf()

IFTIME

IFTIME(*times,days_of_week,days_of_month,months*?[*true*][:*false*])

Conditional: Returns the data following ? if true, otherwise the data following :

times
> Time ranges, in 24-hour format

days_of_week
> Days of the week (mon, tue, wed, thu, fri, sat, sun)

days_of_month
> Days of the month (1-31)

months
> Months (jan, feb, mar, apr, etc.)

See Also

GotoIfTime()

ISNULL

ISNULL(*data*)

Returns 1 if *data* is blank or 0 otherwise.

See Also

LEN(), EXISTS()

KEYPADHASH

KEYPADHASH(*string*)

Hashes the letters in *string* into the equivalent keypad numbers.

```
; Calculate the hashes of the authors' last names.  So, the
; corresponding values would be 623736, 76484, and 82663443536.
exten => 123,1,Set(lastname1=${KEYPADHASH(Madsen)})
exten => 123,n,Set(lastname2=${KEYPADHASH(Smith)})
exten => 123,n,Set(lastname3=${KEYPADHASH(VanMeggelen)})
```

See Also

Directory()

LANGUAGE
Accesses the channel language

LANGUAGE()

Gets or sets the channel's language.

This function is deprecated in favor of CHANNEL(*language*).

See Also

CHANNEL()

LEN
Calculates the string length

LEN(*string*)

Returns the length of *string*.

MATH
Mathematical calculations

MATH(*number1op number2*[,*type_of_result*])

Performs mathematical functions.

```
exten => 123,1,Set(value1=${MATH(1+2)})
```

MD5
Calculates MD5 digest

MD5(*data*)

Computes the MD5 digest of *data*.

See Also

SHA1()

MUSICCLASS

MUSICCLASS()

 This function has been deprecated in favor of CHANNEL(*musicclass*).

Reads or sets the music-on-hold class.

See Also

CHANNEL()

QUEUE_MEMBER_COUNT

QUEUE_MEMBER_COUNT(*queuename*)

Counts the number of members answering a queue.

See Also

QUEUE_MEMBER_LIST()

QUEUE_MEMBER_LIST

QUEUE_MEMBER_LIST(*queuename*)

Returns a list of interfaces in a queue.

See Also

QUEUE_MEMBER_COUNT()

QUEUE_WAITING_COUNT

QUEUE_WAITING_COUNT(*queuename*)

Counts number of calls currently waiting in a queue.

QUEUEAGENTCOUNT

QUEUEAGENTCOUNT(*queuename*)

This function has been deprecated in favor of `QUEUE_MEMBER_COUNT()`.

Counts number of agents answering a queue.

See Also

`QUEUE_MEMBER_COUNT()`, `QUEUE_MEMBER_LIST()`

QUOTE
Escapes a string

`QUOTE(string)`

Quotes a given string, escaping embedded quotes as necessary.

See Also

`FILTER()`

RAND
Random number

`RAND([min][|max])`

Chooses a random number within a range.

`RAND()` randomly picks an integer between *min* and *max*, inclusive, and returns that integer. If *min* is not specified, it defaults to 0. If *max* is not specified, it defaults to the C constant `INT_MAX`, which is 2,147,483,647 on 32-bit platforms. Note that `INT_MAX` is quite a bit larger on 64-bit platforms.

REALTIME
Retrieves real-time data

`REALTIME(family|fieldmatch[|value[||delim1[|delim2]]])`

Real-time read/write functions. Use the above syntax for a read and the following syntax for a write:

```
REALTIME(family|fieldmatch|value|field)
```

REGEX
Compares based upon a regular expression

`REGEX("regular expression" data)`

Matches based upon a regular expression.

SET

SET(*varname*=[*value*])

SET assigns a value to a channel variable. It is frequently used to set values that contain the character |, since that character is normally a delimiter when used with the application Set().

See Also
Set()

SHA1

SHA1(*data*)

Computes the SHA-1 digest of *data*.

See Also
MD5()

SIP_HEADER

SIP_HEADER(*name*[,*number*])

Gets the specified SIP header.

SIPCHANINFO

SIPCHANINFO(*item*)

Gets the specified SIP parameter from the current channel.

Valid items are:

peerip
> The IP address of this SIP peer

recvip
> The source IP address of this SIP peer

from
> The SIP URI from the From: header

uri
> The SIP URI from the Contact: header

useragent

The name of the SIP user agent

peername

The name of the SIP peer

t38passthrough

1 if T38 is offered or enabled in this channel, otherwise 0

SIPPEER

SIPPEER(*peername*[|*item*])

Gets SIP peer information.

Valid items are:

ip

The IP address of this peer. If *item* is not specified, the IP address will be given.

mailbox

This peer's configured mailbox.

context

The peer's configured context.

expire

The epoch time of the next registration expiration.

dynamic

Does this device register with Asterisk? (yes/no)

callerid_name

The Caller ID name configured on this peer.

callerid_number

The Caller ID number configured on this peer.

status

This peer's status (if qualify=yes).

regexten

This peer's registration extension, if configured.

limit

This peer's call limit.

curcalls

Current number of calls. Only available if call-limit is set on this peer.

language

Default language for this peer.

accountcode

Account code for this peer.

---

useragent

> The name of the SIP user agent.

codecs

> The configured codecs for this peer.

codec[*x*]

> Preferred codec index number *x* (beginning with zero).

See Also

IAXPEER()

SORT
Sorts a list

SORT(*key1*:*val1*[...][,*keyN*:*valN*])

Sorts a list of key/vals into a list of keys, based upon the vals which can be any real number (float).

SPEECH
Retrieves info on speech recognition results

SPEECH(*argument*)

Gets information about speech recognition results.

SPEECH_ENGINE
Modifies speech engine property

SPEECH_ENGINE(*name*)=*value*)

Changes a specific attribute of the speech engine.

SPEECH_GRAMMAR
Retrieves speech grammar information

SPEECH_GRAMMAR(*result number*)

Gets the matched grammar of a result if available.

SPEECH_SCORE
Retrieves speech recognition confidence score

SPEECH_SCORE(*result number*)

Gets the confidence score of a result.

SPEECH_TEXT

SPEECH_TEXT(*result number*)

Gets the recognized text of a result.

SPRINTF

SPRINTF(*format*|*arg1*[|...*argN*])

Formats a variable or set of variables according to a format string.

The most common case for the use of SPRINTF is to zero-pad a number to a certain length:

```
; Returns 00123
exten => 123,1,Set(padfive=${SPRINTF(%05d,${EXTEN})})
```

Most of the format options listed in the manpage for `sprintf(3)` are also implemented in this dialplan function.

See Also

STRFTIME()

STAT

STAT(*flag*,*filename*)

Does a check on the specified file.

flag may be one of the following options:

e

> Returns 1 if the file exists; 0 otherwise

s

> Returns the size of the file, in bytes

f

> Returns 1 if the path referenced is a regular file (and not a directory, symlink, socket, or device) or 0 otherwise

d

> Returns 1 if the path referenced is a directory (and not a regular file, symlink, socket, or device) or 0 otherwise

M

> Returns the epoch time when the file contents were last modified

C

> Returns the epoch time when the file inode was last modified

m

Returns the permissions mode of the file (as an octal number)

STRFTIME

Formats the date and time

STRFTIME([epoch][|[timezone][|format]])

Returns the current date/time in a specified format.

STRFTIME passes the *epoch* and *format* arguments directly to the underlying strftime (3) C library call, so check out that manpage for more information. The *timezone* parameter should be the name of a directory/file in */usr/share/zoneinfo* (e.g., *America/Chicago* or *America/New_York*).

See Also

STRPTIME()

STRPTIME

Converts a string into a date and time

STRPTIME(datetime|timezone|format)

Returns the epoch of the arbitrary date/time string structured as described in the format.

The purpose of this function is to take a formatted date/time and convert it back into seconds since the epoch (January 1st, 1970, at midnight GMT), so that you may do calculations with it, or simply convert it into some other date/time format.

STRPTIME passes the string and format directly to the underlying C library call strptime (3), so check out that manpage for more information. The *timezone* parameter should be the name of a directory/file in */usr/share/zoneinfo* (e.g., *America/Chicago* or *America/New_York*).

See Also

STRFTIME()

TIMEOUT

Accesses channel timeout values

TIMEOUT(timeouttype)

Gets or sets timeouts on the channel.

The timeouts that can be manipulated are:

absolute

The absolute maximum amount of time permitted for a call. A setting of **0** disables the timeout.

---

digit

The maximum amount of time permitted between digits when the user is typing in an extension. When this timeout expires, after the user has started to type in an extension, the extension will be considered complete, and will be interpreted. Note that if an extension typed in is valid, it will not have to timeout to be tested, so typically at the expiry of this timeout, the extension will be considered invalid (and thus control will be passed to the i extension, or if it doesn't exist the call will be terminated). The default timeout is five seconds.

response

The maximum amount of time permitted after falling through a series of priorities for a channel in which the user may begin typing an extension. If the user does not type an extension in this amount of time, control will pass to the t extension if it exists, and if not the call will be terminated. The default timeout is 10 seconds.

TXTCIDNAME

DNS lookup

TXTCIDNAME(*number*)

Looks up a caller name via DNS.

URIDECODE

Decodes a URI

URIDECODE(*data*)

Decodes a URI-encoded string according to RFC 2396.

See Also

URIENCODE()

URIENCODE

Encodes a URI

URIENCODE(*data*)

Encodes a string to URI-safe encoding according to RFC 2396.

See Also

URIDECODE()

VMCOUNT

VMCOUNT(*mailbox*[@*context*][|*folder*])

Counts the voicemail in a specified mailbox.

Asterisk Manager Interface Actions

The following is a list of actions you can perform using the Manager Interface, or AMI. See Chapter 14 for more information on the AMI.

AbsoluteTimeout

Sets the sAbsoluteTimeout on a channel

Hangs up a channel after a certain time.

Parameters

Channel
: [required] The name of the channel on which to set the absolute timeout.

Timeout
: [required] The maximum duration of the call, in seconds.

ActionID
: [optional] An identifier that can be used to identify the response to this action.

Notes

Asterisk will acknowledge the timeout setting with a `Timeout Set` message.

Privilege

`call, all`

Example

```
Action: AbsoluteTimeout
    Channel: SIP/testphone-10210698
    Timeout: 15
    ActionID: 12345

    Response: Success
    Message: Timeout Set
    ActionID: 12345
```

AgentCallbackLogin Sets an agent as logged in to the queue system in callback mode

Logs the specified agent in to the Asterisk queue system in callback mode. When a call is distributed to this agent, it will ring the specified extension.

Parameters

Agent
> [required] Agent ID of the agent to log in to the system, as specified in *agents.conf*.

Exten
> [required] Extension to use for callback.

Context
> [optional] Context to use for callback.

AckCall
> [optional] Set to **true** to require an acknowledgement (the agent pressing the # key) to accept the call when agent is called back.

WrapupTime
> [optional] The minimum amount of time after disconnecting before the agent will receive a new call.

ActionID
> [optional] An identifier which can be used to identify the response to this action.

Privilege

agent, all

Example

```
Action: AgentCallbackLogin
Agent: 1001
Exten: 201
Context: Lab
ActionID: 24242424

Response: Success
Message: Agent logged in
ActionID: 24242424

Event: Agentcallbacklogin
Privilege: agent,all
Agent: 1001
Loginchan: 201@Lab
```

Notes

The AgentCallbackLogin action (along with the AgentCallbackLogin() application) has been deprecated. It is suggested you use the QueueAdd action instead. See *doc/queues-with-callback-members.txt* in the Asterisk source code for more information.

AgentLogoff

Logs off the specified agent for the queue system.

Parameters

Agent
: [required] Agent ID of the agent to log off.

Soft
: [optional] Set to **true** to not hangup existing calls.

ActionID
: [optional] An identifier which can be used to identify the response to this action.

Privilege

agent, all

Example

```
Action: AgentLogoff
Agent: 1001
Soft: true
ActionID: blahblahblah

Response: Success
Message: Agent logged out
ActionID: blahblahblah

Event: Agentcallbacklogoff
Privilege: agent,all
Agent: 1001
Reason: CommandLogoff
Loginchan: 201@Lab
Logintime: 5698
```

Agents

This action lists information about all configured agents.

Privilege

agent, all

Example

```
Action: Agents
ActionID: mylistofagents
```

```
Response: Success
Message: Agents will follow
ActionID: mylistofagents

Event: Agents
Agent: 1001
Name: Jared Smith
Status: AGENT_IDLE
LoggedInChan: 201@Lab
LoggedInTime: 1173237646
TalkingTo: n/a
ActionID: mylistofagents

Event: Agents
Agent: 1002
Name: Leif Madsen
Status: AGENT_LOGGEDOFF
LoggedInChan: n/a
LoggedInTime: 0
TalkingTo: n/a
ActionID: mylistofagents

Event: Agents
Agent: 1003
Name: Jim VanMeggelen
Status: AGENT_LOGGEDOFF
LoggedInChan: n/a
LoggedInTime: 0
TalkingTo: n/a
ActionID: mylistofagents

Event: AgentsComplete
ActionID: mylistofagents
```

ChangeMonitor — Changes monitoring filename of a channel

The `ChangeMonitor` action may be used to change the file started by a previous `Monitor` action. The following parameters may be used to control this.

Parameters

`Channel`
> [required] Used to specify the channel to record.

`File`
> [required] The new filename in which the monitored channel will be recorded.

`ActionID`
> [optional] An identifier that can be used to identify the response to this action.

Privilege

call, all

Example

```
Action: ChangeMonitor
Channel: SIP/linksys-084c63c0
File: new-test-recording
ActionID: 555544443333

Response: Success
ActionID: 555544443333
Message: Changed monitor filename
```

Command

Runs an Asterisk CLI command as if it had been run from the CLI.

Parameters

Command
> [required] Asterisk CLI command to run.

ActionID
> [optional] An action identifier that can be used to identify the response from Asterisk.

Privilege

command, all

Example

```
Action: Command
Command: core show version
ActionID: 0123456789abcdef

Response: Follows
Privilege: Command
ActionID: 0123456789abcdef
Asterisk SVN-branch-1.4-r55869 built by jsmith @ hockey on a ppc running Linux
  on 2007-02-21 16:55:26 UTC
--END COMMAND--
```

DBGet

This action retrieves a value from the AstDB database.

Parameters

Family
> [required] The AstDB key family from which to retrieve the value

Key
> [required] The name of the AstDB key.

ActionID
> [optional] An identifier that can be used to identify the response to this action.

Privilege

system, all

Example

```
Action: DBGet
Family: testfamily
Key: mykey
ActionID: 01234-astdb-43210

Response: Success
Message: Result will follow
ActionID: 01234-astdb-43210

Event: DBGetResponse
Family: testfamily
Key: mykey
Val: 42
ActionID: 01234-astdb-43210
```

DBPut   **Puts DB entry**

Sets a key value in the AstDB database.

Parameters

Family
> [required]The AstDB key family in which to set the value.

Key
> [required] The name of the AstDB key.

Val
> [required]The value to assign to the key.

ActionID
> [optional] An identifier that can be used to identify the response to this action.

Privilege

system, all

Example

```
Action: DBPut
Family: testfamily
Key: mykey
Val: 42
ActionID: testing123

Response: Success
Message: Updated database successfully
ActionID: testing123
```

Events

Enables or disables sending of events to this manager connection.

Parameters

EventMask
> [required] Set to on if all events should be sent, off if events should not be sent, or system,call,log to select which type of events should be sent to this manager connection.

ActionID
> [optional] An identifier which can be used to identify the response to this action.

Privilege

Example

```
Action: Events
EventMask: off
ActionID: 2938416

Response: Events Off
ActionID: 2938416

Action: Events
EventMask: log,call
ActionID: blah1234

Response: Events On
ActionID: blah1234
```

ExtensionState

This command reports the extension state for the given extension. If the extension has a hint, this will report the status of the device connected to the extension.

Parameters

Exten
> [required] The name of the extension to check.

Context
> [required] The name of the context that contains the extension.

ActionId
> [optional] An action identifier that can be used to identify this manager transaction.

Privilege

call, all

Example

```
Action: ExtensionState
Exten: 200
Context: lab
ActionID: 54321

Response: Success
ActionID: 54321
Message: Extension Status
Exten: 200
Context: lab
Hint: SIP/testphone
Status: 0
```

Notes

The following are the possible extension states:

-2
> Extension removed

-1
> Extension hint not found

0
> Idle

1
> In use

2
> Busy

GetConfig

Retrieves configuration

Retrieves the data from an Asterisk configuration file.

Parameters

`Filename`
: [required]Name of the configuration file to retrieve.

`ActionID`
: [optional] An identifier that can be used to identify the response to this action.

Privilege

`config, all`

Example

```
Action: GetConfig
Filename: musiconhold.conf
ActionID: 09235012

Response: Success
ActionID: 09235012
Category-000000: default
Line-000000-000000: mode=files
Line-000000-000001: directory=/var/lib/asterisk/moh
Line-000000-000002: random=yes
```

GetVar

Gets the value of a local channel variable or global variable

Parameters

`Channel`
: [optional] The name of the channel from which to retrieve the variable value.

`Variable`
: [required] Variable name.

`ActionID`
: [optional] An identifier that can be used to identify the response to this action.

Privilege

`call, all`

Example

```
Action: GetVar
Channel: SIP/linksys2-1020e2b0
Variable: SIPUSERAGENT
ActionID: abcd1234

Response: Success
Variable: SIPUSERAGENT
Value: Linksys/SPA962-5.1.5
```

```
ActionID: abcd1234

Action: GetVar
Variable: TRUNKMSD

Response: Success
Variable: TRUNKMSD
Value: 1
```

Hangup

Hangs up channel

Hangs up the specified channel.

Parameters

Channel
> [optional] The channel name to be hung up

ActionID
> [optional] An identifier that can be used to identify the response to this action

Privilege

call, all

Example

```
Action: Hangup
Channel: SIP/labrat-8d3a

Response: Success
Message: Channel Hungup

Event: Hangup
Privilege: call,all
Channel: SIP/labrat-8d3a
Uniqueid: 1173448206.0
Cause: 0
Cause-txt: Unknown
```

IAXNetstats

Shows IAX statistics

Shows a summary of network statistics for the IAX2 channel driver.

Privilege

Example

```
Action: IAXNetstats

IAX2/216.207.245.8:4569-1 608 -1 0 -1 -1 0 -1 1 288 508 10 1 3 0 0
```

IAXPeers

Lists all IAX2 peers and their current status.

Privilege

Example

```
Action: IAXPeers

Name/Username    Host                 Mask             Port    Status
jared/jared      192.168.0.71    (S)  255.255.255.255  4569    UNREACHABLE
jaredsmith       192.168.0.72    (S)  255.255.255.255  4569    OK (43 ms)
arrivaltel/8017  172.20.95.2     (S)  255.255.255.255  4569    Unmonitored
sokol/jsmith     172.17.122.217  (S)  255.255.255.255  4569    OK (48 ms)
demo/asterisk    216.207.245.47  (S)  255.255.255.255  4569    Unmonitored
5 iax2 peers [2 online, 1 offline, 2 unmonitored]
```

ListCommands

Lists the action name and synopsis for every Asterisk Manager Interface action.

Privilege

Example

```
Action: ListCommands

Response: Success
AbsoluteTimeout: Set Absolute Timeout (Priv: call,all)
AgentCallbackLogin: Sets an agent as logged in by callback (Priv: agent,all)
AgentLogoff: Sets an agent as no longer logged in (Priv: agent,all)
. . .
ZapTransfer: Transfer Zap Channel (Priv: <none>)
```

Logoff

Logs off this manager session.

Privilege

Example

```
Action: Logoff

Response: Goodbye
Message: Thanks for all the fish.
```

MailboxCount

Retrieves the number of messages for the specified voice mailbox.

Privilege

call, all

Example

```
Action: MailboxCount
Mailbox: 100@lab
ActionID: 54321abcde

Response: Success
ActionID: 54321abcde
Message: Mailbox Message Count
Mailbox: 100@lab
NewMessages: 2
OldMessages: 0
```

MailboxStatus

Checks the status for the specified voicemail box.

Parameters

Mailbox

[required] The full mailbox ID, including mailbox and context (*box context*).

ActionID

[optional] A unique identifier that can be used to identify responses to this manager command.

Privilege

call, all

Example

```
Action: MailboxStatus
Mailbox: 100@lab
```

```
ActionID: abcdef0123456789

Response: Success
ActionID: abcdef0123456789
Message: Mailbox Status
Mailbox: 100@lab
Waiting: 1
```

MeetmeMute

Mutes a particular user in a MeetMe conference bridge.

Parameters

Meetme
> [required] The MeetMe conference bridge number.

Usernum
> [required] The user number in the specified bridge.

ActionID
> [optional] A unique identifier to help you identify responses to this command.

Privilege

`call, all`

Example

```
Action: MeetmeMute
Meetme: 104
Usernum: 1
ActionID: 5432154321

Response: Success
ActionID: 5432154321
Message: User muted

Event: MeetmeMute
Privilege: call,all
Channel: SIP/linksys2-10211dc0
Uniqueid: 1174008176.3
Meetme: 104
Usernum: 1
Status: on
```

Notes

To find the Usernum number for a particular caller, watch the Asterisk Manager Interface when a new member joins a conference bridge. When it happens, you'll see an event like this:

```
Event: MeetmeJoin
Privilege: call,all
```

```
Channel: SIP/linksys2-10211dc0
Uniqueid: 1174008176.3
Meetme: 104
Usernum: 1
```

MeetMeUnmute

Unmutes the specified user in a MeetMe conference bridge.

Parameters

Meetme
> [required] The MeetMe conference bridge number.

Usernum
> [required] The user number in the specified bridge.

ActionID
> [optional] A unique identifier to help you identify responses to this command.

Privilege

`call, all`

Example

```
Action: MeetmeUnmute
Meetme: 104
Usernum: 1
ActionID: abcdefghijklmnop

Response: Success
ActionID: abcdefghijklmnop
Message: User unmuted

Event: MeetmeMute
Privilege: call,all
Channel: SIP/linksys2-10211dc0
Uniqueid: 1174008176.3
Meetme: 104
Usernum: 1
Status: off
```

Monitor

Records the audio on a channel to the specified file.

Parameters

Channel
> [required] Specifes the channel to be recorded.

File
:   [optional] The name of the file in which to record the channel. The path defaults to the Asterisk monitor spool directory, which is usually */var/spool/asterisk/monitor*. If no filename is specified, the filename will be the name of the channel, with slashes replaced with dashes.

Format
:   [optional] The audio format in which to record the channel. Defaults to wav.

Mix
:   [optional] A Boolean flag specifying whether or not Asterisk should mix the inbound and outbound audio from the channel in to a single file.

ActionID
:   [optional] An identifier that can be used to identify the response to this action.

Privilege

call, all

Example

```
Action: Monitor
Channel: SIP/linksys2-10216e38
Filename: test-recording
Format: gsm
Mix: true

Response: Success
Message: Started monitoring channel
```

Originate

Generates an outbound call from Asterisk, and connect the channel to a context/extension/priority combination or dialplan application.

Parameters

Channel
:   [required] Channel name to call. Once the called channel has answered, the control of the call will be passed to the specified Exten/Context/Priority or Application.

Exten
:   [optional] Extension to use (requires Context and Priority).

Context
:   [optional] Context to use (requires Exten and Priority).

Priority
:   [optional] Priority to use (requires Exten and Context).

Application
:   [optional] Application to use.

Data

[optional] Data to pass as parameters to the application (requires `Application`).

Timeout

[optional] How long to wait for call to be answered (in ms).

CallerID

[optional] Caller ID to be set on the outgoing channel.

Variable

[optional] Channel variable to set. Multiple variable headers are allowed.

Account

[optional] Account code.

Async

[optional] Set to `true` for asynchronous origination. Asynchronous origination allows you to originate one or more calls without waiting for an immediate response.

ActionID

[optional] An identifier that can be used to identify the response to this action.

Privilege

`call, all`

Example

```
Action: Originate
Channel: SIP/linksys2
Context: lab
Exten: 201
Priority: 1
CallerID:

Response: Success
Message: Originate successfully queued

Action: Originate
Application: MusicOnHold
Data: default
Channel: SIP/linksys2

Response: Success
Message: Originate successfully queued
```

Park   Parks a channel

Parks the specified channel in the parking lot.

Parameters

Channel

[required] Channel name to park.

`Channel2`

[required] Channel to announce park info to (and return the call to if the parking times out).

`Timeout`

[optional] Number of milliseconds to wait before callback.

`ActionID`

[optional] An identifier which can be used to identify the response to this action.

Privilege

`call, all`

Example

```
Action: Park
Channel: SIP/linksys-10228fb0
Channel2: SIP/linksys2-10231520
Timeout: 45
ActionID: parking-test-01

Response: Success
ActionID: parking-test-01
Message: Park successful
```

Notes

The call parking lot is configured in *features.conf* in the Asterisk configuration directory.

ParkedCalls

Lists parked calls

Lists any calls that are parked in the call parking lot.

Privilege

Example

```
Action: ParkedCalls
ActionID: 0982350175

Response: Success
ActionID: 0982350175
Message: Parked calls will follow

Event: ParkedCall
Exten: 701
Channel: SIP/linksys2-101f98a8
From: SIP/linksys2-101f98a8
Timeout: 26
CallerID: linksys2
CallerIDName: linksys2
```

```
ActionID: 0982350175

Event: ParkedCallsComplete
ActionID: 0982350175
```

Notes

The call parking lot is configured in *features.conf* in the Asterisk configuration directory.

PauseMonitor

Pauses the monitoring (recording) of a channel if it is being monitored.

Parameters

Channel
> [required] The channel identifier of the channel that is currently being monitored.

ActionID
> [optional] An identifier that can be used to identify the response to this action.

Privilege

call, all

Example

```
Action: PauseMonitor
Channel: SIP/linksys2-10212040
ActionID: 987987987987


Response: Success
ActionID: 987987987987
Message: Paused monitoring of the channel
```

Ping

Queries the Asterisk server to make sure it is still responding. Asterisk will respond with a Pong response. This command can also be used to keep the manager connection from timing out.

Example

```
Action: Ping

Response: Pong
```

PlayDTMF

Plays a DTMF digit on the specified channel.

Parameters

Channel
: [required] The identifier for the channel on which to send the DTMF digit.

Digit
: [required] The DTMF digit to play on the channel.

ActionID
: [optional] An identifier that can be used to identify the response to this action.

Privilege

call, all

Example

```
Action: PlayDTMF
Channel: Local/201@lab-157a,1
Digit: 9

Response: Success
Message: DTMF successfully queued
```

QueueAdd

Adds a queue member to a call queue.

Parameters

Queue
: [required] The name of the queue.

Interface
: [required] The name of the member to add to the queue. This will be a technology and resource, such as SIP/Jane or Local/203@lab/n. Agents (as defined in *agents.conf*) can also be added by using the Agent/1234 syntax.

MemberName
: [optional] This is a human-readable alias for the interface, and will appear in the queue statistics and queue logs.

Penalty
: [optional] A numerical penalty to apply to this queue member. Asterisk will distribute calls to members with higher penalties only after attempting to distribute the call to all members with a lower penalty.

Paused

[optional] Whether or not the member should be initially paused.

ActionID

[optional] An action identifier that you can use to identify the response to this manager transaction.

Privilege

agent, all

Example

```
Action: QueueAdd
Queue: myqueue
Interface: SIP/testphone
MemberName: Jared Smith
Penalty: 2
Paused: no
ActionID: 4242424242

Response: Success
ActionID: 4242424242
Message: Added interface to queue

Event: QueueMemberAdded
Privilege: agent,all
Queue: myqueue
Location: SIP/testphone
MemberName: Jared Smith
Membership: dynamic
Penalty: 2
CallsTaken: 0
LastCall: 0
Status: 1
Paused: 0
```

QueuePause   Pauses or unpauses a member in a call queue

Pauses or unpauses a member in a call queue.

Parameters

Interface

[required] The name of the interface to pause or unpause.

Paused

[required] Whether or not the interface should be paused. Set to **true** to pause the member, or **false** to unpause the member.

Queue

> [optional] The name of the queue in which to pause or unpause this member. If not specified, the member will be paused or unpaused in all the queues it is a member of.

ActionID

> [optional] An identifier that can be used to identify the response to this action.

Privilege

agent, all

Example

```
Action: QueuePause
Interface: SIP/testphone
Paused: true
Queue: myqueue

Response: Success
Message: Interface paused successfully

Event: QueueMemberPaused
Privilege: agent,all
Queue: myqueue
Location: SIP/testphone
MemberName: Jared Smith
Paused: 1

Action: QueuePause
Interface: SIP/testphone
Paused: false

Response: Success
Message: Interface unpaused successfully

Event: QueueMemberPaused
Privilege: agent,all
Queue: myqueue
Location: SIP/testphone
MemberName: Jared Smith
Paused: 0
```

QueueRemove
Removes interface from queue

Removes interface from queue.

Parameters

Queue

> [required] Which queue to remove the member from.

Interface

[required] The interface (member) to remove from the specified queue.

ActionID

[optional] An identifier that can be used to identify the response to this action.

Privilege

agent, all

Example

```
Action: QueueRemove
Queue: myqueue
Interface: SIP/testphone

Response: Success
Message: Removed interface from queue

Event: QueueMemberRemoved
Privilege: agent,all
Queue: myqueue
Location: SIP/testphone
MemberName: Jared Smith
```

QueueStatus   Checks queue status

Checks the status of one or more queues.

Parameters

Queue

[optional] If specified, limits the response to the status of the specified queue.

Member

[optional] An action identifier that you can use to identify the response to this manager transaction.

ActionID

[optional] An identifier that can be used to identify the response to this action.

Privilege

Example

```
Action: QueueStatus
Queue: inbound-queue
ActionID: 11223344556677889900

Response: Success
ActionID: 11223344556677889900
Message: Queue status will follow
```

```
Event: QueueParams
Queue: inbound-queue
Max: 0
Calls: 1
Holdtime: 99
Completed: 540
Abandoned: 51
ServiceLevel: 60
ServicelevelPerf: 50.4
Weight: 0
ActionID: 11223344556677889900

Event: QueueMember
Queue: inbound-queue
Location: Local/4020@agents/n
Membership: dynamic
Penalty: 2
CallsTaken: 25
LastCall: 1175563440
Status: 2
Paused: 0
ActionID: 11223344556677889900

Event: QueueEntry
Queue: inbound-queue
Position: 1
Channel: Zap/25-1
CallerID: 8012317154
CallerIDName: JOHN Q PUBLIC
Wait: 377
ActionID: 11223344556677889900

Event: QueueStatusComplete
ActionID: 11223344556677889900
```

Queues

Shows the call queues along with the queue members, callers, and basic queue statistics.

Privilege

Example

```
Action: Queues

inbound-queue has 0 calls (max unlimited) in 'rrmemory' strategy (81s holdtime),
  W:0, C:542, A:51, SL:50.4% within 60s
    Members:
        Local/4020@agents/n with penalty 2 (dynamic) (Unknown) has taken
          27 calls (last was 124 secs ago)
    No Callers
```

Notes

This manager command gives the same output as the `show queues` command from the Asterisk command-line interface. However, the output of this command is somewhat hard to parse programmatically. You may want to use the `QueueStatus` command instead.

Redirect Redirects (transfers) a channel

Redirects a channel to a new context, extension, and priority in the dialplan.

Parameters

Channel
: [required] Channel to redirect.

ExtraChannel
: [optional] Channel identifier of the second call leg to transfer.

ActionID
: [optional] An identifier that can be used to identify the response to this action.

Exten
: [required] Extension in the dialplan to transfer to.

Context
: [required] Context to transfer to.

Priority
: [required] Priority to transfer to.

Privilege

`call, all`

Example

```
Action: Redirect
Channel: SIP/linksys2-10201e90
Context: lab
Exten: 500
Priority: 1
ActionID: 010123234545

Response: Success
ActionID: 010123234545
Message: Redirect successful
```

Lists the currently configured SIP peers along with their status.

Parameters

ActionID

[optional] An action identifier that you can use to identify the response to this manager transaction.

Privilege

system, all

Example

```
Action: SIPPeers
ActionID: 555444333222111

Response: Success
ActionID: 555444333222111
Message: Peer status list will follow

Event: PeerEntry
ActionID: 555444333222111
Channeltype: SIP
ObjectName: labrat
ChanObjectType: peer
IPaddress: 10.0.0.75
IPport: 5060
Dynamic: no
Natsupport: no
VideoSupport: no
ACL: no
Status: OK (318 ms)
RealtimeDevice: no

Event: PeerEntry
ActionID: 555444333222111
Channeltype: SIP
ObjectName: guineapig
ChanObjectType: peer
IPaddress: 172.18.227.72
IPport: 5060
Dynamic: no
Natsupport: no
VideoSupport: no
ACL: no
Status: Unmonitored
RealtimeDevice: no

Event: PeerEntry
```

```
ActionID: 555444333222111
Channeltype: SIP
ObjectName: another
ChanObjectType: peer
IPaddress: 172.18.227.73
IPport: 5060
Dynamic: yes
Natsupport: no
VideoSupport: no
ACL: no
Status: Unmonitored
RealtimeDevice: no

Event: PeerlistComplete
ListItems: 7
ActionID: 555444333222111
```

SIPShowPeer

Shows detailed information about a configured SIP peer.

Parameters

Peer

> [required] The name of the SIP peer.

ActionID

> [optional] An action identifier that you can use to identify the response to this manager transaction.

Privilege

system, all

Example

```
Action: SIPShowPeer
Peer: linksys2
ActionID: 9988776655

Response: Success
ActionID: 9988776655
Channeltype: SIP
ObjectName: linksys2
ChanObjectType: peer
SecretExist: Y
MD5SecretExist: N
Context: lab
Language:
AMAflags: Unknown
CID-CallingPres: Presentation Allowed, Not Screened
Callgroup:
Pickupgroup:
VoiceMailbox:
```

```
TransferMode: open
LastMsgsSent: -1
Call-limit: 0
MaxCallBR: 384 kbps
Dynamic: Y
Callerid: "Linksys #2" <555>
RegExpire: 2516 seconds
SIP-AuthInsecure: no
SIP-NatSupport: RFC3581
ACL: N
SIP-CanReinvite: Y
SIP-PromiscRedir: N
SIP-UserPhone: N
SIP-VideoSupport: N
SIP-DTMFmode: rfc2833
SIPLastMsg: 0
ToHost:
Address-IP: 192.168.5.71
Address-Port: 5061
Default-addr-IP: 0.0.0.0
Default-addr-port: 5056
Default-Username: linksys2
RegExtension: 6100
Codecs: 0x4 (ulaw)
CodecOrder: ulaw
Status: Unmonitored
SIP-Useragent: Linksys/SPA962-5.1.5
Reg-Contact : sip:linksys2@192.168.5.71:5061
```

SetCDRUserField

Sets the UserField setting for the CDR record on the specified channel.

Parameters

Channel
> [required] The channel on which to set the CDR UserField.

UserField
> [required] The value to assign to the UserField in the CDR record.

ActionID
> [optional] An identifier that can be used to identify the response to this action.

Privilege

call, all

Example

```
Action: SetCDRUserField
Channel: SIP/test-10225140
UserField: abcdefg
```

```
Response: Success
Message: CDR Userfield Set
```

SetVar

Sets a global or channel variable.

Parameters

Channel

> [optional] Channel on which to set the variable. If not set, the variable will be set as a global variable.

Variable

> [required] Variable name.

Value

> [required] Value.

ActionID

> [optional] An identifier that can be used to identify the response to this action.

Privilege

call, all

Example

```
Action: SetVar
Channel: SIP/linksys2-10225140
Variable: MyOwnChannelVariable
Value: 42

Response: Success
Message: Variable Set

Action: SetVar
Variable: MyOwnGlobalVariable
Value: 25

Response: Success
Message: Variable Set
```

Status

Lists the status of one or more channels, showing details about their current state.

Parameters

Channel

> [optional] Limits the status output to the specified channel.

ActionID

> [optional] An action identifier that you can use to identify the response to this manager transaction.

Privilege

`call, all`

Example

```
Action: Status
Channel: SIP/test-10225140
ActionID: 101010101010101

Response: Success
ActionID: 101010101010101
Message: Channel status will follow

Event: Status
Privilege: Call
Channel: SIP/test-10225140
CallerID: "Bob Jones" <501>
CallerIDNum: 501
CallerIDName: "Bob Jones"
Account:
State: Up
Context: lab
Extension: 201
Priority: 1
Seconds: 865
Link: Local/200@lab-4d13,1
Uniqueid: 1177550165.0
ActionID: 101010101010101
Event: StatusComplete
ActionID: 101010101010101
```

StopMonitor

Stops the recording of a channel

Stops a previously started monitoring (recording) on a channel.

Parameters

Channel

> [required] The name of the channel to stop monitoring.

ActionID

> [optional] A unique identifier to help you identify responses to this command.

Privilege

`call, all`

Example

```
Action: StopMonitor
Channel: SIP/linksys2-10216e38

Response: Success
Message: Stopped monitoring channel
```

UnpauseMonitor

Unpauses the monitoring (recording) of the specified channel.

Parameters

Channel
: [required] The name of the channel on which to unpause the monitoring.

ActionID
: [optional] A unique identifier to help you identify responses to this command.

Privilege

call, all

Example

```
Action: UnpauseMonitor
Channel: SIP/linksys2-10212040
ActionID: 282828282828282

Response: Success
ActionID: 282828282828282
Message: Unpaused monitoring of the channel
```

UpdateConfig

Dynamically updates an Asterisk configuration file.

Parameters

SrcFilename
: [required] The filename of the configuration file from which to read the current information.

DstFilename
: [required] The filename of the configuration file to be written.

Reload
: [optional] Specifies whether or not a reload should take place after the configuration update, or the name of a specific module that should be reloaded.

`Action-`*`XXXXXX`*

> [required] An action to take. Can be one of `NewCat`, `RenameCat`, `DelCat`, `Update`, `Delete`, or `Append`.

`Cat-`*`XXXXXX`*

> [required] The name of the category to operate on.

`Var-`*`XXXXXX`*

> [optional] The name of the variable to operate on.

`Value-`*`XXXXXX`*

> [optional] The value of the variable to operate on.

`Match-`*`XXXXXX`*

> [optional] If set, an extra parameter that must be matched on the line.

`ActionID`

> [optional] An identifier that can be used to identify the response to this action.

Privilege

`config, all`

Example

```
Action: UpdateConfig
SrcFilename: sip.conf
DstFilename: test.conf
Action-000000: update
Cat-000000: linksys
Var-000000: mailbox
Value-000000: 101@lab

Response: Success
```

Notes

Note that the first set of parameters should be numbered `000000`, the second `000001`, and so on. This allows you to update many different configuration values at the same time. It should also be noted that the Asterisk GUI uses this its primary mechanism for updating the configuration of Asterisk.

UserEvent

Sends an arbitrary event to the Asterisk Manager Interface.

Parameters

`UserEvent`

> [required] The name of the arbitrary event to send.

Header

[optional] The name and value of an arbitrary parameter to your event. You may add as many additional headers (along with their values) to your event.

ActionID

[optional] An identifier which can be used to identify the response to this action.

Privilege

user, all

Example

```
Action: UserEvent
Blah: one
SomethingElse: two
ActionID: 63346

Event: UserEvent
Privilege: user,all
UserEvent:
Action: UserEvent
Blah: one
SomethingElse: two
ActionID: 63346
```

WaitEvent   Waits for an event to occur

After calling this action, Asterisk will send you a Success response as soon as another event is queued by the Asterisk Manager Interface. Once WaitEvent has been called on an HTTP manager session, events will be generated and queued.

Parameters

Timeout

[optional] Maximum time to wait for events.

ActionID

[optional] An identifier that can be used to identify the response to this action.

Privilege

Example

```
Action: WaitEvent
Timeout: 30

Action: Ping

Response: Success
Message: Waiting for Event...
```

```
Event: WaitEventComplete

Response: Pong
```

ZapDNDoff

Toggles the do not disturb state on the specified Zap channel to off.

Parameters

ZapChannel
> [required] The number of the Zap channel on which to turn off the do not disturb status.

ActionID
> [optional] An identifier that can be used to identify the response to this action.

Privilege

Example

```
Action: ZapDNDoff
ZapChannel: 1
ActionID: 01234567899876543210

Response: Success
ActionID: 01234567899876543210
Message: DND Disabled
```

ZapDNDon

Toggles the do not disturb state on the specified Zap channel to on.

Parameters

ZapChannel
> [required] The number of the Zap channel on which to turn on the Do Not Disturb status.

ActionID
> [optional] An identifier that can be used to identify the response to this action.

Privilege

Example

```
Action: ZapDNDon
ZapChannel: 1
ActionID: 98765432100123456789

Response: Success
ActionID: 98765432100123456789
Message: DND Enabled
```

ZapDialOffhook   Dials over Zap channel while off-hook

Dials the specified number on the Zap channel while the phone is off-hook.

Parameters

ZapChannel
> [required] The Zap channel on which to dial the number.

Number
> [required] The number to dial.

ActionID
> [optional] A unique identifier to help you identify responses to this command.

Privilege

Example

```
Action: ZapDialOffhook
ZapChannel: 1
Number: 543215432154321
ActionID: 5676

Response: Success
ActionID: 5676
Message: ZapDialOffhook
```

ZapHangup   Hangs up Zap channel

Hangs up the specified Zap channel.

Parameters

ZapChannel
> [required] The Zap channel to hang up.

ActionID
> [optional] A unique identifier to help you identify responses to this command.

Privilege

Example

```
Action: ZapHangup
ZapChannel: 1-1
ActionID: 98237892

Response: Success
ActionID: 98237892
Message: ZapHangup
```

ZapRestart

Completly restarts the Zaptel channels, terminating any calls in progress.

Privilege

Example

```
Action: ZapRestart

Response: Success
Message: ZapRestart: Success
```

ZapShowChannels

Shows the status of all the Zap channels.

Parameters

ActionID
[optional] An action identifier that can be used to identify the response from Asterisk.

Privilege

Example

```
Action: ZapShowChannels
ActionID: 9999999999

Response: Success
ActionID: 9999999999
Message: Zapata channel status will follow
```

```
Event: ZapShowChannels
Channel: 1
Signalling: FXO Kewlstart
Context: incoming
DND: Disabled
Alarm: No Alarm
ActionID: 9999999999

Event: ZapShowChannels
Channel: 4
Signalling: FXS Kewlstart
Context: incoming
DND: Disabled
Alarm: No Alarm
ActionID: 9999999999

Event: ZapShowChannelsComplete
ActionID: 9999999999
```

ZapTransfer

Transfers a Zap channel.

Privilege

Example

```
Action: ZapTransfer
ZapChannel: 1
ActionID: 4242

Response: Success
Message: ZapTransfer
ActionID: 4242
```

An Example of func_odbc

This appendix contains the examples from the hot-desking feature in the "Getting Funky with func_odbc: Hot-Desking" section in Chapter 12. It may be useful to follow along with this example in Chapter 12 as it can be useful to see where the code is going while reading along with the explanation.

Hot-Desking (extensions.conf)

Dialplan code

```
; Hot Desking Feature
[hotdesk]
; Hot Desk Login
exten => _11XX,1,NoOp()
exten => _11XX,n,Set(E=${EXTEN})
exten => _11XX,n,Verbose(1|Hot Desk Extension ${E} is changing status)
exten => _11XX,n,Verbose(1|Checking current status of extension ${E})
exten => _11XX,n,Set(${E}_STATUS=${HOTDESK_INFO(status,${E})})
exten => _11XX,n,Set(${E}_PIN=${HOTDESK_INFO(pin,${E})})
exten => _11XX,n,GotoIf($[${ISNULL(${${E}_STATUS})}]?invalid_user)

exten => _11XX,n,GotoIf($[${${E}_STATUS} = 1]?logout,1:login,1)

exten => login,1,NoOp()
exten => login,n,Set(PIN_TRIES=0)
exten => login,n,Set(MAX_PIN_TRIES=3)
exten => login,n(get_pin),NoOp()
exten => login,n,Set(PIN_TRIES=$[${PIN_TRIES} + 1])
exten => login,n,Read(PIN_ENTERED|enter-password|${LEN(${${E}_PIN})})
exten => login,n,GotoIf($[${PIN_ENTERED} = ${${E}_PIN}]?valid_login,1)
exten => login,n,Playback(invalid-pin,1)
exten => login,n,GotoIf($[${PIN_TRIES} <= ${MAX_PIN_TRIES}]?get_pin:login_fail,1)

exten => valid_login,1,NoOp()
exten => valid_login,n,Set(LOCATION=${CUT(CHANNEL,/,2)})
exten => valid_login,n,Set(LOCATION=${CUT(LOCATION,-,1)})
exten => valid_login,n,Set(ARRAY(USERS_LOGGED_IN)=${HOTDESK_CHECK_PHONE_LOGINS
(${LOCATION})})
exten => valid_login,n,GotoIf($[${USERS_LOGGED_IN} > 0]?logout_login,1)
```

```
exten => valid_login,n(set_login_status),NoOp()
exten => valid_login,n,Set(HOTDESK_STATUS(${E})=1\,${LOCATION})

exten => valid_login,n,GotoIf($[${ODBCROWS} < 1]?error,1)
exten => valid_login,n,Playback(agent-loginok)
exten => valid_login,n,Hangup()

exten => logout_login,1,NoOp()
exten => logout_login,n,Set(ROW_COUNTER=0)
exten => logout_login,n,While($[${ROW_COUNTER} < ${USERS_LOGGED_IN}])
exten => logout_login,n,Set(WHO=${HOTDESK_LOGGED_IN_USER(${LOCATION},${ROW_COUNTER})})
exten => logout_login,n,Set(HOTDESK_STATUS(${WHO})=0)
exten => logout_login,n,Set(ROW_COUNTER=$[${ROW_COUNTER} + 1])
exten => logout_login,n,EndWhile()
exten => logout_login,n,Goto(valid_login,set_login_status)

exten => logout,1,NoOp()
exten => logout,n,Set(HOTDESK_STATUS(${E})=0)
exten => logout,n,GotoIf($[${ODBCROWS} < 1]?error,1)
exten => logout,n,Playback(silence/1&agent-loggedoff)
exten => logout,n,Hangup()

exten => login_fail,1,NoOp()
exten => login_fail,n,Playback(silence/1&login-fail)
exten => login_fail,n,Hangup()

exten => error,1,NoOp()
exten => error,n,Playback(silence/1&connection-failed)
exten => error,n,Hangup()

exten => invalid_user,1,NoOp()
exten => invalid_user,n,Verbose(1|Hot Desk extension ${E} does not exist)
exten => invalid_user,n,Playback(silence/2&invalid)
exten => invalid_user,n,Hangup()

include => hotdesk_outbound

[hotdesk_outbound]
exten => _X.,1,NoOp()
exten => _X.,n,Set(LOCATION=${CUT(CHANNEL,/,2)})
exten => _X.,n,Set(LOCATION=${CUT(LOCATION,-,1)})
exten => _X.,n,Set(WHO=${HOTDESK_PHONE_STATUS(${LOCATION})})
exten => _X.,n,GotoIf($[${ISNULL(${WHO})}]?no_outgoing,1)
exten => _X.,n,Set(${WHO}_CID_NAME=${HOTDESK_INFO(cid_name,${WHO})})
exten => _X.,n,Set(${WHO}_CID_NUMBER=${HOTDESK_INFO(cid_number,${WHO})})
exten => _X.,n,Set(${WHO}_CONTEXT=${HOTDESK_INFO(context,${WHO})})
exten => _X.,n,Goto(${${WHO}_CONTEXT},${EXTEN},1)

[international]
exten => _011.,1,NoOp()
exten => _011.,n,Set(E=${EXTEN})
exten => _011.,n,Goto(outgoing,call,1)
```

```
exten => i,1,NoOp()
exten => i,n,Playback(silence/2&sorry-cant-let-you-do-that2)
exten => i,n,Hangup()

include => longdistance

[longdistance]
exten => _1NXXNXXXXXX,1,NoOp()
exten => _1NXXNXXXXXX,n,Set(E=${EXTEN})
exten => _1NXXNXXXXXX,n,Goto(outgoing,call,1)

exten => _NXXNXXXXXX,1,Goto(1${EXTEN},1)

exten => i,1,NoOp()
exten => i,n,Playback(silence/2&sorry-cant-let-you-do-that2)
exten => i,n,Hangup()

include => local

[local]
exten => _416NXXXXXX,1,NoOp()
exten => _416NXXXXXX,n,Set(E=${EXTEN})
exten => _416NXXXXXX,n,Goto(outgoing,call,1)

exten => i,1,NoOp()
exten => i,n,Playback(silence/2&sorry-cant-let-you-do-that2)
exten => i,n,Hangup()

[outgoing]
exten => call,1,NoOp()
exten => call,n,Set(CALLERID(name)=${${WHO}_CID_NAME})
exten => call,n,Set(CALLERID(number)=${${WHO}_CID_NUMBER})
exten => call,n,Dial(SIP/service_provider/${E})
exten => call,n,Playback(silence/2&pls-try-call-later)
exten => call,n,Hangup()

[hotdesk_phones]
exten => _11XX,1,NoOp()
exten => _11XX,n,Set(E=${EXTEN})
exten => _11XX,n,Set(LOCATION=${HOTDESK_LOCATION(${E})})
exten => _11XX,n,GotoIf($[${ISNULL(${LOCATION})}]?voicemail,1)
exten => _11XX,n,Dial(SIP/${LOCATION},30)
exten => _11XX,n,Goto(voicemail,1)

exten => voicemail,1,NoOp()
exten => voicemail,n,Voicemail(${E}@hotdesk,u)
exten => voicemail,n,Hangup()
```

See Also

Hot-Desking (*sip.conf*), Hot-Desking (*func_odbc.conf*), Chapter 5, Chapter 6, Read(), CUT, While(), ISNULL, VoiceMail(), CALLERID, Dial(), GotoIf()

Hot-Desking (func_odbc.conf)   Custom dialplan functions

```
[INFO]
prefix=HOTDESK
dsn=asterisk
read=SELECT ${ARG1} FROM ast_hotdesk WHERE extension = '${ARG2}'

[STATUS]
prefix=HOTDESK
dsn=asterisk
write=UPDATE ast_hotdesk SET status = '${VAL1}', location = '${VAL2}' WHERE extension
= '${ARG1}'

[CHECK_PHONE_LOGINS]
prefix=HOTDESK
dsn=asterisk
read=SELECT COUNT(status) FROM ast_hotdesk WHERE status = '1' AND location = '${ARG1}'

[LOGGED_IN_USER]
prefix=HOTDESK
dsn=asterisk
read=SELECT extension FROM ast_hotdesk WHERE status = '1' AND location = '${ARG1}'
ORDER BY id LIMIT '1' OFFSET '${ARG2}'

[PHONE_STATUS]
prefix=HOTDESK
dsn=asterisk
read=SELECT extension FROM ast_hotdesk WHERE location = '${ARG1}' AND status = '1'
```

See Also

Hot-Desking (*extensions.conf*), Hot-Desking (*sip.conf*), *res_odbc.conf*

Hot-Desking (sip.conf)   Two sample phone configurations and sample service provider configuration

```
; HOT DESK USERS
[desk_1]
type=friend
host=dynamic
secret=my_special_secret
context=hotdesk
qualify=yes

[desk_2]
type=friend
host=dynamic
secret=my_special_secret
```

```
context=hotdesk
qualify=yes

; END HOT DESK USERS
```

See Also

Hot-Desking (*extensions.conf*), Hot-Desking (*func_odbc.conf*), Chapter 4

Index

Symbols

! (bang), matching characters with, 138
!= operator, 147
$ (dollar sign), using expressions, 145
% (remainder of sign), 147
& (ampersand)
   and boolean operator, 147
   dialing multiple channels, 131
' (single quotes)
   using the makerequest function, 259
* (asterisk), 70
   GotoIfTime( ) function), 153
   multiplication sign, 147
* (wildcard), 70
*** termcap support not found, 52
+ (plus sign), 147
, (commas), using Set( ), 282
- (minus sign), 147
. (period), matching characters with, 138
/ (forward slash)
   integer division sign, 147
   using Dial( ), 130
7960 (Cisco) telephone, 93–95
: (regular expression operator), 147
< (less than) operator, 147
<= operator, 147
= (equal sign) comparison operator, 147
=> (extensions), 122
> (greater than) comparison operator, 147
>= operator, 147
[ ] (square brackets), 337
   contexts and, 120
   DUNDi peers, defining, 311
   editing the iax.conf file, 109

Manager interface and, 228
\ (backslash), using Set( ), 282
^ (caret), in regular expressions, 147
_ (underscores), using pattern-matching, 137
{ } (curly braces)
   functions and, 148
   variables and, 135
| (pipe), 337
   as a separator, 124
   boolean operator, 146
   mailboxes, creating, 155
   Set( ) application and, 282
μlaw, 177

A

AADK (Asterisk Appliance Developers Kit, 246
AbsoluteTimeout (AMI action), 515
AC (alternating current), 168
accountcode
   CSV value, 293
   parameter (IAX), 338
   SIP parameter, 358
ActiveRecord, 237
adaptors (telephony), 33
AddQueueMember( ) application, 368
Adhearsion, 231–243
   code, distributing and reusing, 239
   dialplans and, 232
   installing, 233–234
   Micromenus, using, 241
adsi parameter (IAX), 338
adsi.conf file, 462
ADSIProg( ) application, 368
adtranvofr.conf file, 463

We'd like to hear your suggestions for improving our indexes. Send email to *index@oreilly.com*.

M

MAC (Media Access Control), 309
Mac OS X, installing Ruby/RubyGems, 234
Macro( ) application, 158, 404
MacroExclusive( ) application, 406
MacroExit( ) application, 406
MacroIf( ) application, 406
macros, 157–160
    arguments, using, 159
${MACRO_CONTEXT} variable, 158
${MACRO_EXTEN} variable, 158
${MACRO_PRIORITY} variable, 158
Madsen, Leif, 8
mailbox parameter
    IAX, 348
    SIP, 363
MailboxCount (AMI action), 526
mailboxdetail parameter (IAX), 341
mailboxes, creating, 154
MailboxExists( ) application, 407
MailboxStatus (AMI action), 526
mailing lists (Asterisk), 7
make clean, 46, 49
make config, 47, 50, 54
make distclean, 49
make progdocs, 49
make program, 48
make samples command, installing Asterisk,
    48
make update, 49
make webvmail, 49
Makefiles, 42
    arguments, 49
makerequest function, 259
Manager interface
    commands, 228
Manager Interface (Asterisk), 227–243, 248,
    249
    HTTP, commands over, 251
manager.conf file, 227, 268, 475
    Asterisk GUI, setting up, 250
master config files, 92
Master.csv, 62
matchexterniplocally SIP parameter, 354
MATH dialplan function, 505
maxcallbitrate SIP parameter, 363
maxexpiry SIP parameter, 354
maxjitterbuffer parameter (IAX), 341
maxjitterinterps parameter (IAX), 341

maxregexpire parameter (IAX), 341
MD5 dialplan function, 505
MD5 hashing, 188
md5secret SIP parameter, 363
media, 81
Media Access Control (MAC), 309
Media Gateway Control Protocol (MGCP),
    192
medium option (bandwidth parameter), 339
medium systems, choosing processors for, 18
MeetMe( ) application, 164, 407
meetme.conf file, 164, 476
meetme/ directory, 60
MeetMeAdmin( ) application, 409
MeetMeCount( ) application, 165, 411
MeetmeMute (AMI action), 527
MeetMeUnmute (AMI action), 528
menuselect, 42, 50
message waiting indication (MWI), 154, 348
metrics, 270
MGCP (Media Gateway Control Protocol),
    192
mgcp.conf file, 476
Micromenus, integrating desk phones with,
    241
Microsoft
    DHCP environments, 86
    NetMeeting client, 191
    Ruby/RubyGems, installing, 234
Milliwatt( ) application, 411
minexpiry SIP parameter, 355
minregexpire parameter (IAX), 341
minus sign (-), 147
MixMonitor( ) application, 412
modem.conf file, 477
modems, 19
modprobe, 76
modules.conf file, 59, 268, 462
mohinterpret parameter (IAX), 342
mohinterpret SIP parameter, 363
mohmp3/ directory, 60
mohsuggest parameter (IAX), 342
mohsuggest SIP parameter, 363
Molex connectors, 75
Monitor (AMI action), 528
Monitor( ) application, 412
monitor/ directory, 60
months argument (GotoIfTime( ) function),
    152

Moore, Geffrey, 8
MorseCode( ) application, 413
motherboard, choosing, 18–20
Mozilla (see Firefox browser)
MP3 (Moving Picture Experts Group Audio Layer 3) codec, 196
MP3Player( ) application, 414
MP3s, 299
MPLS (Multiprotocol Label Switching), 199
multiplication sign (*), 147
Multiprotocol Label Switching (MPLS), 199
Music on Hold, 299–302
MUSICCLASS dialplan function, 506
musicclass SIP parameter, 363
MusicOnHold( ) application, 414
musiconhold.conf file, 270, 477
MWI (message waiting indication), 154, 348
MySQL, 160, 207, 263
    CDRs, storing, 294

N

/n (blank line feed), 223
-n flag, 311
N, matching characters with, 138
names (functions), 148
NANP (North America Number Plan), 139
NAT (Network Address Translation), 185, 190
nat SIP parameter, 364
NATed firewalls, 187
NBScat( ) application, 414
ncursers, 38
NetMeeting client (Microsoft), 191
Network Address Translation (NAT), 185, 190
Network Configuration screen, 65
Network Interface Card (NIC), 19
network transformers, 169
newt-devel package, 39
NIC (Network Interface Card), 19
NoCDR( ) application, 415
nochecksums parameter (IAX), 342
non-root users, 295–298
nonces, 190
NoOP (AGI), 453
NoOp( ) application, 415
Nortel, 1
North America Number Plan (NANP), 139
notifyhold SIP parameter, 355

notifymimetype SIP parameter, 355
notifyringing SIP parameter, 355
${NUMBER} variable, 310
Nutter, Charles, 242
Nyquist's theorem, 176

O

ob_implicit_flush(false) command, 219
ODBC connector, 263
    installing and configuring, 265–267
    voicemail, 286–291, 315
ogg-Vorbis format, 300
OHCI USB controller chip, 46, 53
open architecture, 321
Open Settlement Protocol (OSP), 477
OpenH323 Gatekeeper, 192
OpenSER (SIP Express Router), 82
OpenSSL, 38
OpenWRT, 17
operating systems, debugging AGI scripts from, 223
operator option, creating mailboxes, 155
operators, 146
Originate (AMI action), 529
OS79XX.TXT file, 94
[Osaka] definition, 112
OSP (Open Settlement Protocol), 477
osp.conf file, 477
oss.conf file, 477
outbound dialing, 141
outgoing/ directory, 60
outkey parameter (IAX), 348

P

packet-based transmissions, 186
packet-switched networks, 184
packetization, 186
Page( ) application, 416
Park (AMI action), 530
Park( ) application, 417
ParkAndAnnounce( ) application, 417
ParkedCall( ) application, 418
ParkedCalls (AMI action), 531
parkext setting (features.conf file), 163
parkingtime setting (features.conf file), 163
parkpos setting (features.conf file), 163
parser generators, 38
password-protected voicemail boxes, 153

About the Authors

Jim Van Meggelen is President and CTO of Core Telecom Innovations, a Canadian-based provider of open source telephony solutions. He has over 15 years of enterprise telecom experience, for such companies as Nortel, Williams, and Telus, and has extensive knowledge of both legacy and VoIP equipment from manufacturers such as Nortel, Cisco, and Avaya.

Jim was the architect of two of the world's largest managed enterprise voice networks, each solution serving roughly 20,000 users in more than 1,000 communities across Canada and providing telecommunications in five different languages through six time zones, administered completely from a central location. These networks pioneered the use of extensive automation and database control in a branch voice network—functionalities not generally available in proprietary telecommunications systems. Jim has now moved on from the world of proprietary telecom, and is committed to open source telephony.

Jim is one of the principal contributors to the Asterisk Documentation Project. He enjoys teaching, public speaking, improvisational acting, and writing.

Leif Madsen is a graduate of the Telecommunications Technology program from the Sheridan Institute of Technology and CEO of LeifMadsen Enterprise, Incorporated, a documentation and consulting firm specializing in Asterisk. He was one of the first Digium Certified Asterisk Professionals (dCAP), and assists with the Astricon conferences and trainings organized by IPsando, LLC.

Leif first took an interest in Asterisk while attempting to find a voice conferencing solution for himself and his friends. After someone suggested trying Asterisk, the obsession began. Wanting to contribute and be involved with the community, and noticing the lack of Asterisk documentation, he co-founded the Asterisk Documentation Project.

Our look is the result of reader comments, our own experimentation, and feedback from distribution channels. Distinctive covers complement our distinctive approach to technical topics, breathing personality and life into potentially dry subjects.

Jared Smith is one of those rare individuals whose beloved hobby is the same as his profession. The son of a computer store owner, Jared wrote his first computer program at the age of seven on his Commodore 64. The obvious choice of major for this geek-in-embryo was Computer Engineering, and Jared received his Bachelor of Science degree with a minor in Computer Science from Utah State University. He now has more than a decade of professional systems administration and programming experience in the simulation, market research, and web analytics industries. As a key architect of one of the world's largest Asterisk installations, Jared has a wealth of hands-on telephony and VoIP knowledge, which he shares through users groups and various public speaking engagements. He is an active member of the Asterisk community and a co-founder of the Asterisk Documentation Project.

Jared is active in his community, donating Asterisk services to local schools and serving in his church. The greatest joy in Jared's life comes from spending time with his children, Caleb and Sydney Jo, and his wife, Jenny.

Colophon

The animals on the cover of *Asterisk: The Future of Telephony* are starfish. Starfish are classified as *Asteroidea*. They are a group of echinoderms, spiny-skinned invertebrates found only in the sea. Most starfish have fivefold symmetry (arms or rays in multiples of five), though some species can have four or nine arms. But all starfish are radially symmetrical: they have arms or rays branching out from a central body disc. There are over 1,500 species of starfish.

Starfish live on the floor of the sea and in tidal pools, clinging to rocks and moving (slowly) using a water-based vascular system to manipulate their hundreds of tiny, tube-like legs, called *podia*. A small bulb or *ampulla* at the top of the tube contracts, expelling water and expanding the starfish's leg. The ampulla relaxes, and the leg retracts. Starfish use muscles to bend their legs, but it is the flow of water pressure that keeps the feet moving. At the tip of each leg, starfish have suction cups that allow them to pry open clam, oyster, or mussel shells. Many starfish can push their stomachs out through their mouths in order to digest their prey in its shell. Starfish are carnivores; they eat coral, fish, and snails, as well as bivalves.

Starfish can flex and rearrange their arms to fit into small places as they move over the ocean floor. At the end of each arm, they have eyespots, primitive sensors that detect light and help the starfish determine direction. Starfish also have the ability to regenerate a missing limb. Some species can even regrow a complete, new starfish from a severed arm.

The cover image is from the Dover Pictorial Archive. The cover font is Adobe ITC Garamond. The text font is Linotype Birka; the heading font is Adobe Myriad Condensed; and the code font is LucasFont's TheSans Mono Condense.

Try the online edition
free for 45 days

Get the information you need when you need it, with Safari Books Online. Safari
Books Online contains the complete version of the print book in your hands plus
thousands of titles from the best technical publishers, with sample code ready to
cut and paste into your applications.

Safari is designed for people in a hurry to get the answers they need so they can
get the job done. You can find what you need in the morning, and put it to work in
the afternoon. As simple as cut, paste, and program.

To try out Safari and the online edition of the above title FREE for 45 days,
go to www.oreilly.com/go/safarienabled and enter the coupon code YVTITYG.

To see the complete Safari Library visit:
safari.oreilly.com

70502

Related Titles from O'Reilly

Networking

Our books are available at most retail and online bookstores.

To order direct: 1-800-998-9938 • order@oreilly.com • www.oreilly.com

Online editions of most O'Reilly titles are available by subscription at *safari.oreilly.com*

The O'Reilly Advantage

Stay Current and Save Money

Order books online:
www.oreilly.com/order_new

Questions about our products or your order:
order@oreilly.com

Join our email lists: Sign up to get topic specific email announcements or new books, conferences, special offers and technology news
elists@oreilly.com

For book content technical questions:
booktech@oreilly.com

To submit new book proposals to our editors:
proposals@oreilly.com

Contact us:
O'Reilly Media, Inc.
1005 Gravenstein Highway N.
Sebastopol, CA U.S.A. 95472
707-827-7000 or
800-998-9938
www.oreilly.com

Did you know that if you register your O'Reilly books, you'll get automatic notification and upgrade discounts on new editions?

And that's not all! Once you've registered your books you can:

» Win free books, T-shirts and O'Reilly Gear

» Get special offers available only to registered O'Reilly customers

» Get free catalogs announcing all our new titles (US and UK Only)

Registering is easy! Just go to www.oreilly.com/go/register